Marina Owners Workshop Manual

Bruce Gilmour

Models covered:

Marina 1700 Saloons and Estates, including L and HL versions
Covers manual and automatic transmission versions

ISBN 0 85696 808 0

Printed in England *(526-9N2)*

Haynes Publishing Group
Sparkford Nr Yeovil
Somerset BA22 7JJ England

Haynes Publications, Inc
861 Lawrence Drive
Newbury Park
California 91320 USA

Acknowledgements

Thanks are due to Castrol Limited who supplied lubrication data and to the Champion Sparking Plug Company for the illustrations showing the various spark plug conditions. The bodywork repair photographs used in this manual were provided by Lloyds Industries Limited who supply 'Turtle Wax', 'Dupli-color Holts' and other Holts range products.

The Section in Chapter 10 dealing with the suppression of radio interference was originated by Mr I P Davey and was first published in Motor magazine.

Lastly, thanks are due to all those people at Sparkford who helped in the production of this manual, particularly Brian Horsfall and Les Brazier who carried out the mechanical work and took the photographs respectively, Stanley Randolph who planned the layout of each page and Chris Rogers who edited the text.

About this manual

Its aim

The aim of this manual is to help you get the best value from your car. It can do so in several ways. It can help you decide what work must be done (even should you choose to get it done by a garage), provide information on routine maintenance and servicing, and give a logical course of action and diagnosis when random faults occur. However, it is hoped that you will make use of the manual by tackling the work yourself. On simpler jobs it may be even quicker than booking the car into a garage, and having to go there twice, to leave and collect it. Perhaps most important, a lot of money can be saved by avoiding the costs the garage must charge to cover its labour and overheads.

The manual has drawings and descriptions to show the function of the various components so that their layout can be understood. Then the tasks are described and photographed in a step-by-step sequence so that even a novice can do the work.

Its arrangement

The manual is divided into twelve Chapters, each covering a logical sub-division of the vehicle. The Chapters are each divided into consecutively numbered Sections and the Sections into paragraphs (or sub-sections), with decimal numbers following on from the Section they are in, eg, 5.1, 5.2, 5.3 etc.

It is freely illustrated, especially in those parts where there is a detailed sequence of operations to be carried out. There are two forms of illustration; figures and photographs. The figures are numbered in sequence with decimal numbers, according to their position in the Chapter; eg, Fig. 6.4 is the 4th drawing/illustration in Chapter 6. Photographs are numbered (either individually or in related groups) the same as Section or sub-section of the text where the operation they show is described.

There is an alphabetical index at the back of the manual as well as a contents list at the front.

References to the 'left' or 'right' of the vehicle are in the sense of a person in a seat facing towards the front of the vehicle.

Unless otherwise stated, nuts and bolts are removed by turning anti-clockwise and tightened by turning clockwise.

Vehicle manufacturers continually make changes to specifications and recommendations, and these, when notified, are incorporated into our manuals at the earliest opportunity.

Whilst every care is taken to ensure that the information in this manual is correct, no liability can be accepted by the authors or publishers for loss, damage or injury caused by any errors in, or omissions, from the information given.

Introduction to the Marina 1700

There are two approaches to building a car. A conventional design can be used, and from it can be expected the reliability and dependability that must come from the thorough development of a well proven layout. The unconventional car should give some startling advantages, but in return some penalties must be accepted. In British Leyland the ordinary mass produced cars are made by BL Cars Limited. Their unconventional cars such as the Mini, the Maxi and more recently the Allegro, have transverse engines driving the front wheels, and unusual suspension, such as hydrolastic and hydragas. These cars are renowned for their road holding and the large space inside with small overall dimensions. The conventional car is usually cheaper to build and to repair. Its different layout suits some owners who do not take to the transverse engined layout. The Marina is aimed at this large market. The dealer and agency loyalty in the United Kingdom was built originally when Austin and Morris were separate, and rivals. Whilst the Maxi and Allegro used the Austin agency on the home market, the Marina was launched as a Morris.

The Marina was introduced in 1971 with a 1.3 litre or 1.8 litre B series engine. In 1975, the 1.8 litre Mk I was replaced by the Mk 2 which had a much improved suspension.

The Marina 1700 introduced in September 1978 has basically the same body and suspension except for a few minor improvements, but is equipped with a completely new 1695 cc engine. The B series engine, which has pushrod operated valves, is replaced by a new O series engine with an overhead camshaft.

Automatic transmission is available as an optional extra on L and HL models.

The Marina is simple and straightforward to work on. All the major components have been in use in earlier models so have a long development period behind them resulting in good reliability.

Contents

The Morris Marina 1700 HL Estate

The Morris Marina 1700 L Saloon

Quick reference chart

Dimensions

1700 4-door saloon:	
Overall height	4ft $8\frac{1}{8}$ in (1.425 m)
Overall width	5 ft $6\frac{47}{64}$ in (1.659 m)
Overall length	14 ft $0\frac{15}{16}$ in (4.291 m)
Ground clearance (kerbside weight)	0ft 6 in (153 mm)
Turning circle (kerb to kerb)	33ft 6 in (10.21 m)
Track:	
Front	4ft 4 in (1.321 m)
Rear	4ft 4 in (1.321 m)
1700 5-door estate:	
Overall height	4ft $8\frac{35}{64}$ in (1.436 m)
Overall width	5ft $6\frac{47}{64}$ in (1.659 m)
Overall length	14ft $1\frac{61}{64}$ in (4.317 m)
Ground clearance (kerbside weight)	0ft $6\frac{1}{4}$ in (160 mm)
Turning circle (kerb to kerb)	33ft 6 in (10.21 m)
Track:	
Front	4ft 4 in (1.321 m)
Rear	4ft 4 in (1.321 m)

Kerb weights (full fuel tank)

1700 4-door saloon:	
Manual	2072 lb (940 kg)
Automatic	2110 lb (958 kg)
1700 5-door estate:	
Manual	2195 lb (995 kg)
Automatic	2235 lb (1003 kg)

Capacities

Fuel tank	11.5 gal (52 litres)
Engine sump (with filter)	7 pints (4 litres)
Gearbox (manual)	1.5 pints (0.85 litres)
Automatic transmission (top-up from MIN to MAX on dipstick)	1 pint (0.57 litres)
Rear axle	1.25 pints (0.71 litres)
Cooling system (with heater)	10.5 pints (6 litres)

Load capacities

Maximum load on towing hitch	75 to 100 lb (34 to 35 kg)
Maximum payload:	
Saloon	875 lb (397 kg)
Estate	900 lb (408 kg)
Maximum roof rack load	100 lb (45 kg)

Buying spare parts and vehicle identification numbers

Buying spare parts

Spare parts are available from many sources, for example: BL dealers, other garages and accessory shops, and motor factors. Our advice regarding spare part sources is as follows:

Officially appointed BL garages – This is the best source of parts which are peculiar to your car and are otherwise not generally available (eg, complete cylinder heads, internal gearbox components, badges, interior trim etc). It is also the only place at which you should have repairs carried out if your car is still under warranty – non-BL components may invalidate the warranty. To be sure of obtaining the correct parts it will always be necessary to give the storeman your car's vehicle identification number, and if possible, to take the old part along for positive identification. It obviously makes good sense to go straight to the specialists on your car for this type of part for they are best equipped to supply you.

Other garages and accessory shops – These are often very good places to buy materials and components needed for the maintenance of your car (eg, spark plugs, bulbs, fan belts, oils and greases, filler paste etc). They also sell general accessories, usually have convenient opening hours, charge reasonable prices and can often be found not far from home.

Motor factors – Good factors will stock all the more important components which wear out relatively quickly (eg, clutch components, pistons, valves, exhaust systems, brake cylinders/pipes/hoses/seals/shoes and pads etc). Motor factors will often provide new or reconditioned components on a part exchange basis – this can save a considerable amount of money.

Vehicle identification numbers

The car number is located on a metal plate fixed to the bonnet lock platform.

The engine number is stamped on a plate fixed to the cylinder block on the right-hand side of the engine.

The manual gearbox number is on a label attached to the right-hand side of the casing. On automatic transmission models, the number is stamped on the left-hand side of the casing.

The rear axle number is stamped on the outside face of the differential casing joint flange.

Tools and working facilities

Introduction

A selection of good tools is a fundamental requirement for anyone contemplating the maintenance and repair of a motor vehicle. For the owner who does not possess any, their purchase will prove a considerable expense, offsetting some of the savings made by doing-it-yourself. However, provided that the tools purchased meet the relevant national safety standards and are of good quality, they will last for many years and prove an extremely worthwhile investment.

To help the average owner to decide which tools are needed to carry out the various tasks detailed in this manual, we have compiled three lists of tools under the following headings: *Maintenance and minor repair, Repair and overhaul,* and *Special.* The newcomer to practical mechanics should start off with the *Maintenance and minor repair* tool kit and confine himself to the simpler jobs around the vehicle. Then, as his confidence and experience grow, he can undertake more difficult tasks, buying extra tools as, and when, they are needed. In this way, a *Maintenance and minor repair* tool kit can be built-up into a *Repair and overhaul* tool kit over a considerable period of time without any major cash outlays. The experienced do-it-yourselfer will have a tool kit good enough for most repair and overhaul procedures and will add tools from the *Special* category when he feels the expense is justified by the amount of use to which these tools will be put.

It is obviously not possible to cover the subject of tools fully here. For those who wish to learn more about tools and their use there is a book entitled *How to Choose and Use Car Tools* available from the publishers of this manual.

Maintenance and minor repair tool kit

The tools given in this list should be considered as a minimum requirement if routine maintenance, servicing and minor repair operations are to be undertaken. We recommend the purchase of combination spanners (ring one end, open-ended the other); although more expensive than open-ended ones, they do give the advantages of both types of spanner.

Combination spanners - 10, 11, 12, 13, 14 & 17 mm
Adjustable spanner - 9 inch
Engine sump/gearbox/rear axle drain plug key
Spark plug spanner (with rubber insert)
Spark plug gap adjustment tool
Set of feeler gauges
Brake adjuster spanner
Brake bleed nipple spanner
Screwdriver - 4 in long x 1/4 in dia (flat blade)
Screwdriver - 4 in long x 1/4 in dia (cross blade)
Combination pliers - 6 inch
Hacksaw (junior)
Tyre pump
Tyre pressure gauge
Grease gun
Oil can
Fine emery cloth (1 sheet)
Wire brush (small)
Funnel (medium size)

Repair and overhaul tool kit

These tools are virtually essential for anyone undertaking any major repairs to a motor vehicle, and are additional to those given in the *Maintenance and minor repair* list. Included in this list is a comprehensive set of sockets. Although these are expensive they will be found invaluable as they are so versatile - particularly if various drives are included in the set. We recommend the ½ in square-drive type, as this can be used with most proprietary torque wrenches. If you cannot afford a socket set, even bought piecemeal, then inexpensive tubular box spanners are a useful alternative.

The tools in this list will occasionally need to be supplemented by tools from the *Special* list.

Sockets (or box spanners) to cover range in previous list
Reversible ratchet drive (for use with sockets)
Extension piece, 10 inch (for use with sockets)
Universal joint (for use with sockets)
Torque wrench (for use with sockets)
'Mole' wrench - 8 inch
Ball pein hammer
Soft-faced hammer, plastic or rubber
Screwdriver - 6 in long x 5/16 in dia (flat blade)
Screwdriver - 2 in long x 5/16 in square (flat blade)
Screwdriver - 1 1/2 in long x 1/4 in dia (cross blade)
Screwdriver - 3 in long x 1/8 in dia (electricians)
Pliers - electricians side cutters
Pliers - needle nosed
Pliers - circlip (internal and external)
Cold chisel - 1/2 inch
Scriber
Scraper
Centre punch
Pin punch
Hacksaw
Valve grinding tool
Steel rule/straight-edge
Allen keys (inc. splined/Torx type if necessary)
Selection of files
Wire brush (large)
Axle-stands
Jack (strong trolley or hydraulic type)

Special tools

The tools in this list are those which are not used regularly, are expensive to buy, or which need to be used in accordance with their manufacturers' instructions. Unless relatively difficult mechanical jobs are undertaken frequently, it will not be economic to buy many of these tools. Where this is the case, you could consider clubbing together with friends (or joining a motorists' club) to make a joint purchase, or borrowing the tools against a deposit from a local garage or tool hire specialist.

The following list contains only those tools and instruments freely available to the public, and not those special tools produced by the vehicle manufacturer specifically for its dealer network. You will find occasional references to these manufacturers' special tools in the text of this manual. Generally, an alternative method of doing the job without the vehicle manufacturers' special tool is given. However, sometimes, there is no alternative to using them. Where this is the case and the relevant tool cannot be bought or borrowed, you will have to entrust the work to a franchised garage.

Valve spring compressor (where applicable)
Piston ring compressor
Balljoint separator
Universal hub/bearing puller
Impact screwdriver
Micrometer and/or vernier gauge
Dial gauge
Stroboscopic timing light

Dwell angle meter/tachometer
Universal electrical multi-meter
Cylinder compression gauge
Lifting tackle
Trolley jack
Light with extension lead

Buying tools

For practically all tools, a tool factor is the best source since he will have a very comprehensive range compared with the average garage or accessory shop. Having said that, accessory shops often offer excellent quality tools at discount prices, so it pays to shop around.

There are plenty of good tools around at reasonable prices, but always aim to purchase items which meet the relevant national safety standards. If in doubt, ask the proprietor or manager of the shop for advice before making a purchase.

Care and maintenance of tools

Having purchased a reasonable tool kit, it is necessary to keep the tools in a clean serviceable condition. After use, always wipe off any dirt, grease and metal particles using a clean, dry cloth, before putting the tools away. Never leave them lying around after they have been used. A simple tool rack on the garage or workshop wall, for items such as screwdrivers and pliers is a good idea. Store all normal wrenches and sockets in a metal box. Any measuring instruments, gauges, meters, etc, must be carefully stored where they cannot be damaged or become rusty.

Take a little care when tools are used. Hammer heads inevitably become marked and screwdrivers lose the keen edge on their blades from time to time. A little timely attention with emery cloth or a file will soon restore items like this to a good serviceable finish.

Working facilities

Not to be forgotten when discussing tools, is the workshop itself. If anything more than routine maintenance is to be carried out, some form of suitable working area becomes essential.

It is appreciated that many an owner mechanic is forced by circumstances to remove an engine or similar item, without the benefit of a garage or workshop. Having done this, any repairs should always be done under the cover of a roof.

Wherever possible, any dismantling should be done on a clean, flat workbench or table at a suitable working height.

Any workbench needs a vice: one with a jaw opening of 4 in (100 mm) is suitable for most jobs. As mentioned previously, some clean dry storage space is also required for tools, as well as for lubricants, cleaning fluids, touch-up paints and so on, which become necessary.

Another item which may be required, and which has a much more general usage, is an electric drill with a chuck capacity of at least $5/16$ in (8 mm). This, together with a good range of twist drills, is virtually essential for fitting accessories such as mirrors and reversing lights.

Last, but not least, always keep a supply of old newspapers and clean, lint-free rags available, and try to keep any working area as clean as possible.

Jacking and towing

Jacking points

To change a wheel in an emergency, use the jack supplied with the car. Ensure that the roadwheel nuts are slackened before jacking up the car, that the handbrake is applied and that the diagonally opposite wheel to the one being removed is suitably chocked. Make sure that the jack is standing on a firm surface and that the peg on the jack head locates in the hole of the jacking point.

The jack supplied with the car is not suitable for use when raising the car for maintenance or repair operations. For this work use a trolley, hydraulic or screw type jack located under the front chassis members or the rear axle. Always supplement the jack with axle-stands or other suitable supports before crawling under the car.

Towing

If the car is being towed, attach the tow rope to the towing eyes located forward of each front tie-bar on the chassis members. If the car is equipped with automatic transmission, the distance towed must not exceed 30 miles (48 km), nor the speed exceed 30 mph (48 kmh), otherwise serious damage to the transmission may result. If these limits are likely to be exceeded, or there are unusual noises coming from the transmission, disconnect and remove the propeller shaft.

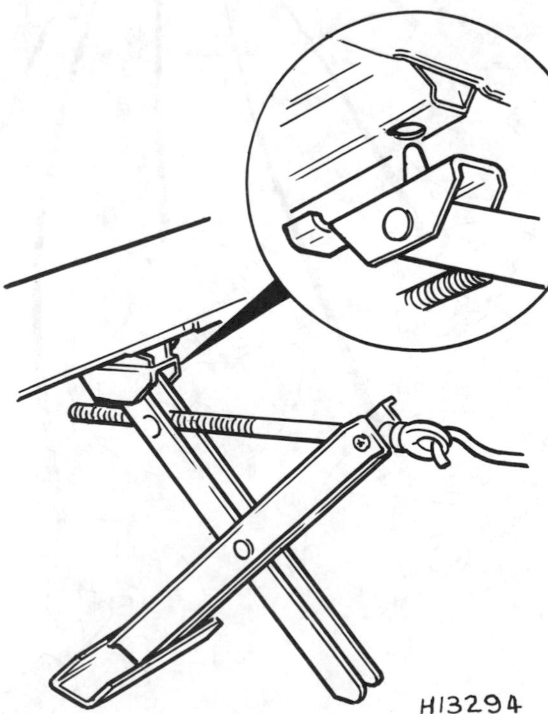

Jacking up the car

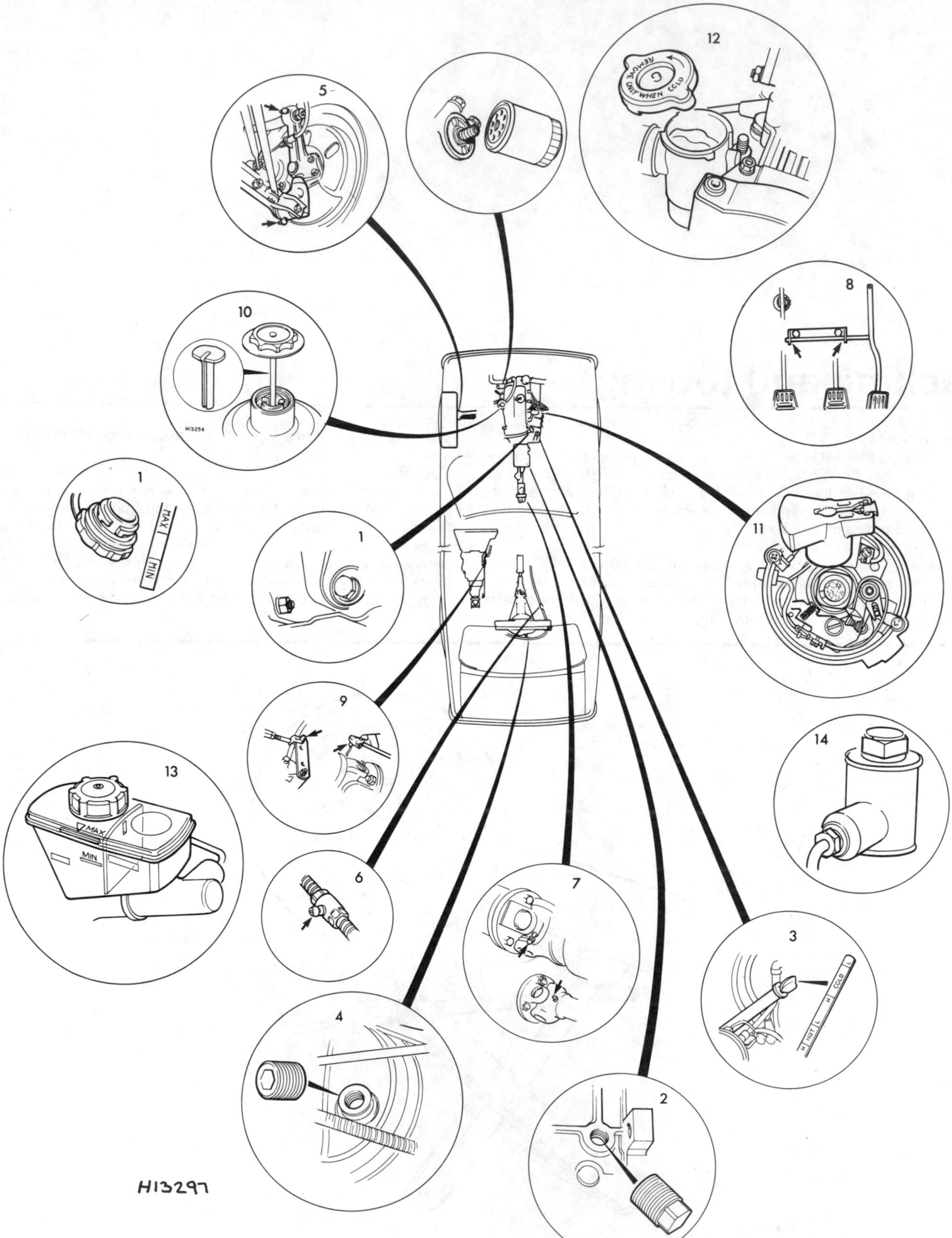

Lubrication chart

See page 11 for Recommended Lubricants and Fluids

Recommended lubricants and fluids

Component	Lubricant type or specification	Castrol product
1 Engine	Multigrade engine oil	**GTX**
2 Manual gearbox	Hypoid gear oil EP 90 (top-up) or EP 80 (refill)	**Hypoy B EP90 or Hypoy Light**
3 Automatic transmission	Automatic transmission fluid, BL type G	**TQF**
4 Rear axle	Hypoid gear oil, EP 90	**Hypoy B EP90**
5 Front suspension	Multi-purpose lithium-based grease	**LM Grease**
6 Handbrake cable	Multi-purpose lithium-based grease	**LM Grease**
7 Propeller shaft	Multi-purpose lithium-based grease	**LM Grease**
8 Pedal pivots	Engine oil	**GTX**
9 Selector linkage (automatic transmission)	Engine oil	**GTX**
10 Carburettor dashpot	Engine oil	**GTX**
11 Distributor	Engine oil/multi-purpose grease	**GTX/LM Grease**
12 Cooling system	Anti-freeze to BS 3151 or 3152	**Anti-freeze**
13 Brake hydraulic system	Hydraulic fluid to SAE J1703	**Universal Brake and Clutch Fluid**
14 Clutch hydraulic system	Hydraulic fluid to SAE J1703	**Universal Brake and Clutch Fluid**
Steering rack	Hypoid gear oil, EP 90	**Hypoy**
Front shock absorbers	Armstrong Super Shock Absorber Fluid	

Note: *The above are general recommendations only. Different operating territories and conditions require different lubricants. If in doubt, consult the operator's handbook supplied with the vehicle, or a BL dealer.*

Safety first!

Professional motor mechanics are trained in safe working procedures. However enthusiastic you may be about getting on with the job in hand, do take the time to ensure that your safety is not put at risk. A moment's lack of attention can result in an accident, as can failure to observe certain elementary precautions.

There will always be new ways of having accidents, and the following points do not pretend to be a comprehensive list of all dangers; they are intended rather to make you aware of the risks and to encourage a safety-conscious approach to all work you carry out on your vehicle.

Essential DOs and DON'Ts

DON'T rely on a single jack when working underneath the vehicle. Always use reliable additional means of support, such as axle stands, securely placed under a part of the vehicle that you know will not give way.

DON'T attempt to loosen or tighten high-torque nuts (e.g. wheel hub nuts) while the vehicle is on a jack; it may be pulled off.

DON'T start the engine without first ascertaining that the transmission is in neutral (or 'Park' where applicable) and the parking brake applied.

DON'T suddenly remove the filler cap from a hot cooling system – cover it with a cloth and release the pressure gradually first, or you may get scalded by escaping coolant.

DON'T attempt to drain oil until you are sure it has cooled sufficiently to avoid scalding you.

DON'T grasp any part of the engine, exhaust or catalytic converter without first ascertaining that it is sufficiently cool to avoid burning you.

DON'T allow brake fluid or antifreeze to contact vehicle paintwork.

DON'T syphon toxic liquids such as fuel, brake fluid or antifreeze by mouth, or allow them to remain on your skin.

DON'T inhale dust – it may be injurious to health (see *Asbestos* below).

DON'T allow any spilt oil or grease to remain on the floor – wipe it up straight away, before someone slips on it.

DON'T use ill-fitting spanners or other tools which may slip and cause injury.

DON'T attempt to lift a heavy component which may be beyond your capability – get assistance.

DON'T rush to finish a job, or take unverified short cuts.

DON'T allow children or animals in or around an unattended vehicle.

DO wear eye protection when using power tools such as drill, sander, bench grinder etc, and when working under the vehicle.

DO use a barrier cream on your hands prior to undertaking dirty jobs – it will protect your skin from infection as well as making the dirt easier to remove afterwards; but make sure your hands aren't left slippery. Note that long-term contact with used engine oil can be a health hazard.

DO keep loose clothing (cuffs, tie etc) and long hair well out of the way of moving mechanical parts.

DO remove rings, wristwatch etc, before working on the vehicle – especially the electrical system.

DO ensure that any lifting tackle used has a safe working load rating adequate for the job.

DO keep your work area tidy – it is only too easy to fall over articles left lying around.

DO get someone to check periodically that all is well, when working alone on the vehicle.

DO carry out work in a logical sequence and check that everything is correctly assembled and tightened afterwards.

DO remember that your vehicle's safety affects that of yourself and others. If in doubt on any point, get specialist advice.

IF, in spite of following these precautions, you are unfortunate enough to injure yourself, seek medical attention as soon as possible.

Asbestos

Certain friction, insulating, sealing, and other products – such as brake linings, brake bands, clutch linings, torque converters, gaskets, etc – contain asbestos. *Extreme care must be taken to avoid inhalation of dust from such products since it is hazardous to health.* If in doubt, assume that they *do* contain asbestos.

Fire

Remember at all times that petrol (gasoline) is highly flammable. Never smoke, or have any kind of naked flame around, when working on the vehicle. But the risk does not end there – a spark caused by an electrical short-circuit, by two metal surfaces contacting each other, by careless use of tools, or even by static electricity built up in your body under certain conditions, can ignite petrol vapour, which in a confined space is highly explosive.

Always disconnect the battery earth (ground) terminal before working on any part of the fuel or electrical system, and never risk spilling fuel on to a hot engine or exhaust.

It is recommended that a fire extinguisher of a type suitable for fuel and electrical fires is kept handy in the garage or workplace at all times. Never try to extinguish a fuel or electrical fire with water.

Note: *Any reference to a 'torch' appearing in this manual should always be taken to mean a hand-held battery-operated electric lamp or flashlight. It does NOT mean a welding/gas torch or blowlamp.*

Fumes

Certain fumes are highly toxic and can quickly cause unconsciousness and even death if inhaled to any extent. Petrol (gasoline) vapour comes into this category, as do the vapours from certain solvents such as trichloroethylene. Any draining or pouring of such volatile fluids should be done in a well ventilated area.

When using cleaning fluids and solvents, read the instructions carefully. Never use materials from unmarked containers – they may give off poisonous vapours.

Never run the engine of a motor vehicle in an enclosed space such as a garage. Exhaust fumes contain carbon monoxide which is extremely poisonous; if you need to run the engine, always do so in the open air or at least have the rear of the vehicle outside the workplace.

If you are fortunate enough to have the use of an inspection pit, never drain or pour petrol, and never run the engine, while the vehicle is standing over it; the fumes, being heavier than air, will concentrate in the pit with possibly lethal results.

The battery

Never cause a spark, or allow a naked light, near the vehicle's battery. It will normally be giving off a certain amount of hydrogen gas, which is highly explosive.

Always disconnect the battery earth (ground) terminal before working on the fuel or electrical systems.

If possible, loosen the filler plugs or cover when charging the battery from an external source. Do not charge at an excessive rate or the battery may burst.

Take care when topping up and when carrying the battery. The acid electrolyte, even when diluted, is very corrosive and should not be allowed to contact the eyes or skin.

If you ever need to prepare electrolyte yourself, always add the acid slowly to the water, and never the other way round. Protect against splashes by wearing rubber gloves and goggles.

When jump starting a car using a booster battery, for negative earth (ground) vehicles, connect the jump leads in the following sequence: First connect one jump lead between the positive (+) terminals of the two batteries. Then connect the other jump lead first to the negative (–) terminal of the booster battery, and then to a good earthing (ground) point on the vehicle to be started, at least 18 in (45 cm) from the battery if possible. Ensure that hands and jump leads are clear of any moving parts, and that the two vehicles do not touch. Disconnect the leads in the reverse order.

Mains electricity and electrical equipment

When using an electric power tool, inspection light etc, always ensure that the appliance is correctly connected to its plug and that, where necessary, it is properly earthed (grounded). Do not use such appliances in damp conditions and, again, beware of creating a spark or applying excessive heat in the vicinity of fuel or fuel vapour. Also ensure that the appliances meet the relevant national safety standards.

Ignition HT voltage

A severe electric shock can result from touching certain parts of the ignition system, such as the HT leads, when the engine is running or being cranked, particularly if components are damp or the insulation is defective. Where an electronic ignition system is fitted, the HT voltage is much higher and could prove fatal.

Routine maintenance

Maintenance is essential for ensuring safety and desirable for the purpose of getting the best in terms of performance and economy from your car. Over the years the need for periodic lubrication, oiling, greasing and so on, has been drastically reduced if not totally eliminated. This has unfortunately tended to lead some owners to think that because no such action is required, components either no longer exist, or will last forever. This is a serious delusion. It follows therefore that the largest initial element of maintenance is visual examination. This may lead to repairs or renewals.

The maintenance instructions are those recommended by the manufacturers. They are supplemented by additional maintenance tasks which, from practical experience, need to be carried out.

Every 250 miles (400 km) or weekly – whichever comes first

1 Remove the dipstick and check the engine oil level which should be up to the MAX mark. Top up the oil in the sump with the recommended oil. On no account allow the oil to fall below the MIN mark on the dipstick. The distance between MAX and MIN marks corresponds to approximately 1.5 pints (0.85 litre) (photo).

2 Check the battery electrolyte level and top up as necessary with distilled water. Make sure that the top of the battery is always kept clean and free of moisture. See Chapter 10 (photo).

3 Check the level of coolant in the translucent expansion tank. This should be maintained at the required level mark by adding coolant via the cap. If the expansion tank is empty, top up the cooling system via the filler cap on the thermostat housing as described in Chapter 2, Section 4, and check for leaks. **Note:** *If the engine is hot, place a rag over the expansion tank pressure cap and release the pressure cap slowly to avoid injury from escaping steam.*

4 Check the tyre pressures with an accurate gauge and adjust as necessary. Make sure that the tyre walls and treads are free of damage. Remember that the tyre tread should have a minimum of 1 millimetre depth across three quarters of the total width of the tread.

5 Refill the windscreen washer container. Add an anti-freezing solution satchet in cold weather to prevent freezing (do not use ordinary anti-freeze). Check that the jets operate correctly (photos).

6 Remove the wheel trims and check all wheel nuts for tightness but take care not to overtighten.

Every 6000 miles (10 000 km) or 6 months

Complete the service items in the weekly service check as applicable, plus:

1 Run the engine until it is hot and then place a container of 8 pints (4.55 litres) under the engine sump drain plug located on the right-hand side at the rear of the sump. Remove the drain plug and its copper sealing washer (photo). Allow the oil to drain out for 10 minutes. Whilst this is being done, unscrew the old oil filter cartridge located on the left-hand side of the engine and discard. Smear the rubber seal on a new cartridge with a little oil and refit it to the filter head. Screw it on and tighten hand tight only (photo). Check the drain plug copper sealing washer and if damaged, fit a new one. Refit the drain plug and sealing washer. Refill the engine with 7 pints (4.0 litres) of the recommended engine oil and clean off any oil which may have been spilt over the engine or its components. Run the engine and check the oil level.

Note: *The interval between oil changes should be reduced in very hot or dusty conditions or during cold weather with much slow or stop/start driving.*

2 Top up the carburettor damper with engine oil. Unscrew the cap, then carefully lift the piston and damper to the top of its travel, taking care not to dislodge the retainer. If the retainer comes out, remove the air cleaner, then hold up the piston and push the retainer into position. Top up the retainer recess with oil and push the damper down until the cap touches the top of the suction chamber. This topping-up procedure should be repeated until oil can just be seen at the bottom of the retainer recess with the piston down. Tighten the cap.

3 Check the carburettor adjustment as described in Chapter 3.

4 Carefully examine the cooling and heater systems for signs of leaks. Make sure that all hose clips are tight and that none of the hoses

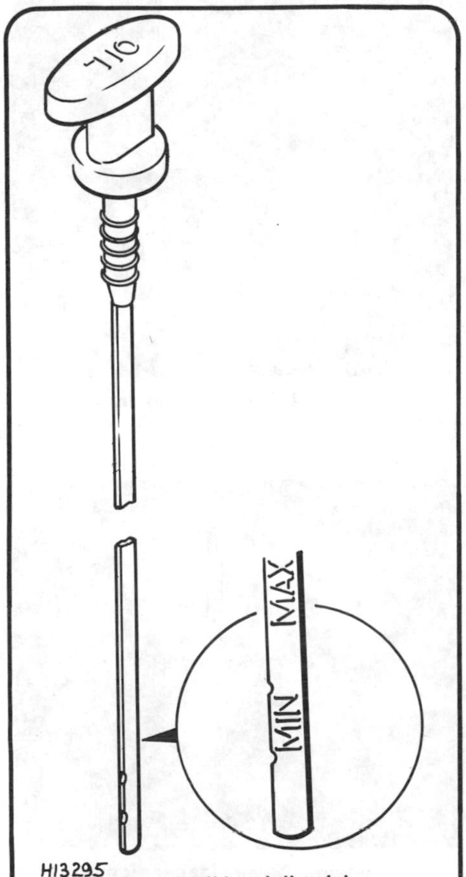

H13295 Engine oil level dipstick

Topping-up the engine oil level

Topping-up the battery with distilled water

Fill the windscreen washer reservoir

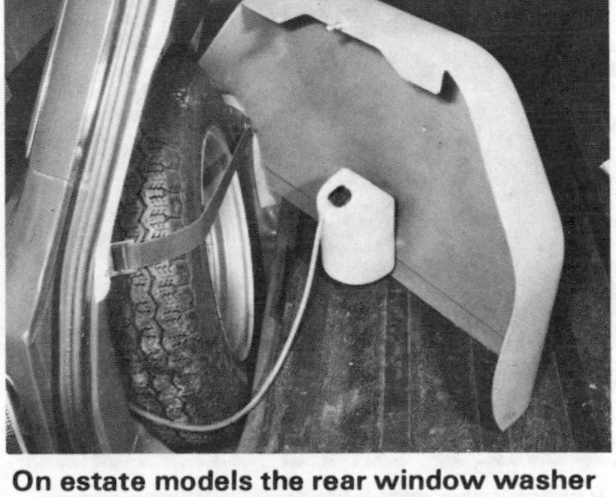
On estate models the rear window washer reservoir is mounted on the inside of the left-hand panel in the boot

The oil sump drain plug

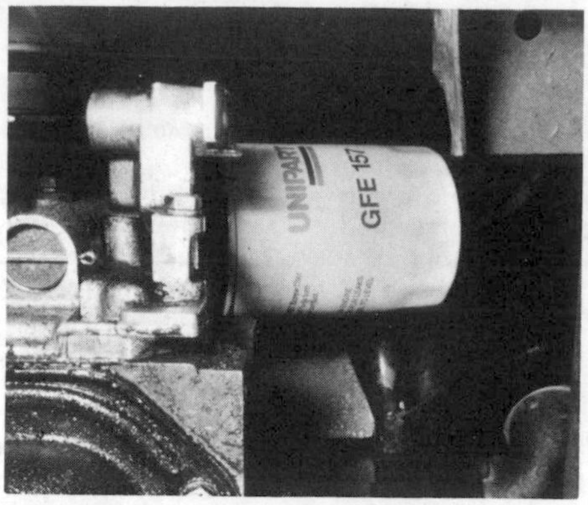

The oil filter is accessible from underneath the car

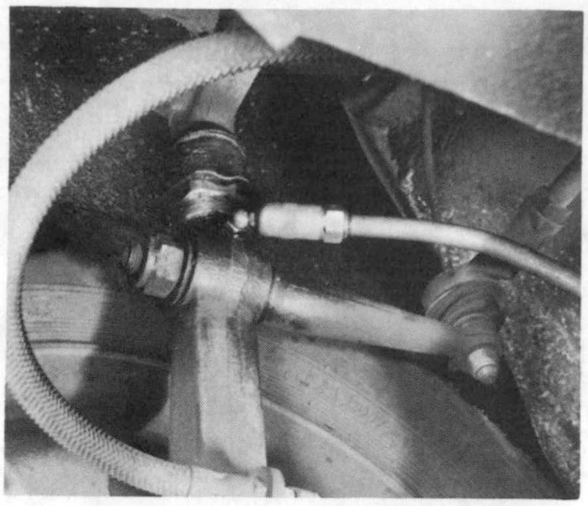

Grease the top nipple on the swivel pin ...

... and the bottom nipple

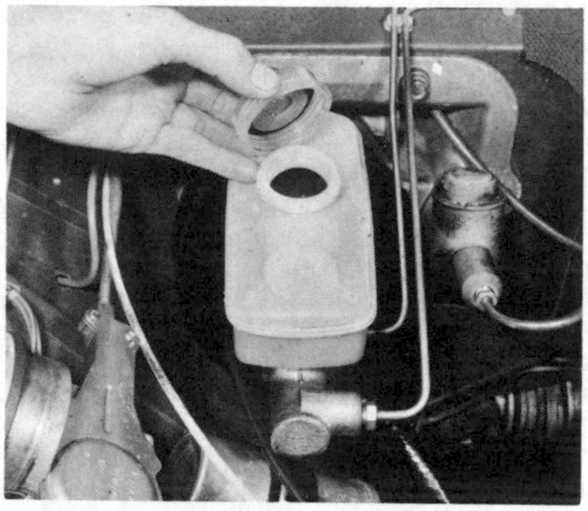

Top-up the brake fluid reservoir

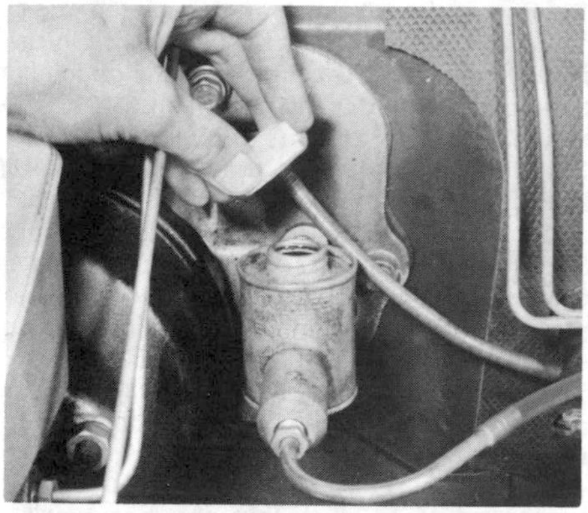

Check the level of fluid in the clutch fluid reservoir

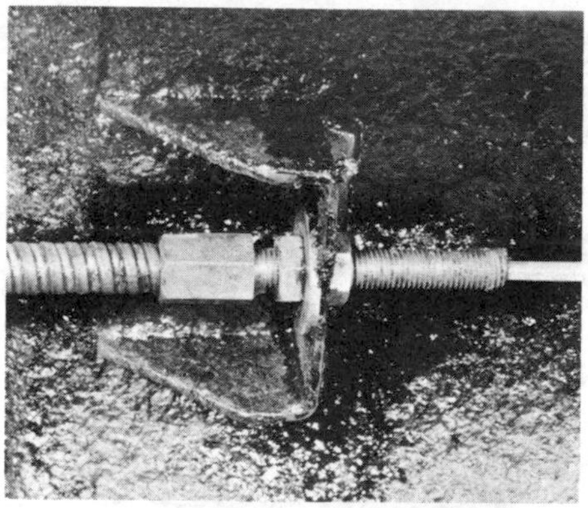

The handbrake cable adjuster

Lubricate the propeller shaft

Gearbox level/filler plug and drain plug

Remove the rear axle level/filler plug and check the oil level

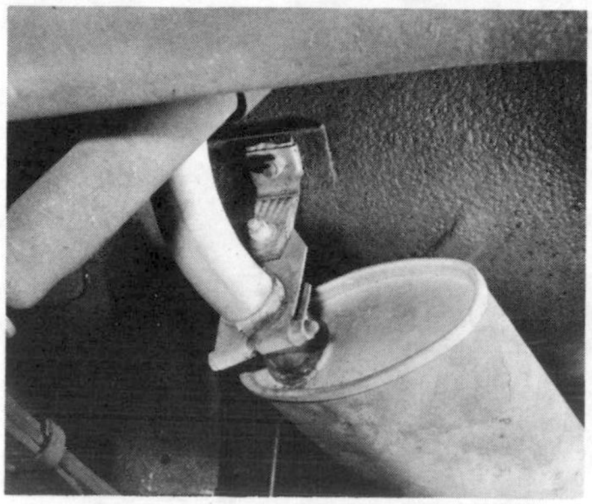

Check the security of the exhaust supports

Renewing the air cleaner element

have cracked or perished. Do not attempt to repair a leaking hose, always fit a new item. Generally inspect the exterior of the engine for signs of water leaks or stains. The method of repair will depend on its location. This check is particularly important before filling the cooling system with anti-freeze as it has a greater searching action than pure water and is bound to find any weak spots.
5 Examine the condition of the alternator/water pump drivebelt. Check the belt tension and adjust if necessary. Refer to Chapter 10, Section 9.
6 Lubricate the accelerator control linkage cable and pedal fulcrum with a little engine oil.
7 Inspect the steering rack rubber boots for signs of leakage or deterioration which, if evident, must be rectified as described in Chapter 11.
8 Lubricate the top and bottom nipples on each of the front swivel joints with several strokes of a grease gun filled with multi-purpose lithium grease (photos).
9 Inspect all steering balljoints for signs of wear, leaking rubber boots, and securing nuts for tightness. If a balljoint rubber boot has failed, the whole assembly must be renewed. Refer to Chapter 11.
10 Check the front wheel alignment. For this, special equipment is necessary, therefore this job should be left to the local BL garage. See Chapter 11.
11 Wipe the top of the brake and clutch master cylinder and unscrew the caps. Check the level of hydraulic fluid in the reservoirs and top up, if necessary, to the marks on the exterior of the reservoir with the recommended fluid. Make sure the cap breather vent is clean and then refit the cap (photos). Take care not to spill any hydraulic fluid on the paintwork as it acts as a solvent.
12 Check the adjustment of the handbrake and footbrake. If travel is excessive, refer to Chapter 9 and check the footbrake adjustment and then the handbrake if its travel is still excessive (photo).
13 Refer to Chapter 9 and inspect the brake linings and pads for wear and the front discs and rear drums for scoring.
14 Carefully examine all brake hydraulic pipes and unions for signs of leakage. Check flexible hoses are not in contact with any body or mechanical component when the steering is turned through both locks.
15 Lubricate all moving parts of the handbrake system with engine oil.
16 Lubricate the propellor shaft front universal joint with 3 to 4 strokes of a grease gun filled with multi-purpose lithium grease (photo).
17 Refer to Chapter 4 and remove the spark plugs. Clean, adjust if necessary, and refit.
18 Refer to Chapter 4 and clean and adjust the distributor contact breaker points.
19 Lift off the rotor arm and lightly grease the cam surface. Apply a few drops of thin oil to the pad in the top of the cam spindle. Apply a few drops of oil through the hole in the contact breaker base plate to lubricate the advance mechanism.
20 Refer to Chapter 4 and check the ignition timing. Adjust if necessary.
21 Inspect the shock absorbers for signs of leakage.
22 Wipe the area around the gearbox level/filler plug. Unscrew the plug and check the level of oil which should be up to the bottom of the threads. Top up if necessary using the recommended oil and refit the plug. Wipe away any spilled oil (photo).
23 Wipe the area around the rear axle level/filler plug. Unscrew the plug and check the level of oil which should be up to the bottom of the threads. Top up if necessary using the recommended oil and refit the plug. Wipe away any spilled oil (photo).
24 *Automatic transmission:* With the vehicle standing on level ground, apply the handbrake and move the selector to the P position. Start the engine and allow to run at idle speed for a minimum of 2 minutes. With the engine still running, withdraw the dipstick from the filler tube to be found at the rear of the engine compartment. Wipe the dipstick and quickly fit and withdraw the dipstick again. Check the level of oil and top up if necessary with the recommended oil. Take great care not to overfill.
25 Generally check the operation of all lights and electrical equipment. Renew any blown bulbs with bulbs of the same wattage rating and rectify any electrical equipment fault. See Chapter 10.
26 Check the battery electrolyte specific gravity as described in Chapter 10. Clean the battery terminals and smear them with vaseline (petroleum jelly) to prevent corrosion.
27 Check the alignment of the headlights and adjust if necessary. See Chapter 10.
28 Check the condition of the windscreen wiper blades and fit new items if the blade end has frayed, softened or perished. They should be renewed every 12 months.
29 Check the exhaust system for damage and signs of leakage. If badly corroded, the system must be renewed. Check all exhaust mountings for security (photo).
30 Carefully examine all clutch and fuel lines and unions for signs of leakage, and flexible hoses for signs of perishing. Check the tightness of all unions and renew any faulty lines or hoses.
31 Lubricate all door, bonnet and boot lid locks, hinges and controls with light oil.
32 Inspect the seat belts for damage to the webbing. Check that all seat and seat belt mountings are tight.
33 Make sure that the rear view mirror is firm in its mounting and is not crazed or cracked.

Every 12 000 miles (20 000 km) or 12 months

Complete the service items in the 6000 mile service check as applicable plus:
1 To fit a new air cleaner element, unscrew the wing nut and lift away the cover, body and element which should be discarded. Wipe out the container and fit a new element (photo). Refit the cover. Make sure that the sealing ring between the air cleaner body and carburettor is not damaged or perished. Refit the air cleaner to the carburettor and secure with the fibre washer and nut. Disconnect the intake ducts from the air temperature control valve. Press the valve plate and check that it returns to its original position when released. Refit the intake ducts.
2 Fit new spark plugs.
3 Inspect the ignition HT leads for cracks or deterioration. Renew as necessary.
4 On early models renew the oil filler cap and filter assembly on the top of the camshaft cover. On later models, equipped with an oil filler tube bolted to the side of the cylinder block, the cap on the camshaft cover contains a wire mesh filter. This filter does not require renewal but must be thoroughly washed in paraffin and shaken dry.
5 Refer to Chapter 11 and adjust the front wheel bearing endfloat.
6 Balance the roadwheels to eliminate any vibration in the steering. This must be done on specialist equipment.

Every 24 000 miles (40 000 km) or 2 years

Complete the service items in the 6000 and 12 000 mile service check as applicable plus:

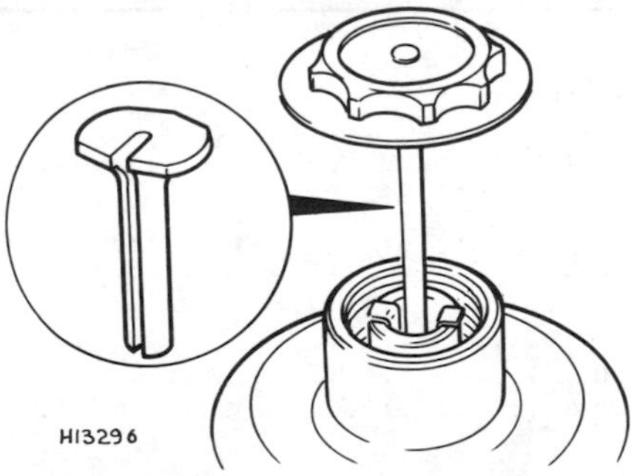

Topping-up the carburettor damper

1 Remove the timing belt cover and check the condition and tension of the timing belt. Refer to Chapter 1, Section 11. Adjust or renew as necessary.
2 Fit new rubber grommets on the timing belt cover.
3 Completely drain the brake hydraulic system, renew all system seals and refill with clean, fresh fluid. Refer to Chapter 9.

Every 48 000 miles (80 000 km) or 4 years

1 Refer to Chapter 1, Section 11 and renew the timing belt.
2 Refer to Chapter 11, Section 23 and renew the brake servo unit air filter.

Chapter 1 Engine

Contents

Specifications

Engine (general)

Type	17V, overhead camshaft
Number of cylinders	4
Bore	3·325 in (84·45 mm)
Stroke	2·984 in (75·8 mm)
Capacity	103·73 cu in (1700 cc)
Firing order	1 – 3 – 4 – 2
Valve operation	Overhead camshaft
Compression ratio	9·0 : 1
Torque at 3700 rpm	99 lbf ft (13·69 kgf m)

Crankshaft

Main journal diameter	2·1262 to 2·1270 in (54·005 to 54·026 mm)
Crankpin journal diameter	1·8754 to 1·8759 in (47·635 to 47·647 mm)
Crankshaft endthrust	Taken on thrust washers at centre bearing
Crankshaft endfloat	0·001 to 0·005 in (0·025 to 0·14 mm)

Main bearings

Number and type	5, steel-backed thin wall type
Width:	
Front, centre and rear	1·125 in (28·57 mm)
Intermediate	0·760 in (19·3 mm)
Diametrical clearance	0·001 to 0·003 in (0·025 to 0·077 mm)
Undersizes	0·010 in (0·254 mm)
	0·020 in (0·508 mm)
	0·030 in (0·762 mm)
	0·040 in (1·016 mm)

Connecting rods

Type	Horizontal, split big-end, plain small-end offset
Length between centres	5·86 in (149 mm)

Big-end bearings

Type	Steel-backed thin wall
Width	0·775 to 0·785 in (19·68 to 19·94 mm)
Diametrical clearance	0·0015 to 0·0032 in (0·038 to 0·081 mm)
Undersizes	0·010 in (0·254 mm)
	0·020 in (0·508 mm)
	0·030 in (0·762 mm)
	0·040 in (1·016 mm)

Gudgeon pin

Type	Press in connecting rod
Fit in piston	Hand push at 16°C (60°F)
Diameter	0·8125 to 0·8127 in (20·638 to 20·634 mm)

Pistons

Type	Duotherm, solid skirt with combustion chamber in crown
Clearance in cylinder:	
Below oil control groove	0·0008 to 0·0051 in (0·02 to 0·13 mm)
Bottom of skirt	0·0004 to 0·0023 in (0·01 to 0·06 mm)
Number of rings	3 (2 compression, 1 oil control)
Width of ring grooves:	
Top and second	0·070 to 0·071 in (1·78 to 1·80 mm)
Oil control	0·157 to 0·158 in (4·00 to 4·02 mm)
Gudgeon pin bore	0·8128 to 0·8130 in (20·646 to 20·651 mm)
Offset from centre	0·059 in (1·5 mm)

Piston rings

Compression rings	
Type:	
Top	Plain, chrome faced
Second	Stepped scraper
Ring to groove clearance	0·0015 to 0·0027 in (0·04 to 0·07 mm)
Fitted gap	0·012 to 0·020 in (0·3 to 0·5 mm)
Oil control ring	
Type	Two chrome faced rings with butted expander
Fitted gap	0·015 to 0·055 in (0·33 to 1·4 mm)

Camshaft

Journal diameters	1·888 to 1·889 in (47·96 to 47·97 mm)
Number of bearings	3
Bearing type	Direct in cylinder head and cover
Diametrical clearance	0·0017 to 0·0037 in (0·043 to 0·094 mm)
Endthrust	Taken on rear cover
Endfloat	0·003 to 0·007 in (0·07 to 0·18 mm)
Drive	Toothed belt from crankshaft sprocket

Tappets

Type	Bucket with flat base
Outside diameter	1·2491 to 1·2498 in (31·729 to 31·745 mm)
Adjustment	Selective shim

Valves

Face angle	45° 30′
Seat angle	45°
Head diameter:	
Inlet	1·575 in (40 mm)
Exhaust	1·339 in (34 mm)
Stem diameter:	
Inlet	0·2917 to 0·2921 in (7·41 to 7·42 mm)
Exhaust	0·2909 to 0·2917 in (7·39 to 7·41 mm)
Stem to guide clearance:	
Inlet	0·001 to 0·002 in (0·027 to 0·053 mm)
Exhaust	0·0015 to 0·0028 in (0·04 to 0·073 mm)
Cam lift	0·375 in (9·525 mm)

Valve guides

Length	1·532 in (38·90 mm)
Outside diameter	0·474 to 0·475 in (12·04 to 12·06 mm)
Inside diameter	0·293 to 0·2937 in (7·45 to 7·46 mm)
Fitted height above head	0·394 in (10 mm)
Interference fit in head	0·0015 to 0·003 in (0·04 to 0·09 mm)

Valve springs

Free length	1·646 in (41·81 mm)
Fitted length	1·375 in (34·92 mm)
Load at fitted length	44·5 lbf (20 kgf)
Number of working coils	4·5

Valve timing

Inlet valve:	
Opens	15° BTDC
Closes	45° ABDC
Exhaust valve:	
Opens	50° BBDC
Closes	10° ATDC
Tappet clearance (standard)	0·012 in (0·30 mm)

Lubrication

System type	Wet sump, pressure fed
Pressure:	
Running	60 to 90 lbf/in^2 (4·2 to 6·3 kgf/cm^2)
Idling	40 to 60 lbf/in^2 (2·8 to 4·2 kgf/cm^2)
Oil pump type	Eccentric rotors, mounted around crankshaft
Relief valve spring:	
Free length	1·525 in (38·7 mm)
Fitted length	0·960 in (24·4 mm)
Load at fitted length	17·4 lbf (7·9 kgf)
Oil filter type	Full flow, disposable cartridge
Bypass valve opens	8 to 12 lbf/in^2 (0·6 to 0·7 kgf/cm^2)
Sump capacity	7 pints (4 litres)

Torque wrench settings

	lbf ft	kgf m
Main bearing cap bolts	75	10·4
Big-end bearing cup nuts	33	4·5
Backplate:		
8 mm bolts	22	3·0
10 mm bolts	37	5·1
Flywheel to crankshaft bolts	42	5·8
Oil pump cover bolts	2	0·3
Oil pump securing bolts	8	1·1
Oil pressure switch	9	1·2
Camshaft sprocket bolt	48	6·6
Crankshaft pulley bolt	62	8·5
Camshaft cover bolts	9	1·2
Cylinder head bolts:		
Pre-tighten	35	4·8
Fully tighten	60	8·3
Manifold to cylinder head bolts	18	2·5
Oil separator to block bolts	22	3·0
Water pump to block	8	1·1
Thermostat housing to cylinder head	8	1·1
Coolant temperature switch	5	0·7
Sump drain plug	27	3·8
Mounting bracket to engine bolts	38	5·2
Mounting bracket to body bolts	26 to 28	3·6 to 3·9
Mounting rubber to bracket nuts	38	5·2
Rear mounting rubber to crossmember nuts	20 to 22	2·7 to 3·0
Restrictor plate to mounting rubber	28 to 35	3·9 to 4·8
Rear crossmember to body	20 to 22	2·7 to 3·0
Exhaust pipe to manifold	21 to 23	2·9 to 3·2

1 General description

The 1700 cc engine is a four cylinder, overhead camshaft type fitted with an SU carburettor.

The valves are mounted vertically in the die-cast alloy cylinder head and run in pressed-in valve guides. The cylinder head acts as a support for the overhead camshaft which is retained in position by the camshaft cover. The camshaft, which is driven by a toothed composite rubber drivebelt from a sprocket on the front end of the crankshaft, operates directly on the bucket type valve tappets. A spring-loaded belt tensioner is fitted to eliminate backlash and prevent slackness of the belt.

The combined crankcase and cylinder block is made of cast iron and houses the pistons and crankshaft. Attached to the underside of the crankcase is a pressed steel sump which acts as a reservoir for the engine oil.

The aluminium pistons have concave tops which act as combustion chambers. The pistons are connected to the forged steel connecting rods by gudgeon pins. Two compression rings and one scraper ring, all located above the gudgeon pin, are fitted.

The crankshaft is supported by five renewable main bearings. Crankshaft endfloat is controlled by four semi-circular thrust washers located on each side of the centre main bearing.

The centrifugal water pump and alternator are driven by a belt from the pulley on the front end of the crankshaft. The distributor is mounted on the right-hand side of the camshaft cover and is driven by a sprocket on the camshaft. The mechanical fuel pump is mounted on the left-hand side of the camshaft cover and is operated by a separate cam on the camshaft. The oil pump is mounted on the front of the crankcase and is driven directly by a key on the crankshaft.

Attached to the rear of the crankshaft by six bolts is the flywheel to which is bolted the diaphragm spring clutch. Mounted on the circumference of the flywheel is the starter ring gear into which the

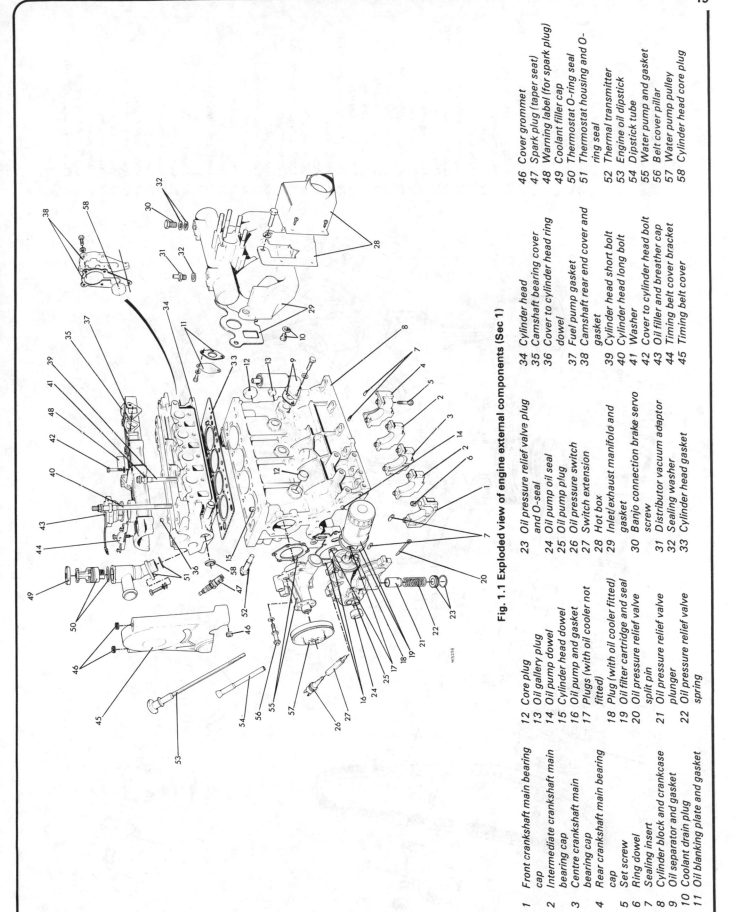

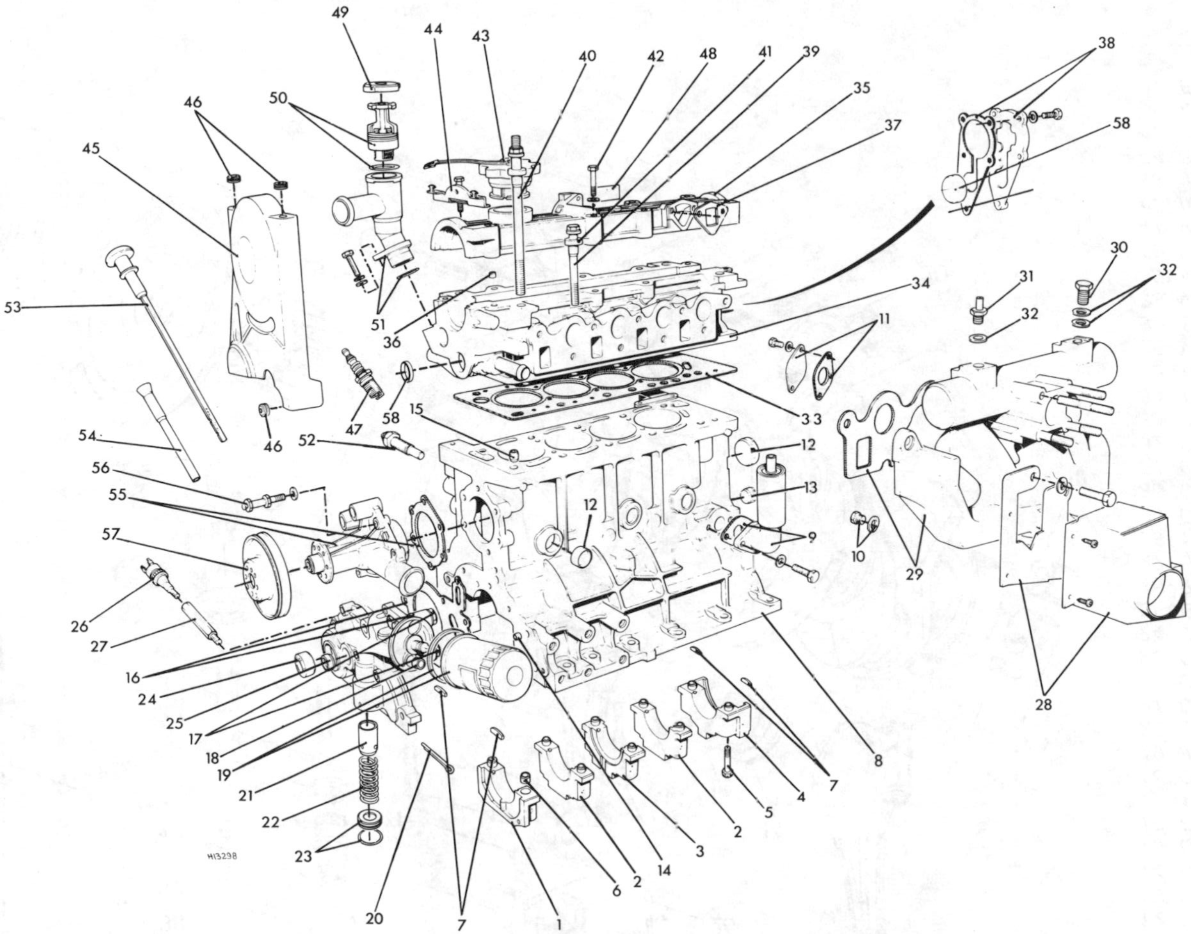

Fig. 1.1 Exploded view of engine external components (Sec 1)

1 Front crankshaft main bearing cap
2 Intermediate crankshaft main bearing cap
3 Centre crankshaft main bearing cap
4 Rear crankshaft main bearing cap
5 Set screw
6 Ring dowel
7 Sealing insert
8 Cylinder block and crankcase
9 Oil separator and gasket
10 Coolant drain plug
11 Oil blanking plate and gasket
12 Core plug
13 Oil gallery plug
14 Oil pump dowel
15 Cylinder head dowel
16 Oil pump and gasket
17 Plugs (with oil cooler not fitted)
18 Plug (with oil cooler fitted)
19 Oil filter cartridge and seal
20 Oil pressure relief valve split pin
21 Oil pressure relief valve plunger
22 Oil pressure relief valve spring
23 Oil pressure relief valve plug and O-seal
24 Oil pump oil seal
25 Oil pump plug
26 Oil pressure switch
27 Switch extension
28 Hot box
29 Inlet/exhaust manifold and gasket
30 Banjo connection brake servo screw
31 Distributor vacuum adaptor
32 Sealing washer
33 Cylinder head gasket
34 Cylinder head
35 Camshaft bearing cover
36 Cover to cylinder head ring dowel
37 Fuel pump gasket
38 Camshaft rear end cover and gasket
39 Cylinder head short bolt
40 Cylinder head long bolt
41 Washer
42 Cover to cylinder head bolt
43 Oil filler and breather cap
44 Timing belt cover bracket
45 Timing belt cover
46 Cover grommet
47 Spark plug (taper seat)
48 Warning label (for spark plug)
49 Coolant filler cap
50 Thermostat O-ring seal
51 Thermostat housing and O-ring seal
52 Thermal transmitter
53 Engine oil dipstick
54 Dipstick tube
55 Water pump and gasket
56 Belt cover pillar
57 Water pump pulley
58 Cylinder head core plug

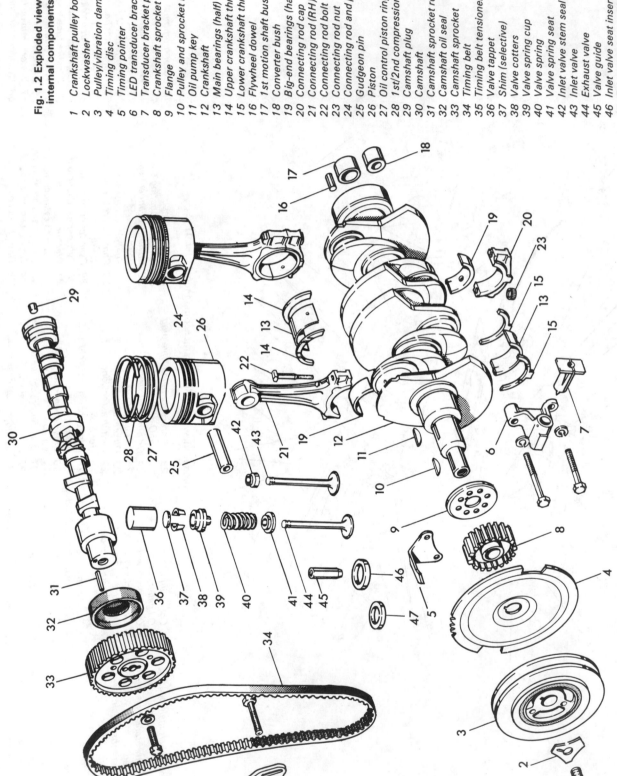

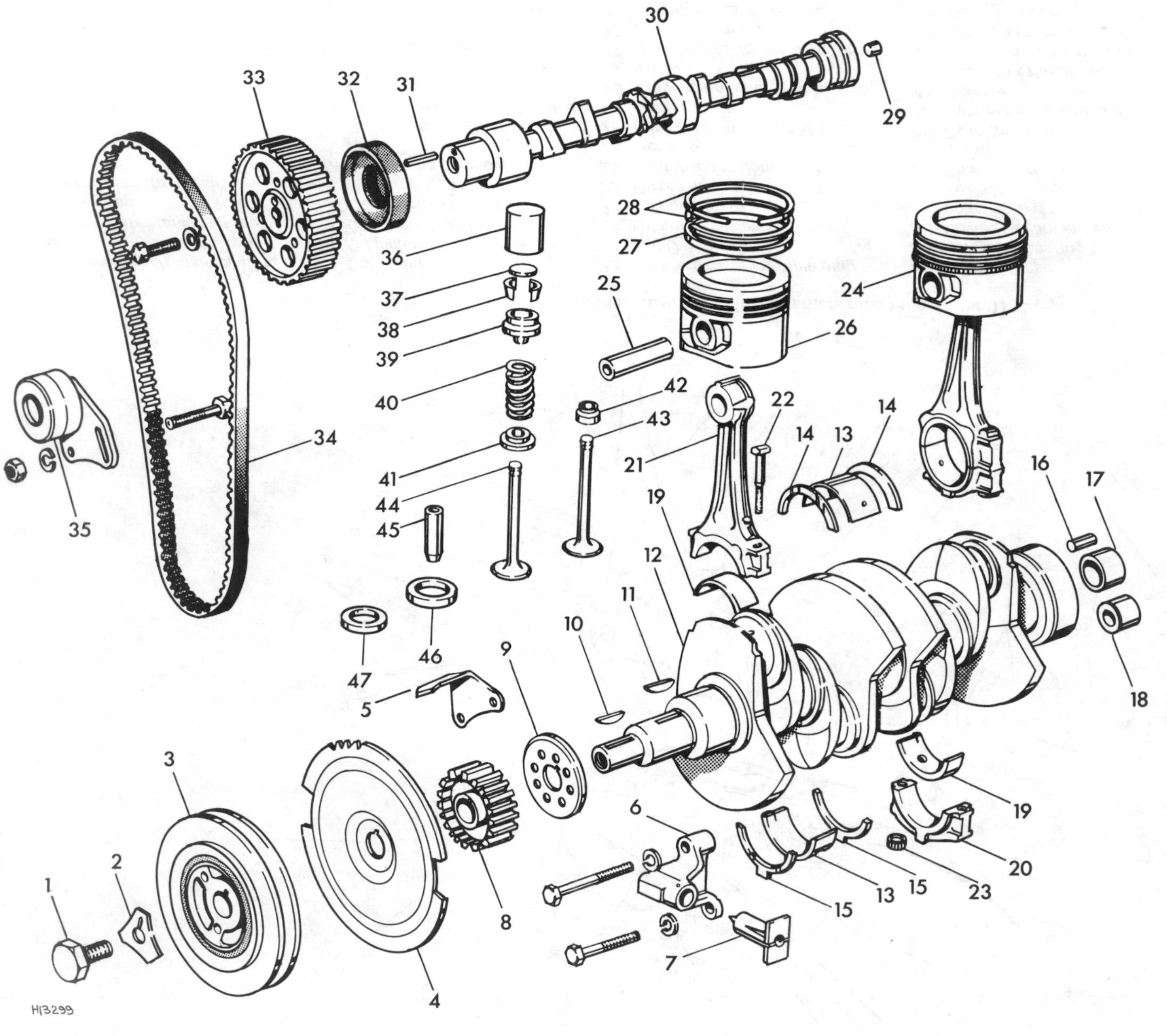

Fig. 1.2 Exploded view of engine internal components (Sec 1)

1 *Crankshaft pulley bolt*
2 *Lockwasher*
3 *Pulley/vibration damper*
4 *Timing disc*
5 *Timing pointer*
6 *LED transducer bracket*
7 *Transducer bracket plug*
8 *Crankshaft sprocket*
9 *Flange*
10 *Pulley and sprocket key*
11 *Oil pump key*
12 *Crankshaft*
13 *Main bearings (half)*
14 *Upper crankshaft thrustwashers*
15 *Lower crankshaft thrustwashers*
16 *Flywheel dowel*
17 *1st motion shaft bush*
18 *Converter bush*
19 *Big-end bearings (half)*
20 *Connecting rod cap*
21 *Connecting rod (RH)*
22 *Connecting rod bolt*
23 *Connecting rod nut*
24 *Connecting rod and piston (LH)*
25 *Gudgeon pin*
26 *Piston*
27 *Oil control piston rings*
28 *1st/2nd compression piston rings*
29 *Camshaft plug*
30 *Camshaft*
31 *Camshaft sprocket roll pin*
32 *Camshaft oil seal*
33 *Camshaft sprocket*
34 *Timing belt*
35 *Timing belt tensioner*
36 *Valve tappet*
37 *Shim (selective)*
38 *Valve cotters*
39 *Valve spring cup*
40 *Valve spring*
41 *Valve spring seat*
42 *Inlet valve stem seal*
43 *Inlet valve*
44 *Exhaust valve*
45 *Valve guide*
46 *Inlet valve seat inserts*
47 *Exhaust valve seat inserts*

starter motor drive gear engages when starting the engine.

The intake/exhaust manifold is mounted on the left-hand side of the cylinder head.

2 Major operations possible with engine in car

The following major operations can be carried out on the engine with it in place in the car:

1 Removal and refitting of camshaft
2 Removal and refitting of cylinder head
3 Removal and refitting of timing belt
4 Removal and refitting of oil pump
5 Removal and refitting of oil sump
6 Removal and refitting of pistons and connecting rods and big-end bearings
7 Removal and refitting of engine mountings
8 Removal and refitting of flywheel
9 Removal and refitting of front and rear crankshaft oil seals

3 Major operations requiring engine removal

The following major operations must be carried out with the engine out of the car and on a bench or the floor:

1 Removal and refitting of the main bearings
2 Removal and refitting of the crankshaft

4 Valve tappet clearances – checking and adjusting

To retain the camshaft in position during the checking operation, BL special tools 18G 1301 and 18G 1302 will be required. If these tools are not available it is possible to make up suitable alternatives as described below and shown in the photos. Read through the entire Section first to familiarize yourself with the procedure and then obtain or make up the special tools before proceeding.

1 Unscrew the distributor cap securing screws, lift off the cap and place it to one side (photo).

2 Mark the relative position of the distributor mounting flange to the camshaft cover as a guide to refitting. Disconnect the vacuum pipe from the vacuum advance unit, and the LT lead at the wiring connector.

3 Undo and remove the two distributor securing nuts and washers and withdraw the distributor (photo).

4 Disconnect the two fuel pipes from the fuel pump (photo). Plug the pipe ends to prevent loss of fuel and dirt ingress.

5 Turn the crankshaft to the 90° BTDC position. This is achieved when the dimple on the rear face of the camshaft sprocket is opposite the pointer on the upper surface of the camshaft cover (photo). At the same time the single notch in the crankshaft timing disc is opposite the timing pointer. The crankshaft can be turned using a socket on the crankshaft pulley bolt or, on cars with manual transmission only, by engaging top gear and moving the car forward. Removal of the spark plugs will enable the crankshaft to be turned with greater ease.

6 Lift away the timing belt cover (photo).

7 Undo and remove the nut securing the brake servo vacuum pipe retaining clip to the cylinder head. Lift off the bracket and move it to one side. Undo and remove the vacuum pipe banjo union on the inlet manifold, taking care not to lose the two washers. Lift the retaining clip off the cylinder head stud and position the vacuum pipe clear of the engine.

8 If BL special tools 18G 1301 and 18G 1302 are not available, make up three camshaft retaining brackets and a rear support plate to the dimensions shown in Fig. 1.3 and 1.4 (photos), using suitable scrap metal and angle iron.

9 Undo and remove the five bolts securing the camshaft rear end cover and lift off the cover and gasket (photo). Fit the home-made rear support plate (or tool 18G 1302) in position and secure with two of the rear end cover retaining bolts. Make sure that the rounded bolt head of the home-made tool, or the tapered projection of the BL tool, fits squarely into the recess at the rear of the camshaft (photos).

10 Slacken evenly and in a side-to-side sequence each of the nine

4.1 Undo and remove the screws and lift off the distributor cap

4.3 Unscrew the two nuts and washers and withdraw the distributor

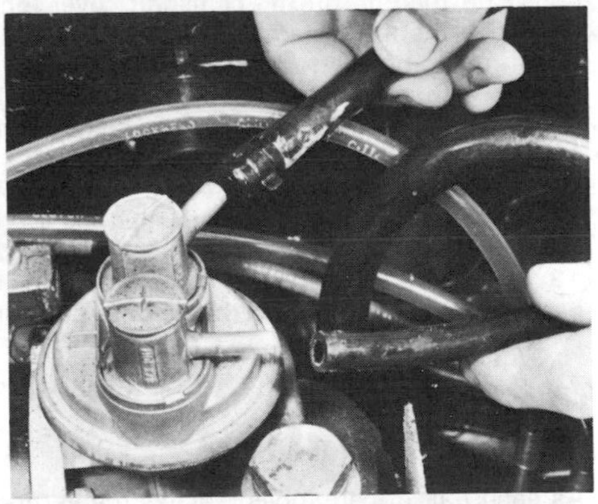

4.4 Disconnect the fuel pipes from the pump

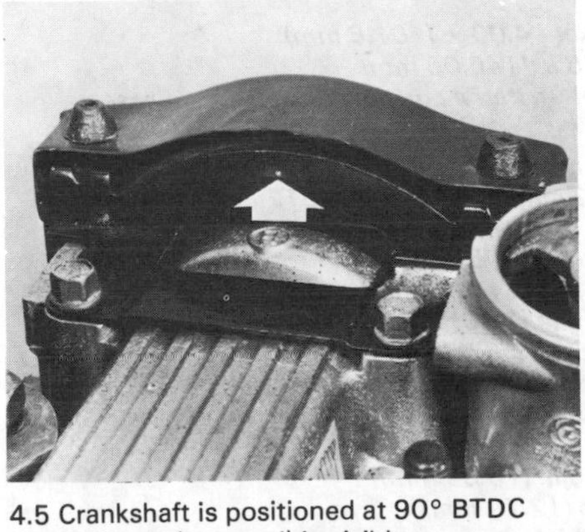

4.5 Crankshaft is positioned at 90° BTDC when dimple (arrowed) is visible

4.6 Lift off the timing belt cover

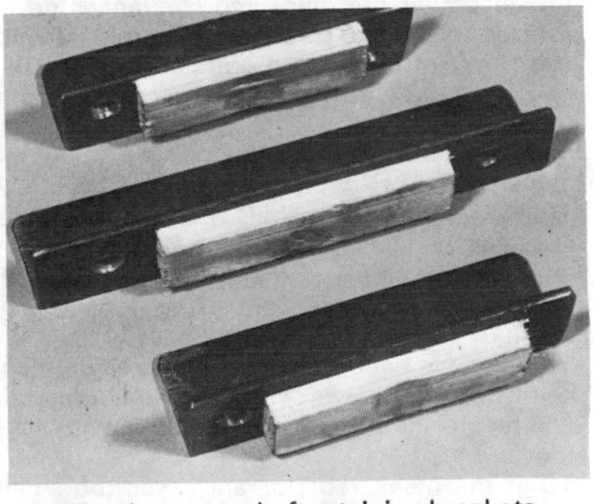

4.8a The three camshaft retaining brackets...

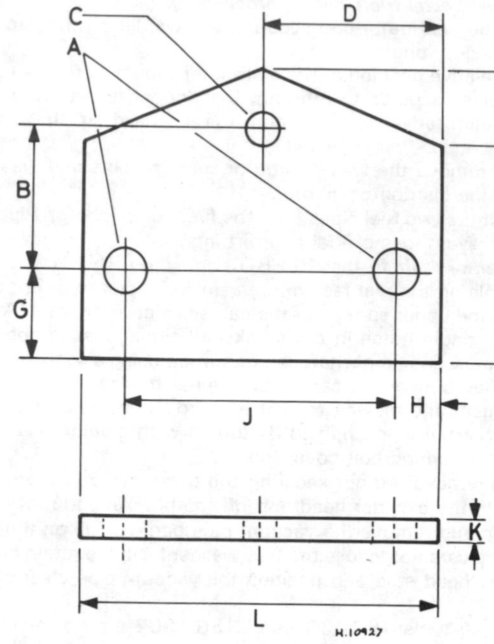

Fig. 1.3 Camshaft rear support plate dimensions (Sec 4)

A = 0.31 in (7.93 mm)	B = 1.00 in (25.4 mm)
C = 0.25 in (6.35 mm) diameter clearance for $\frac{1}{4}$ UNF nut and bolt	
D = 1.25 in (31.75 mm)	H = 0.31 in (7.93 mm)
E = 2.00 in (50.80 mm)	J = 1.87 in (47.62 mm)
F = 1.50 in (38.10 mm)	K = 0.12 in (3.17 mm)
G = 0.62 in (15.87 mm)	L = 2.50 in (63.50 mm)

bolts securing the camshaft cover to the cylinder head. Remove the bolts and then lift off the camshaft cover (photo).

11 Place the three clamps of BL special tool 18G 1302, or the three home-made brackets, in position over the camshaft bearings and secure with the camshaft cover retaining bolts (photo). When in position the clamps or brackets should be exerting a light pressure only on the camshaft bearing journals, sufficient to hold the camshaft firmly in place while still allowing it to be rotated.

12 Release the timing belt tensioner and slip the belt off the camshaft gear (photos).

13 Rotate the camshaft in the normal direction of rotation to bring each pair of cam lobes to the vertical position, ie:

1 and 4
2 and 5
6 and 7
3 and 8

As each pair of cam lobes reaches the vertical position, measure and record the clearance between the heel of the cam lobe and the tappet using feeler gauges (photo).

14 The standard valve tappet clearance is given in the Specifications. Provided that the measured clearances do not exceed this figure, or are not less than 0.008 in (0.20 mm), adjustment is not necessary and the camshaft cover and components can be refitted as described in paragraphs 22 to 30. If however, adjustment is required, proceed as follows.

15 Undo and remove the bolt securing the camshaft gear to the camshaft and withdraw the gear (photo).

16 Remove the camshaft rear support plate and then progressively slacken the camshaft retaining bracket bolts. When the spring tension is released, remove the bolts and lift away the brackets.

17 Lift the camshaft out of its location in the cylinder head and carefully remove the front oil seal.

18 Remove each tappet requiring adjustment in turn, and extract the shim from the valve spring cup (or from inside the tappet). Measure and record the thickness of the shim with a micrometer. Do not mix up the tappets.

19 If the measured clearance was too small, then a smaller shim is required. If the clearance was too large then a larger shim is required, ie:

A + B − C = Shim thickness required
Where: A = Measured clearance
B = Thickness of existing shim
C = Standard tappet clearance

For example:

0.006 in (0.15 mm) + 0.105 in (2.67 mm) − 0.012 in (0.30 in) = 0.099 in (2.52 mm)

Shims are available in the following sizes:

in	mm
0.091	2.32
0.093	2.37
0.095	2.42
0.097	2.47
0.099	2.52
0.101	2.56

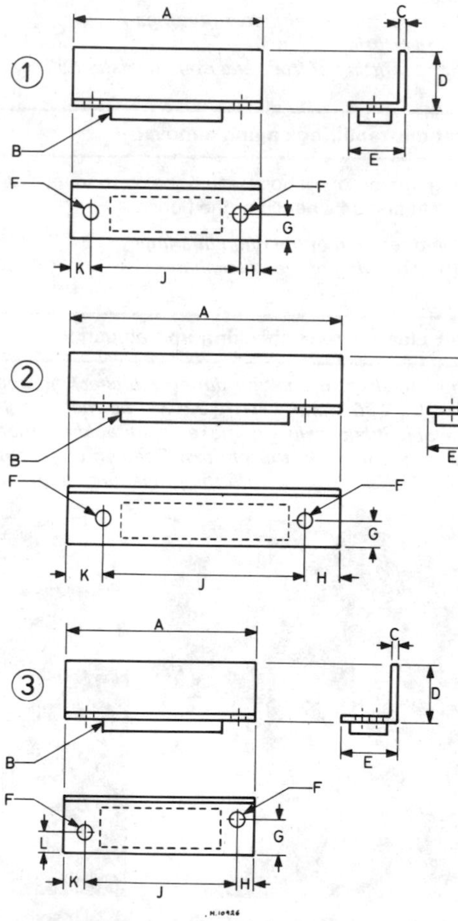

Fig. 1.4 Camshaft retaining bracket dimensions (Sec 4)

1 Front bracket 2 Centre bracket 3 Rear bracket

A Front and rear bracket - 4.00 in (101.6 mm)
Centre bracket - 5.75 in (146.05 mm)
B Plywood or other soft material positioned centrally on base of bracket
C = 0.125 in (3.175 mm) E = 1.187 in (30.162 mm)
D = 1.187 in (30.162 mm) F = 0.312 in (7.937 mm)
G Front and centre brackets - 0.5 in (12.7 mm)
Rear bracket - 0.75 in (19.05 mm)
H Front bracket - 0.437 in (11.112 mm)
Centre bracket - 0.75 in (19.05 mm)
Rear bracket - 0.375 in (9.525 mm)
J Front and rear bracket - 3.125 in (79.37 mm)
Centre bracket - 4.25 in (107.95 mm)
K Front bracket - 0.437 in (11.112 mm)
Centre bracket - 0.75 in (19.05 mm)
Rear bracket - 0.5 in (12.70 mm)
L = 0.437 in (11.112 in)

4.8b ...and rear support plate. These can easily be made from scrap materials

4.9a Remove the camshaft rear end cover...

4.9b ...and fit the rear support plate

4.9c Rear support plate in position (engine removed for clarity)

4.10 Remove the camshaft cover...

4.11 ...and retain the camshaft using the three brackets

4.12a Release the timing belt tensioner...

4.12b ...and slip off the timing belt

4.13 Using feeler gauges to measure the valve tappet clearance

4.15 Remove the camshaft gear retaining bolt and the gear

4.21 Liberally lubricate the camshaft journals when refitting

4.23 Tighten the camshaft cover retaining bolts to the specified torque

0.103	*2.62*
0.105	*2.67*
0.107	*2.72*
0.109	*2.77*
0.111	*2.83*
0.113	*2.87*
0.115	*2.93*
0.117	*2.98*

20 Having selected the required new shims, place them in the spring cup on top of the valve and refit the tappet in its original bore.
21 Liberally lubricate the camshaft bearing journals with engine oil and place the camshaft in position (photo). Lubricate the outer circumference and sealing lip of the oil seal and position it over the camshaft until it is in contact with the register in the cylinder head.
22 Remove all traces of old sealant from the camshaft cover and cylinder head and ensure that the mating faces are clean and dry.
23 Apply a very thin bead, $\frac{1}{16}$ in (1.5 mm) diameter, of RTV silicone sealant to the mating face of the camshaft cover. Refit the cover and the timing belt cover bracket and then tighten the retaining bolts evenly and progressively to the specified torque (photo).
24 Refit the camshaft rear end cover, using a new gasket if necessary.
25 Refit the camshaft gear and fully tighten the retaining bolt.
26 Ensure that the crankshaft is still at the 90° BTDC position and then turn the camshaft until the dimple on the gear is aligned with the pointer on the cover. Refit and tension the timing belt as described in Section 11.
27 Refit the brake servo vacuum pipe banjo union to the inlet manifold. Reposition the retaining clip over the cylinder head stud and secure with the retaining nut.
28 Refit the timing belt cover.
29 Reconnect the two fuel pipes to the fuel pump.
30 Refer to Chapter 4 and refit the distributor.

5 Methods of engine removal

The engine may be removed either on its own or in unit with the gearbox. On models fitted with automatic transmission, it is recommended that the engine be lifted out on its own because of the weight factor.

If the engine only is being removed, it can be lifted out of the engine compartment using a suitable hoist. When removing the engine/gearbox as a unit, it is necessary to raise the front of the car a minimum of 27 in (685 mm) and then remove the assembly by lowering it with a hoist and withdrawing it from under the car.

It is easier if a hydraulic type trolley jack is used in conjunction with two pairs of axle-stands. Overhead lifting tackle will be necessary in both cases.

6 Engine/gearbox assembly – removal

1 The do-it-yourself owner should be able to remove the power unit fairly easily in about four hours provided he has the equipment recommended in Section 5.
2 The sequence of operations listed in this Section is not critical as the position of the person undertaking the work, or the tool in his hand, will determine to a certain extent the order in which the work is tackled. Obviously the power unit cannot be removed until everything is disconnected from it and the following sequence will ensure that nothing is forgotten:
3 Open the bonnet and using a soft pencil, mark the outline position of both the hinges on the bonnet to act as a datum for refitting.
4 With the help of an assistant to take the weight of the bonnet, undo and remove the hinge to bonnet securing bolts and remove the plain and spring washers. There are two bolts to each hinge.
5 Lift away the bonnet and store it in a safe place where it will not be scratched or damaged.
6 Remove the battery as described in Chapter 10, Section 2.
7 Drain the cooling system as described in Chapter 2, Section 2. Undo the securing bolts and remove the fan.
8 Slacken the securing clips and remove the radiator top hose, also the heater and radiator bottom hoses (photo).
9 Undo and remove the bolt that secures the heater pipe to the heat shield support bracket.
10 Remove the distributor vacuum pipe (photo).
11 Slacken the clip that secures the breather hose to the carburettor and detach the hose.
12 Carefully pull the fuel supply hose from the union on the car-

6.8 Remove the bottom hose and heater hoses

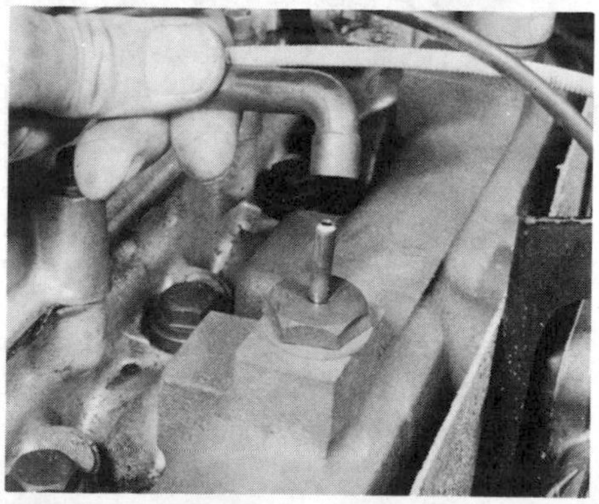

6.10 Disconnect the distributor vacuum pipe from the inlet manifold

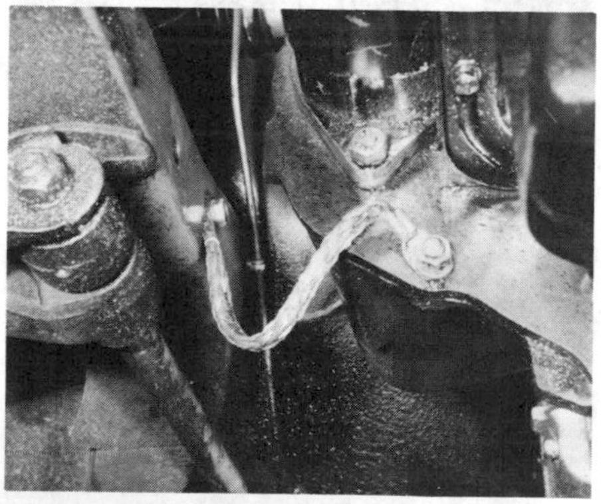

6.26 Disconnect the earth strap

6.34 Remove the gearbox mounting bracket bolts

6.39 Remove the engine mounting bolts

burettor. Plug the end of the hose to prevent the ingress of dirt.
13 Disconnect the air intake ducts and remove the air cleaner. Refer to Chapter 3, Section 6.
14 Slacken the accelerator cable to trunnion clamp nut and withdraw the accelerator inner cable from the trunnion. Press the plastic retainer located on the underside of the abutment bracket and withdraw the cable from the bracket.
15 Using two open-ended spanners, slacken the choke cable nut and withdraw the inner cable. Withdraw the cable from the carburettor.
16 On models with automatic transmission, disconnect the downshift cable from the carburettor.
17 Remove the securing bolt and lift away the carburettor, gaskets and insulator block; then remove the heat shield and support bracket.
18 Slacken the clip that secures the brake servo vacuum hose to the inlet manifold and disconnect the hose. Remove the clip that secures the vacuum hose to the cylinder head.
19 Slacken the clip that secures the fuel supply pipe to the fuel pump. Pull off the hose and plug the end to prevent the ingress of dirt.
20 Make a note of the cable connections to the rear of the starter motor solenoid and detach the cables.
21 Disconnect the electrical wiring from the engine at the engine harness multi-pin connector.
22 Release the HT cable from the ignition coil, mark the spark plug HT cables to ensure refitting in the correct order and disconnect from the spark plugs.
23 Undo the two screws that secure the distributor cap to the distributor body and lift away the distributor cap and HT leads.
24 Chock the rear wheels, raise the front of the car and support it on axle-stands or other suitable supports. The front of the car must be raised high enough (at least 27 in) to allow for removal of the engine/gearbox assembly from under the car.
25 Undo and remove the exhaust manifold to downpipe securing nuts and tie the exhaust pipe to the side on the left-hand wing valance.
26 Undo and remove the bolt that secures the braided earth strap to the body (photo).
27 Undo and remove the bolt and spring washer that secures the speedometer cable retaining clamp to the gearbox housing.
28 Lift away the retaining clamp and withdraw the speedometer cable from the gearbox. Tuck the end of the cable back out of the way so it will not be damaged during subsequent operations.
29 Wipe the area around the top of the clutch master cylinder reservoir. Unscrew the cap, place a thin piece of polythene over the top of the reservoir and then refit the cap. This will prevent hydraulic fluid syphoning out when the clutch hydraulic pipe is disconnected.
30 Wipe the area around the clutch slave cylinder hydraulic pipe union and disconnect the pipe from the slave cylinder. Wrap a piece of polythene round the end of the pipe and tuck it back out of the way.
31 Refer to Chapter 12 and remove the centre console (if fitted), then disconnect the wiring from the reverse lamp switch.
32 Mark the gearbox and propeller shaft mating flanges so that they may be refitted in their original positions. Undo and remove the four self-locking nuts and bolts.
33 Move the propeller shaft to the side and secure it to the torsion bar with wire or rope.
34 Using a jack, support the weight of the gearbox then undo and remove the two bolts, spring and plain washers that secure the gearbox mounting bracket to the underside of the body (photo).
35 Undo and remove the one bolt and spring washer that secure the mounting bracket to the underside of the gearbox. Lift away the mounting.
36 Lower the power unit approximately 3 in (76 mm) and release the gear lever from the gearbox as described in Chapter 6, Section 2, paragraph 15.
37 Remove the front anti-roll bar as described in Chapter 11.
38 Fit lifting hooks, or place a rope sling or chain around the engine, and support its weight using an overhead hoist. Remove the jack from under the gearbox.
39 Undo and remove the bolts that secure the engine mountings to the main body member brackets (photo).
40 Place a low trolley or a piece of board under the power unit and carefully lower the complete power unit down through the engine compartment until it is resting on the trolley or board. Whilst lowering the unit, check to make sure that none of the control cables or wiring get caught by the engine.
41 Detach the hoist from the engine and withdraw the engine/gearbox assembly from underneath the car.

7 Engine removal without gearbox

1 If only the engine is to be removed, leaving the gearbox in position, the following sequence will enable the engine to be removed.
2 Follow the instructions given in Section 6, paragraphs 3 to 23 and 25. On models with automatic transmission, refer to Chapter 6 and drain the fluid into a suitable container.
3 Remove the radiator as described in Chapter 2.
4 Undo and remove the bolts that secure the anti-roll bar bushes, release the anti-roll bar and rotate it to allow removal of the engine.
5 Undo and remove the bolts that secure the sump connecting plate to the underside of the clutch bellhousing.
6 Undo and remove the two nuts, bolts and spring washers that secure the starter motor to the engine backplate and gearbox bellhousing. Note the earth strap fitted on the lower bolt. Lift away the starter motor.
7 Using a jack, support the weight of the gearbox.
8 Fit lifting hooks, or place a rope sling or chain around the engine, and support its weight with an overhead hoist.
9 Undo and remove the remaining nuts, bolts and spring washers that secure the engine to the gearbox bellhousing.
10 Remove the nuts and spring washers that secure the engine mountings to the brackets attached to the cylinder block.
11 Lift the engine clear of the mountings and move it forward until the clutch is clear of the first motion shaft.
12 Check that no controls, cables or pipes have been left connected to the engine and that they are safely tucked to one side where they will not be caught up as the engine is being removed.
13 Continue lifting the engine out of the engine compartment, taking care not to damage the front valance. Draw it forwards, or push the car rearwards, and lower to the ground.

8 Engine/gearbox assembly – gearbox removal

1 Remove the remaining nut, bolt and spring washer that secure the starter motor to the engine backplate and gearbox bellhousing. Lift away the starter motor.
2 Undo and remove the bolts that secure the sump connecting plate to the underside of the bellhousing.
3 Undo and remove the remaining nuts, bolts and spring washers that secure the gearbox to the engine.
4 Carefully withdraw the gearbox from the engine, ensuring that the weight of the gearbox is not allowed to hang on the first motion shaft as it will bend the shaft.

9 Dismantling the engine – general

1 It is best to mount the engine on a dismantling stand, but if one is not available, stand the engine on a strong bench to be at a comfortable working height. It can be dismantled on the floor but it is not easy.
2 During the dismantling process, greatest care should be taken to keep the exposed parts free from dirt. As an aid to achieving this, thoroughly clean down the outside of the engine, removing all traces of oil and congealed dirt.
3 Use paraffin or a proprietary grease solvent. The latter compound will make the job much easier for after the solvent has been applied and allowed to stand for a time, a vigorous jet of water will wash off the solvent with all the grease and dirt. If the dirt is thick and deeply embedded, work the solvent into it with a wire brush.
4 Finally wipe down the exterior of the engine with a rag and only then, when it is quite clean, should the dismantling process begin. As the engine is stripped, clean each part in a bath of paraffin.
5 Never immerse parts with oilways (for example the crankshaft) in paraffin, but to clean, wipe down carefully with a petrol dampened cloth. Oilways can be cleaned out with nylon pipe cleaners. If an air line is available, all parts can be blown dry and the oilways blown through as an added precaution.
6 Re-use of old engine gaskets is false economy and will lead to oil and water leaks, if nothing worse. Always use new gaskets throughout.
7 Do not throw the old gasket away, for it sometimes happens that

an immediate replacement cannot be found and the old gasket is then very useful as a template. Hang up the old gaskets as they are removed.

8 To strip the engine it is best to work from the top down. The underside of the crankcase when supported on wood blocks acts as a firm base. When the stage is reached where the crankshaft and connecting rods have to be removed, the engine can be turned on its side and all other work carried out with it in this position.

9 Whenever possible, refit nuts, bolts and washers finger-tight from wherever they were removed. This helps avoid loss and muddle later. If they cannot be refitted lay them out in such a fashion that it is clear from whence they came.

10 Engine – removing ancillary components

Before basic engine dismantling begins it is necessary to remove the following ancillary components:

Alternator
Distributor
Spark plugs
Fuel pump
Inlet/exhaust manifold
Thermostat housing and thermostat
Coolant temperature sender unit
Water pump
Oil pressure sender unit
Oil separator
Oil filter
Clutch assembly
Oil filler tube (later models)

Some of these items have to be removed for individual servicing or renewal periodically. Details can be found in the appropriate Chapter.

11 Timing belt – checking, removal and refitting

1 The condition and tension of the timing belt must be checked after 24 000 miles (40 000 km) or 2 years and adjusted if necessary. If defective, it must be renewed. A new timing belt must be fitted after 48 000 miles (80 000 km) or 4 years usage.

2 Remove the timing belt cover, fan belt and fan.

3 Examine the belt. If the teeth are worn or show signs of uneven wear, cracking or oil contamination, it must be renewed (photo).

4 Using a spring balance connected to the timing belt at the point in-line with the water pump inlet pipe, check the pull required to align the outside face of the timing belt with the line on the inlet pipe, see Fig. 1.5.

5 The belt tension for a used belt should be 11 lbf (5 kgf) and for a new belt, 13 lbf (6 kgf). If the belt is in good condition but extends beyond the line on the water inlet it must be adjusted as follows:

(a) Slacken the timing belt tensioner securing nuts
(b) Push the tensioner against the belt and tighten the tensioner securing nuts
(c) Recheck the timing belt tension and readjust the tension further if necessary

6 To remove the timing belt, the fan belt must first be removed.

7 Turn the crankshaft to 90° BTDC (the V-notch in the timing disc opposite the timing pointer). **Note:** *It is essential that the crankshaft is set at this position before removing the timing belt. The crankshaft must not be turned while the camshaft is disconnected as this will result in damage to the valves and pistons.*

8 Slacken the timing belt tensioner securing nuts and release the belt tension.

9 Ease the timing belt off the camshaft gear. Withdraw the plug from the LED sensor timing bracket, lift the timing pointer slightly and lift away the timing belt.

10 Refitting the timing belt is the reverse of the removal procedures but the following additional points should be noted:

11 Ensure that the crankshaft is still set at 90° BTDC and that the dimple on the rear face of the camshaft gear is opposite the pointer on the camshaft cover.

12 Adjust the belt tension, see paragraphs 4 and 5.

13 Adjust the alternator drive belt tension, refer to Chapter 10.

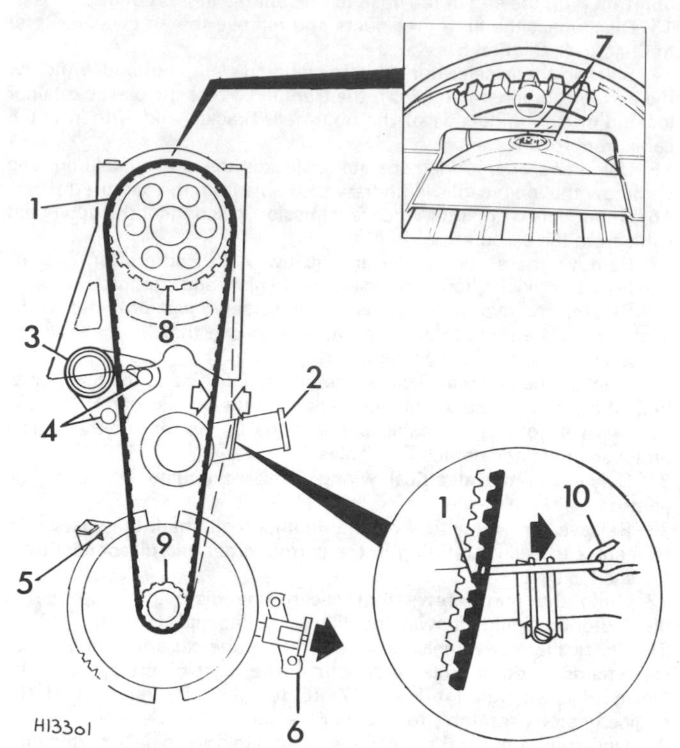

Fig. 1.5 Checking the timing belt tension (Sec 11)

1 Timing belt
2 Water pump inlet
3 Timing belt tensioner
4 Tensioner securing nuts
5 Crankshaft 90° BTDC setting
6 LED timing bracket
7 Camshaft sprocket dimple and camshaft cover pointer
8 Camshaft sprocket
9 Crankshaft sprocket
10 Spring balance

11.3 Examine the timing belt teeth for wear

12 Cylinder head – removal (engine in car)

1 Open the bonnet and mark the outline of the hinges on the bonnet to act as a datum for refitting.

2 With the help of an assistant, undo and remove the hinge to bonnet securing bolts, plain and spring washers. There are two bolts to each hinge.

3 Lift away the bonnet and place it where it will not get scratched or otherwise damaged.

4 Disconnect the earth lead from the battery negative terminal.

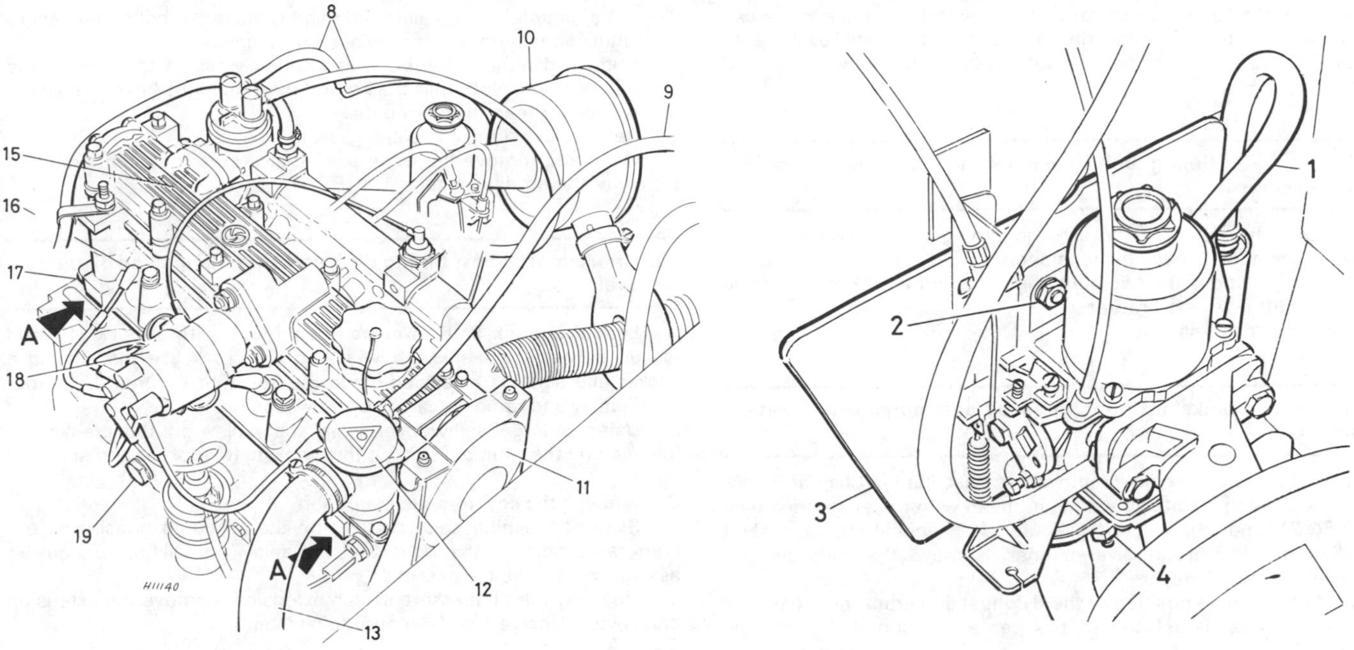

Fig. 1.6 Cylinder head attachments (Sec 12)

1 Crankcase breather hose
2 Carburettor mounting nut
3 Heat shield
4 Exhaust pipe flange nut
8 Fuel lines
9 Coolant air bleed pipe
10 Air cleaner
11 Timing belt cover
12 Coolant filler cap
13 Radiator top hose
15 Camshaft cover
16 Spark plug cap
17 Joint
18 Distributor
19 Cylinder block drain plug
A Cylinder head levering lugs

5 Drain the cooling system, refer to Chapter 2, Section 2.
6 To provide access to the crankshaft pulley securing bolt (required when rotating the crankshaft), remove the fan and radiator as described in Chapter 2, Section 5.
7 Mark the spark plug HT leads to ensure correct refitting and detach the leads from the spark plugs.
8 Disconnect the HT lead to the coil and the earth lead, then undo the screws that secure the distributor cap and remove the cap and spark plug leads.
9 Disconnect the lead from the coolant temperature sender located beneath the thermostat housing.
10 Remove the distributor vacuum pipe.
11 Disconnect the crankcase breather hose from the carburettor.
12 Disconnect the fuel supply hose from the carburettor.
13 Slacken the accelerator cable trunnion securing nut, then press in the accelerator cable retaining clips on the underside of the support bracket and pull the cable through the bracket.

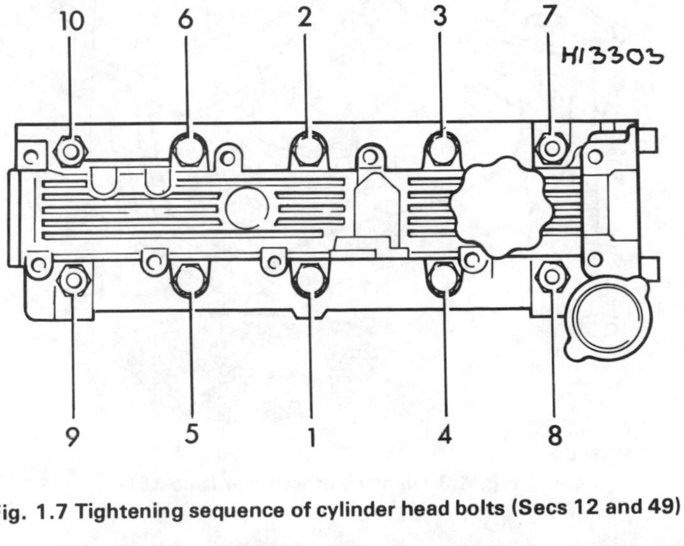

Fig. 1.7 Tightening sequence of cylinder head bolts (Secs 12 and 49)

14 Slacken the nut that secures the choke control cable to the choke lever and withdraw the complete cable from the carburettor.
15 On cars with automatic transmission, disconnect the downshift cable from the carburettor linkage.
16 Disconnect the cold/warm air ducts from the air cleaner. Undo and remove the carburettor securing nuts.
17 Lift away the carburettor and air cleaner as an assembly.
18 Remove the heat shield, support bracket and insulator block from the manifold studs.
19 Remove the clip that secures the vacuum servo unit hose to the cylinder head and disconnect the vacuum hose from the induction manifold.
20 Disconnect the tank fuel supply hose from the fuel pump and plug the end of the hose to prevent ingress of dirt.
21 Disconnect the heater inlet hose from the cylinder head.
22 Undo and remove the six nuts that secure the exhaust twin flange downpipe to the manifold.
23 Remove the timing belt cover by pulling it off its mounting spigots.
24 The crankshaft must be positioned at 90° BTDC before the timing belt is removed. Rotate the crankshaft until the V in the timing disc is in-line with the timing pointer on the crankcase and the dimple on the rear face of the camshaft gear is opposite the pointer on the camshaft cover. **Note:** *The crankshaft must not be rotated whilst the camshaft is disconnected as the pistons and valves may be damaged.*
25 Release the timing belt tensioner by slackening the securing nuts and remove the timing belt from the camshaft gear.
26 Slacken the cylinder head bolts in a progressive manner in the reverse of the tightening sequence shown in Fig. 1.7. Remove the bolts.
27 The cylinder head assembly can now be removed by lifting upwards. To break the seal, it will probably be necessary to lever under the two lugs A in Fig. 1.6. Do not try to prise it apart from the block by inserting a screwdriver between the faces of the cylinder head and the cylinder block as this will result in serious damage to the cylinder head.
28 Remove the cylinder head gasket from the cylinder block.

13 Cylinder head – removal (engine on bench)

The procedure for removing the cylinder head with the engine on the bench is the same as that for removal with the engine in the car, with the exception of disconnecting the controls and services. Refer to Section 12 and follow the operations described in paragraphs 23 to 28 inclusive.

14 Timing belt, timing belt tensioner and LED sensor timing bracket – removal

1 Remove the nuts that secure the timing belt tensioner to the cylinder block and lift away the tensioner.
2 Undo and remove the two bolts that secure the LED sensor timing bracket and lift away the bracket.
3 Remove the timing belt.

15 Flywheel, crankshaft rear oil seal and engine backplate – removal

1 With the clutch assembly removed as described in Chapter 5, lock the flywheel using a suitable wedge in mesh with the starter ring gear (photo 50.2). Undo the bolts that secure the flywheel to the crankshaft in a diagonal and progressive manner. Remove the bolts and the locking plate (Fig. 1.8).
2 Mark the relative position of the flywheel and crankshaft to ensure that the flywheel is refitted in the same position. Lift away the flywheel.
3 Undo and remove the bolts that secure the crankshaft rear oil seal retainer and lift away the retainer.
4 The engine backplate complete with oil seal and gasket can now be lifted away.

16 Sump and oil strainer – removal

1 Remove the drain plug and drain the oil into a suitable container. Refit the drain plug.
2 If the engine is in the car, slacken the front anti-roll bar mountings and rotate the anti-roll bar so that the sump can be removed.

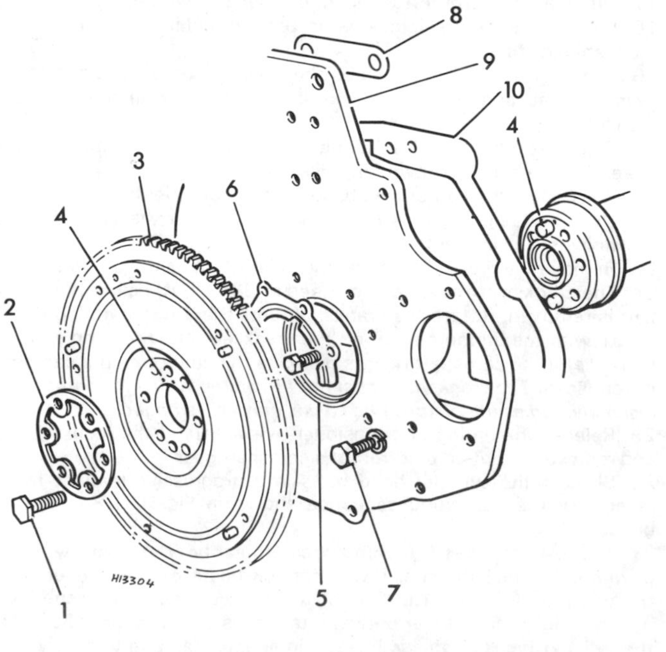

Fig. 1.8 Flywheel and engine backplate removal (Sec 15)

1 Flywheel securing bolts
2 Shaped lockwasher
3 Flywheel
4 Mating marks
5 Crankshaft rear oil seal
6 Oil sealer retainer
7 Backplate securing bolt and spring washer
8 Gasket (upper)
9 Engine backplate
10 Gasket (lower)

3 If the engine is in the car, undo and remove the bolts that secure the sump connecting plate to the gearbox bellhousing.
4 Undo and remove the bolts and spring washers that secure the sump in position. Note that the bolt fitted at the right-hand rear of the sump is longer than the other bolts.
5 Remove the sump and sump gasket.
6 Undo and remove the three bolts that secure the oil strainer and lift away the strainer and gasket (Fig. 1.9).

17 Crankshaft pulley, timing disc, sprocket, flange and oil pump – removal

1 Lock the crankshaft, to prevent it turning, with a block of wood placed between a crankshaft web and the crankcase then, using a socket and suitable extension, undo the bolt that secures the crankshaft pulley and remove the bolt and lockwasher.
2 Using a large screwdriver, ease the pulley off the crankshaft, followed by the timing disc then the crankshaft sprocket and sprocket flange.
3 Remove the key from the crankshaft.
4 Slacken the nine bolts that secure the oil pump housing to the crankcase, remove the bolts and lift away the oil pump housing assembly. Remove the gasket.
5 Unscrew the oil pressure switch extension. Remove the extension and switch. Unscrew and remove the oil filter.

18 Piston, connecting rods and big-end bearings – removal

1 Note that the pistons are marked on the top with the word FRONT or by an arrow or a groove which must face towards the front of the engine.
2 Check that the big-end bearing caps and connecting rods have identification marks. If not, suitably mark them to ensure that the correct end caps are fitted to the correct connecting rods and the correct connecting rods are fitted in their respective cylinder bores.
3 Remove the big-end cap securing nuts and put them to one side in the order in which they were removed.
4 Remove the big-end caps, taking care to keep them in the right order. Ensure that the bearing shells are kept with their respective big-end caps unless the bearings are to be renewed.

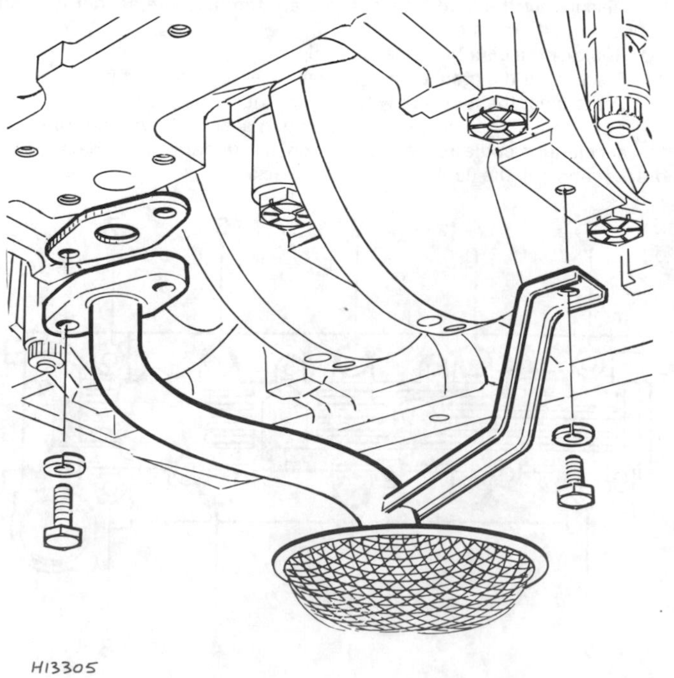

Fig. 1.9 Oil strainer removal (Sec 16)

5 If the big-end caps are difficult to remove, they may be gently tapped with a soft-faced hammer.
6 To remove the shell bearings, press the bearing opposite the groove in both the connecting rod and its cap and the bearing shell will slide out easily.
7 Push the piston and connecting rod assemblies upwards and withdraw them from the top of the cylinder block.

19 Crankshaft and main bearings – removal

1 Note that the main bearing caps Nos 2, 3 and 4 have their numbers, together with arrows, cast on the front face of the caps.
2 Undo by one turn at a time the bolts that secure the main bearing caps.
3 Lift away each main bearing cap and the bottom half of each bearing shell, taking care to keep the bearing shell with the right cap.
4 When removing the front and rear main bearing caps, note the cork sealing joints.
5 When removing the centre main bearing cap, note the bottom semi-circular halves of the thrustwashers, one half located on each side of the main bearing. Lay them, with the centre bearing cap, along the correct side.
6 Slightly rotate the crankshaft to free the upper halves of the bearing shells and thrustwashers which can be extracted and placed over the respective bearing caps.
7 Remove the crankshaft by lifting it from the crankcase.

20 Cylinder head and camshaft assembly – dismantling

1 Remove the securing bolts and lift away the hot air box, inlet/exhaust manifold and the manifold gasket.
2 Unscrew the coolant temperature switch.
3 Undo and remove the bolt that secures the thermostat housing to the cylinder head and lift away the housing complete with thermostat. Remove the O-ring seal.
4 Undo and remove the bolt that secures the camshaft sprocket and pull the sprocket off the camshaft.
5 Undo and remove the five bolts that secure the camshaft end cover and lift away the cover and gasket.
6 Slacken the nine bolts that secure the camshaft bearing cover in a progressive and even sequence until the valve spring tension is released, then remove the bolts, release the oil filler cap and lift away the cover bracket.
7 Lift off the camshaft bearing cover.
8 Remove the camshaft complete with the front oil seal from the cylinder head. Pull the oil seal off the front of the camshaft.
9 Lift out each tappet and shim and lay them out in the correct order, 1 to 8, to ensure that they can be refitted in their original positions.
10 To remove the valves, compress each spring in turn with a universal valve spring compressor until the two halves of the collets can be removed. Release the compressor and lift away the spring top cup, valve spring, oil seal (inlet valves only), valve spring seat and the valve.
11 If, when the valve spring compressor is screwed down the valve spring top cup refuses to free and expose the split collet, do not continue to screw down on the compressor but gently tap the top of the tool directly over the cup with a light hammer. This should free the cup. To avoid the compressor jumping off the valve retaining cup when it is tapped, hold the compressor firmly in position.
12 It is essential that the valves are kept in their correct order unless they are so badly worn or burnt that they are to be renewed. If they are going to be refitted, place them in their correct sequences along with the tappets and shims previously removed. Also keep the valve springs, cups and collets in the same order.

21 Gudgeon pin – removal and refitting

The gudgeon pins are a press fit in the connecting rods and it is important that no damage is caused during removal and refitting. Because of this, should it be necessary to fit new pistons or connecting rods, take the parts along to the local BL garage who will have the special equipment required to carry out this work.

22 Piston rings – removal

1 To remove the piston rings, slide them carefully over the top of the piston, taking care not to scratch the aluminium alloy of the piston. Never slide them off the bottom of the piston skirt.It is very easy to break piston rings if they are pulled off roughly so this operation should be done with extreme caution. It is helpful to use an old 0·020 in (0.5 mm) feeler gauge to facilitate their removal.
2 Lift one end of the piston ring to be removed out of its groove and insert the end of the feeler gauge under it.
3 Turn the feeler gauge slowly round the piston and as the ring comes out of its groove it rests on the land above. It can then be eased off the piston with the feeler gauge stopping it from slipping into any empty grooves.

23 Lubrication system – description

The pressed steel oil sump is attached to the underside of the crankcase and acts as a reservoir for the engine oil. The oil pump draws oil through a strainer located under the oil surface, passes it along a short passage and into the full flow oil filter which is screwed onto the oil pump housing. The freshly filtered oil flows from the filter and enters the main gallery. Five small drillings connect the main gallery to the five main bearings. The oil passes from the main bearings through drillings in the crankshaft to the big-end bearings.

When the crankshaft is rotating, oil is thrown from the hole in each big-end of the connecting rod and splashes the thrust side of the piston and bore.

Further drillings connect the main oil gallery to the overhead camshaft in order to lubricate the bearings, cams and tappets. The oil then drains back to the sump via large drillings in the cylinder head and cylinder block.

A pressure relief valve is incorporated in the oil pump housing to keep the oil pressure within the specified limit.

A disposable oil filler cap and filter assembly is fitted to the top of the camshaft cover on early models. Later versions are equipped with a separate oil filler tube bolted to the side of the cylinder block, with the cap on the camshaft cover which contains a reusable wire mesh filter for crankcase ventilation.

24 Oil filter – removal and refitting

1 The oil filter is a throw-away cartridge type screwed into the left-hand side of the oil pump housing (photo).
2 Simply unscrew the old unit; if it is very tight use a strap wrench to slacken it.
3 Fit a new sealing ring in position and smear it with engine oil, then screw the new filter on, hand-tight only.
4 Run the engine and check for oil leaks.

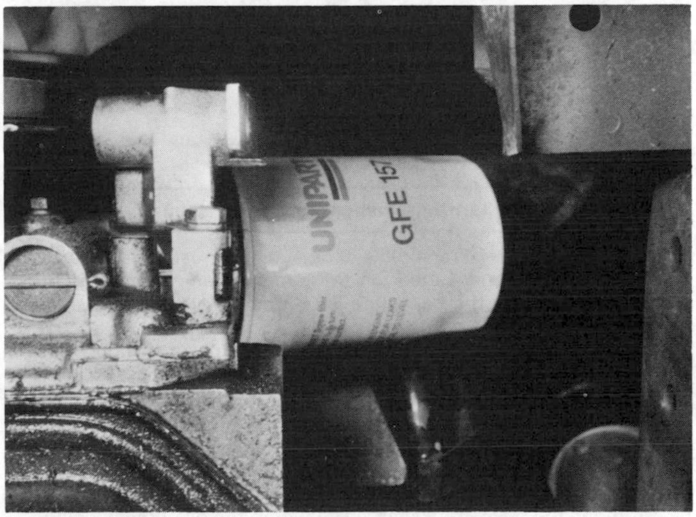

24.1 The oil filter is accessible from under the car

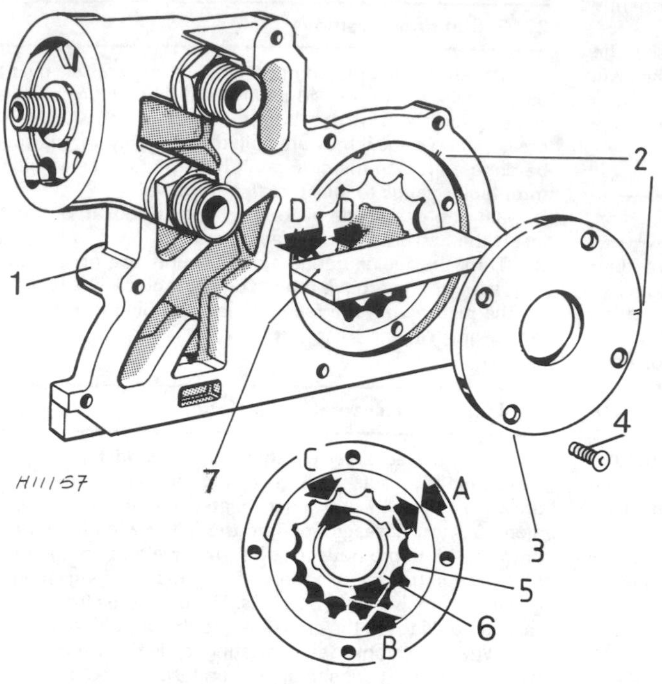

Fig. 1.10 Checking the oil pump for wear (Sec 25)

1	Oil pump housing	5	Outer rotor
2	Alignment marks	6	Inner rotor
3	Backplate	7	Straight-edge
4	Screw		

25.2a Remove the oil pump backplate securing screws ...

25.2b ... and lift off the backplate

25.8 Measure the lobe clearance with a feeler gauge

25.12a Fitting a new seal in the oil pump housing

25.12b The oil seal should be flush with the housing

25 Oil pump – dismantling, inspection and reassembly

The oil pump is not repairable, if defective it must be renewed.

1 Scribe an alignment mark on the pump housing and pump backplate to ensure that the backplate is refitted in its original position at reassembly (Fig. 1.10).

2 Using a 3 mm Allen key, remove the backplate securing screws and lift off the backplate (photos).

3 Remove the outer and inner rotors.

4 Remove the split-pin that retains the pressure relief valve, then remove the plug from the housing by pushing down on the relief valve spring with a screwdriver through the oil return hole. Collect the spring and valve plunger. Discard the plug O-ring seal.

5 Remove the front crankshaft oil seal from the oil pump housing.

6 Thoroughly clean all the component parts then check the rotor endfloat and lobe clearances.

7 Fit the outer rotor in the housing and, using a feeler gauge, check the clearance between the outer rotor and the pump body A in Fig. 1.10. The maximum permissible clearance being 0.004 in (0.10 mm).

8 Fit the inner rotor and measure the lobe clearance B and C in Fig. 1.10. The maximum permissible clearance being 0.002 in (0.05 mm) (photo).

9 Measure the outer rotor to housing face clearance D in Fig. 1.10. The maximum permissible clearance being 0·003 in (0·08 mm).

10 If any of these pump clearances exceed the specified limit renew the pump.

11 Check that the free length of pressure relief valve spring is 1·525 in (38·7 mm). If not, it must be renewed.

12 Reassembly is the reverse of the removal procedure. Always fit a new O-ring on the pressure relief valve plug and a new crankshaft oil seal in the pump housing (photos).

26 Engine components – examination and renovation, general

With the engine stripped down and all parts thoroughly cleaned, it is now time to examine every component for wear. The components listed in the following Sections should be inspected and, where necessary, renovated or renewed.

27 Crankshaft – examination and renovation

1 Inspect the main bearing journals and crankpins. If there are any scratches or score marks then the shaft will need regrinding. Such conditions will nearly always be accompanied by similar deterioration in the matching bearing shells.
2 Each bearing journal should also be round and can be checked with a micrometer or caliper gauge around the periphery at several points. If there is more than 0·001 in (0.025 mm) of ovality, regrinding is necessary.
3 A main BL Agent or motor engineering specialist will be able to decide to what extent regrinding is necessary and also supply the special undersize shell bearing to match whatever may need grinding off.
4 Before taking the crankshaft for regrinding, check also the cylinder bores and pistons as it may be advantageous to have the whole engine done at the same time.
5 Check the condition of the spigot bush in the crankshaft rear flange. Renew it if necessary.

28 Main and big-end bearings – examination and renovation

1 With careful servicing and regular oil and filter changes, bearings will last for a very long time but they can still fail for unforeseen reasons. With big-end bearings the indication is a regular rhythmic loud knocking from the crankcase. The frequency depends on engine speed and is particularly noticeable when the engine is under load. This symptom is accompanied by a fall in oil pressure although this is not normally noticeable unless an oil pressure gauge is fitted. Main bearing failure is usually indicated by serious vibration, particularly at higher engine revolutions, accompanied by a more significant drop in oil pressure and a rumbling noise.
2 Bearing shells in good condition have bearing surfaces with a smooth even matt silver/grey colour all over. Worn bearings will show patches of a different colour where the bearing metal has worn away and exposed the underlay. Damaged bearings will be pitted or scored. It is always well worthwhile fitting new shells as their cost is relatively low. If the crankshaft is in good condition it is merely a question of obtaining another set of standard size. A reground crankshaft will need new bearing shells as a matter of course.

29 Cylinder bores – examination and renovation

1 The cylinder bores must be examined for taper, ovality, scoring and scratches. Start by carefully examining the top of the cylinder bores. If they are at all worn a very slight ridge will be found on the thrust side. This marks the top of the piston ring travel. The owner will have a good indication of the bore wear prior to dismantling the engine or removing the cylinder head. Excessive oil consumption accompanied by blue smoke from the exhaust is a sure sign of worn cylinder bores and piston rings.
2 Measure the diameter of the bore just under the ridge with a micrometer and compare it with the diameter at the bottom of the bore, which is not subject to such wear. If the difference between the two measurements is more than 0·006 in (0·1524 mm) then it will be necessary to fit a ring set or to have the cylinders rebored and fit oversize pistons and rings. If no micrometer is available, remove the rings from one piston and place the piston in each bore in turn about ¾ in (19mm) below the top of the bore. If an 0·010 in (0·254 mm) feeler gauge can be slid between the piston and the cylinder wall on the thrust side of the bore, then remedial action must be taken. Oversize pistons are available in the following sizes:

+ 0·010 in (0·254 mm)
+ 0·020 in (0·508 mm)
+ 0·030 in (0·762 mm)
+ 0·040 in (1·016 mm)

3 These are accurately machined to just below these measurements so as to provide correct running clearances in bores bored out to the exact oversize dimensions.
4 If the bores are slightly worn but not so badly worn as to justify reboring them, special oil control rings can be fitted to the existing pistons which will restore compression and stop the engine burning oil. Several different types are available and the manufacturers instructions concerning their fitting must be followed closely.

30 Pistons and piston rings – examination and renovation

1 If the old pistons are to be refitted, carefully remove the piston rings and then thoroughly clean them. Take particular care to clean out the piston ring grooves. Do not scratch the aluminium in any way. If new rings are to be fitted to the old pistons, then the top ring should be stepped so as to clear the ridge left above the previous top ring. If a normal but oversize new ring is fitted, it will hit the ridge and break, because the new ring will not have worn in the same way as the old.

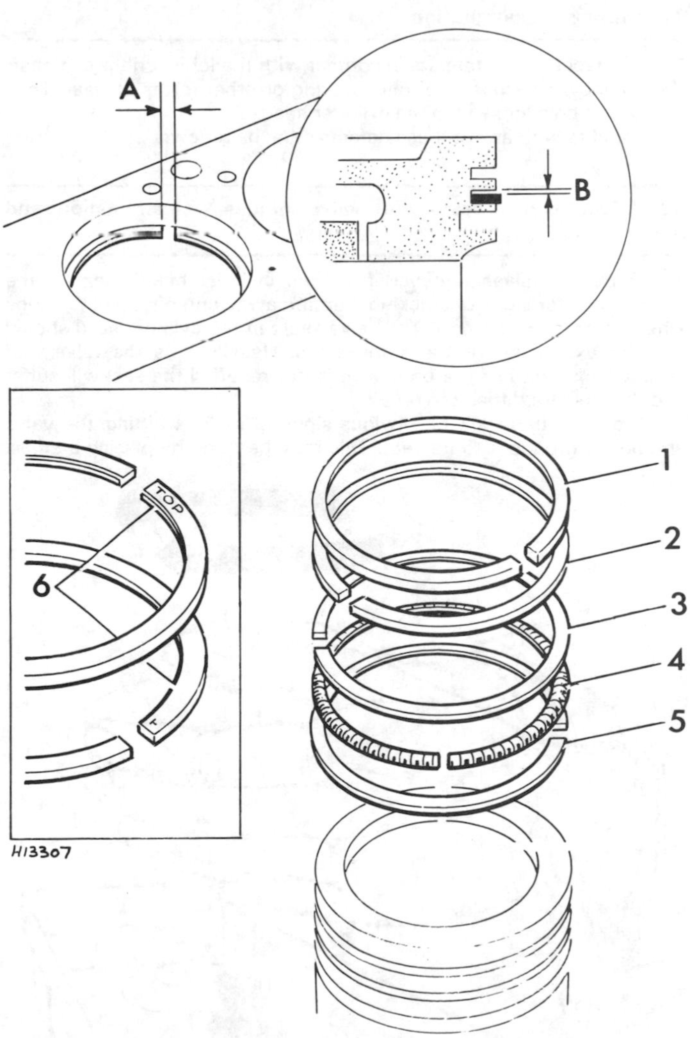

Fig. 1.11 Piston ring measurement (Secs 30 and 39)

A Piston ring gap
B Ring side clearance
1 Top compression ring
2 Stepped compression ring
3 Control ring top rail
4 Control ring expander
5 Control ring bottom rail
6 The compression rings are marked TOP or T

2 Before fitting the rings on the pistons, each should be inserted approximately 3 in (76mm) down the cylinder bore and the gap measured with a feeler gauge (Fig. 1.11). This should be between the limits given in the Specifications at the beginning of this Chapter. It is essential that the gap is measured at the bottom of the ring travel, for if it is measured at the top of a worn bore and gives a perfect fit, it could easily seize at the bottom. If the ring gap is too small, rub down the ends of the ring with a very fine file until the gap is correct when fitted. To keep the rings square in the bore for measurement, line each one up in turn with an old piston in the bore upside down and use the piston to push the ring down about 3 in (76mm). Remove the piston and measure the piston ring gap.

3 The groove clearance of the new rings in old pistons should be within the specified tolerance given in the Specifications. If it is not enough, the rings could stick in the piston grooves causing loss of compression. The ring grooves in the piston in this case will need machining out to accept the new rings

4 Before fitting new rings onto an old piston, clean out the grooves with a piece of broken ring.

5 If new pistons are obtained the rings will be included, so it must be emphasised that the top ring be stepped if fitted to a cylinder bore that has not been rebored or has not had the top ridge removed.

31 Tappets – examination

1 The faces of the tappets in contact with the lobes on the camshaft should show no signs of pitting, scoring or other forms of wear. They should not be a loose fit in the cylinder head.

2 Tappets with any of these defects must be renewed.

32 Valves, valve seats and valve guides – examination and renovation

1 With the valves removed from the cylinder head, examine the valve heads for signs of cracking, burning away and pitting of the edge where it seats in the port. The valve seats in the cylinder head should also be examined for the same signs. Usually it is the valve that deteriorates first but if a bad valve is not rectified the seat will suffer and this is more difficult to repair.

2 Provided there are no obvious signs of serious pitting, the valve should be ground into its seat. This may be done by placing a smear of fine carborundum paste on the edge of the valve and, using a suction type valve holder, grinding the valve in situ. This is done with a semi-rotary action, rotating the handle of the valve holder between the hands and lifting it occasionally to redistribute the traces of paste. As soon as a matt grey unbroken line appears on both the valve and seat the valve is ground in. All traces of carbon should also be cleaned from the head and neck of the valve stem.

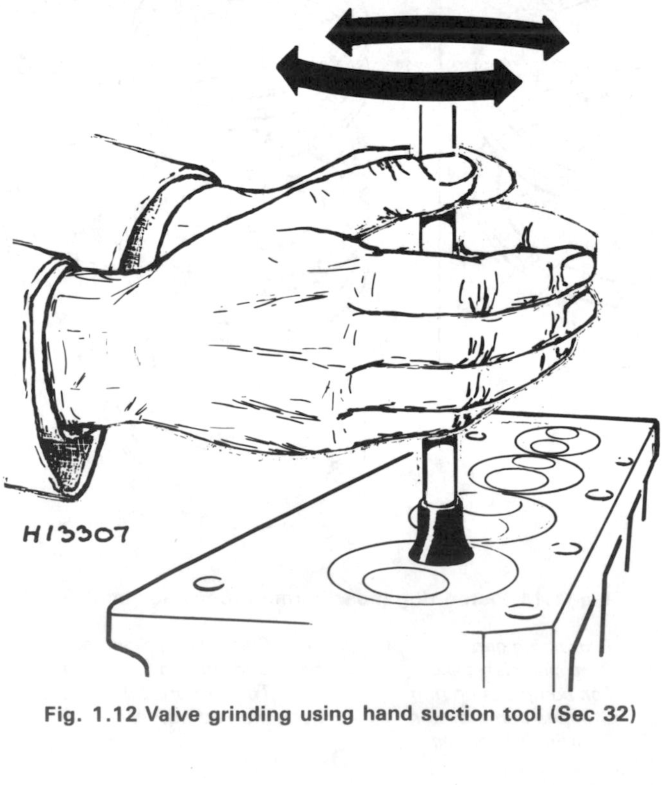

Fig. 1.12 Valve grinding using hand suction tool (Sec 32)

3 If the valve requires renewal it should be ground into the seat in the same way as the old valve.

4 Another form of valve wear can occur on the stem where it runs in the guide in the cylinder head. This can be detected by trying to rock the valve from side to side. If there is any movement at all it is an indication that the valve stem or guide is worn. Check the stem first with a micrometer at points along and around its length and, if they are not within the specified size, new valves will probably solve the problem. If the guides are worn however, they will need renewing. The valve seats will also need recutting to ensure they are concentric with the stems. This work should be given to your BL dealer or local engineering works.

5 When valve seats are badly burnt or pitted, requiring renewal, inserts may be fitted, or replaced if already fitted once before. Once again, this is a specialist task to be carried out by a suitable engineering firm.

6 When all valve grinding is completed it is essential that every trace of grinding paste is removed from the valves and ports in the cylinder head. This should be done by thorough washing in paraffin and blowing out with a jet of air. If particles of carborundum should work their way into the engine they would cause havoc with bearings or cylinder walls.

33 Camshaft and camshaft bearings – examination and renovation

1 Check the camshaft journals and cams for scoring and wear. If there are very slight scoring marks these can be removed with emery cloth or a fine oil stone. The greatest care must be taken to keep the cam profiles smooth.

2 Examine the ignition distributor drivegear for wear or chipping of the gear teeth.

3 Examine the camshaft bearing surfaces in the cylinder head and camshaft bearing cover, if they are scored and worn it means a new cylinder head and camshaft cover will be required.

34 Flywheel starter ring gear – examination and renovation

1 If the teeth on the flywheel starter ring gear are badly worn, or if some are missing, then it will be necessary to remove the ring. This is achieved by splitting the old ring using a hacksaw and cold chisel. Care must be taken not to damage the flywheel during this process.

2 To fit a new ring gear, it will be necessary to heat it gently and evenly with an oxy-acetylene flame until a temperature of approximately 350°C is reached. This is indicated by a grey/brown surface colour. With the ring gear at this temperature, fit it to the flywheel with the bevelled edge of the teeth facing the engine facing end of the flywheel. The ring gear should be either pressed or lightly tapped onto its register and left to cool naturally when the contraction of the metal on cooling will ensure that it is a secure and permanent fit. Great care must be taken not to overheat the ring gear, for if this happens the temper of the ring gear will be lost.

3 An alternative method is to use a high temperature oven to heat the ring.

4 Because of the need of oxy-acetylene equipment or a special oven it is not practical for refitment to take place at home. Take the flywheel and new starter ring to an engineering works willing to do the job.

35 Cylinder head and pistons – decarbonisation

1 This operation can be carried out with the engine either in or out of the car. With the cylinder head off, carefully remove with a wire brush and blunt plastic scraper, all traces of carbon deposits from the combustion spaces and the ports. The valve stems and valve guides should also be freed from any carbon deposits. Wash the combustion spaces and ports down with paraffin and scrape the cylinder head surface free

of any foreign matter with the side of a steel rule or similar article. Take care not to scratch the surfaces.

2 Clean the pistons and top of the cylinder bores. If the pistons are still in the cylinder bores then it is essential that great care is taken to ensure no carbon gets into the bores as this could scratch the cylinder walls or cause damage to the piston and rings. To ensure that this does not happen, first turn the crankshaft so that two of the pistons are at the top of the bores. Place clean non-fluffy rag into the other two bores or seal them off with paper and masking tape. The water and oilways should also be covered with a small piece of masking tape to prevent particles of carbon entering the lubrication system and causing damage to a bearing surface.

3 There are two schools of thought as to how much carbon ought to be removed from the piston crown. One is that a ring of carbon should be left around the edge of the piston and on the cylinder bore wall as an aid to keeping oil consumption low. The other is to remove all traces of carbon during decarbonisation and leave everything clean.

4 If all traces of carbon are to be removed, press a little grease into the gap between the cylinder walls and the two pistons which are to be worked on. With a blunt scraper, carefully scrape away the carbon from the combustion chamber in the piston crown, taking care not to scratch the aluminium. Also scrape away the carbon from the surrounding lip of the cylinder wall. When all carbon has been removed, scrape away the grease which will now be contaminated with carbon particles, taking care not to press it into the bores. To assist prevention of carbon build up the piston crown can be polished with a metal polish. Remove the rags or masking tape from the other two cylinders and turn the crankshaft so that those two pistons which were at the bottom are now at the top. Place a non-fluffy rag into the other two bores or seal them with paper and masking tape. Do not forget the waterways and oilways as well. Proceed as previously described.

5 If a ring of carbon is going to be left around the piston, this can be helped by inserting an old piston ring into the top of the bore to rest on the piston and ensure that carbon is not accidentally removed. Check that there are no particles of carbon in the cylinder bores. Decarbonisation is now complete.

36 Engine reassembly – general

1 To ensure maximum life with minimum trouble from a rebuilt engine, not only must every part be correctly assembled but everything must be spotlessly clean, all the oilways must be clear, locking washers and spring washers must always be fitted where needed and all bearings and other working surfaces must be throughly lubricated during assembly. Before assembly begins, renew any bolts or studs the threads of which are in any way damaged, and whenever possible use new spring washers.

2 Apart from your normal tools, a supply of non-fluffy rag, an oil can filled with engine oil (an empty washing up liquid plastic bottle thoroughly cleaned and washed out, will do), a supply of new spring washers, a set of new gaskets and a torque wrench should be collected together.

37 Crankshaft – refitting

Ensure that the crankcase is thoroughly clean and that all the oilways are clear. A thin twist drill is useful for cleaning them out. If possible, blow them out with compressed air. Treat the crankshaft in the same fashion and then inject engine oil into the crankshaft oilways.

Commence work on rebuilding the engine by refitting the crankshaft and main bearings as follows:

1 Fit the five upper halves of the main bearing shells to their location in the crankcase after wiping the location clean.

2 Note that on the back of each bearing is a tab which engages in locating grooves in either the crankcase or the main bearing cap housings (photo).

3 If new bearings are being fitted, carefully clean away all traces of the protective grease with which they are coated.

4 With the five upper bearing shells securely in place, wipe the lower bearing cap housings and fit the five lower shell bearings to their caps ensuring that the right shell goes into the right cap if the old bearings

37.2 Ensure the tab on the bearing shell locates in the groove

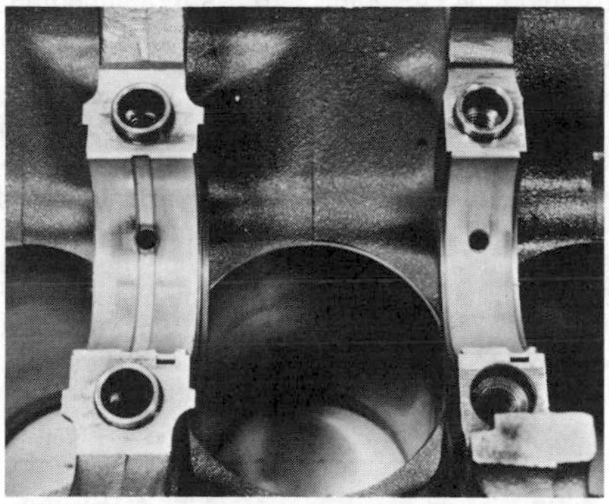

37.4 Fit the bearing shells in the crankcase

37.5 The centre main bearing is recessed on each side to accept the thrustwashers

37.6 The thrustwashers are fitted with the oil grooves facing outwards

37.7a Oil the main bearings

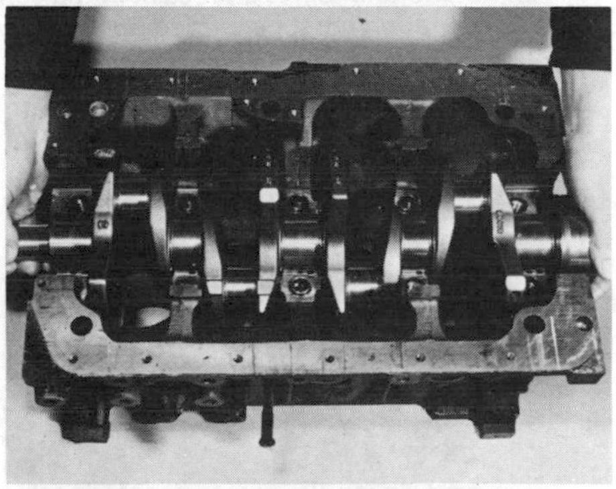

37.7b Lower the crankshaft into position

37.9 Fitting the centre main bearing cap

37.10 Screw in the main bearing cap bolts

37.12 Use a torque wrench to tighten the main bearing cap bolts

are being refitted (photo).
5 Wipe the recesses either side of the centre main bearing which locate the upper halves of the thrustwashers (photo).
6 Introduce the upper halves of the thrustwashers (the halves without tabs) into their grooves either side of the centre main bearing with their oil grooves facing outwards (photo).
7 Generously lubricate the crankshaft journals and the upper and lower main bearing shells and carefully lower the crankshaft into position. Make sure that it is the right way round (photos).
8 Fit the main bearing caps in position ensuring that they locate properly. The mating surfaces must be spotlessly clean or the caps will not seat correctly.
9 When refitting the centre main bearing cap, ensure that the thrustwashers, generously lubricated, are fitted with their oil grooves facing outwards and the locating tab of each washer is in the slot in the bearing cap (photo).
10 Refit the main bearing cap bolts and screw them up finger-tight (photo).
11 Test the crankshaft for freedom of rotation. Should it be very stiff to turn or possess high spots, a most careful inspection must be made, preferably by a skilled mechanic with a micrometer to trace the cause of the trouble. It is very seldom that any trouble of this nature will be experienced when fitting the crankshaft.
12 Tighten the main bearing cap bolts using a torque wrench set to 75 lbf ft (10·4 kgf m) and recheck the crankshaft for freedom of rotation (photo).
13 Using a screwdriver between one crankshaft web and main bearing cap, lever the crankshaft forwards and check the endfloat using feeler gauges. This should be 0·001 to 0·005 in (0·025 to 0·127 mm). If excessive, new thrustwashers must be fitted. Thrustwashers are available in standard size and 0·003 in (0·076 mm) oversize.

38 Pistons and connecting rods – reassembly

As the gudgeon pin is a press fit in the connecting rod (see Section 21), this operation must be carried out by the local BL garage.

39 Piston rings – refitting

1 Check that the piston ring grooves and oilways are thoroughly clean and unblocked. Piston rings must always be fitted over the head of the piston and never from the bottom (Fig. 1.11).
2 Refitment is the exact opposite procedure to removal, see Section 22.
3 An alternative method is to fit the rings by holding them slightly open with the thumbs and both your index fingers. This method requires a steady hand and great care for it is easy to open the ring too much and break it.
4 The special oil control ring requires a special fitting procedure. First fit the bottom rail of the oil control ring to the piston and position it below the bottom groove. Refit the oil control expander into the bottom groove and move the bottom oil control ring rail up into the bottom groove. Fit the top oil control rail into the bottom groove.
5 Ensure the ends of the expander are butting without overlapping.
6 Set the two upper ring gaps 90° to each other.

40 Pistons – refitting

The pistons, complete with connecting rods, can be fitted to the cylinder bores as follows:
1 Wipe the cylinder bores clean with a clean rag.
2 The pistons, complete with connecting rods, must be fitted to their bores from the top of the block.
3 Check that the two upper piston ring gaps are spaced at 90° to each other.
4 Check that the piston is the correct one for the cylinder bore and that the small-end offset of the connecting rod is correct for the bore, see Fig. 1.13.
5 Lubricate the cylinder bore and piston with clean engine oil.
6 Fit a universal piston ring clamp and insert the first piston into the bore, making sure that the front of the piston (marked on top with the word FRONT or an arrow) is towards the front of the engine (photo).
7 Insert the piston into the bore up to the bottom of the piston ring

40.6 The piston is marked with the word FRONT

40.7 Tap the piston into the cylinder bore

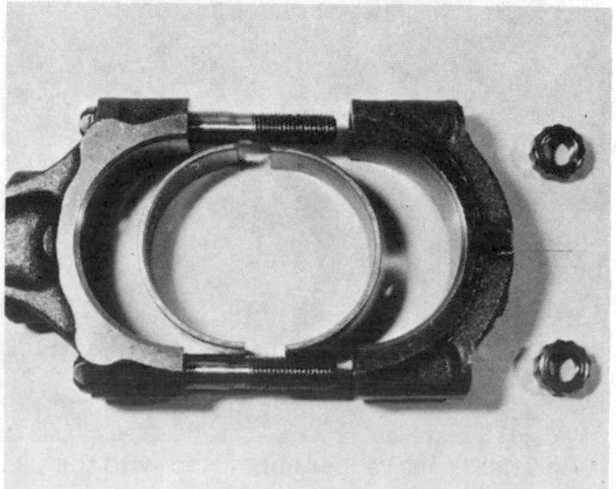
41.1 Connecting rod big-end assembly

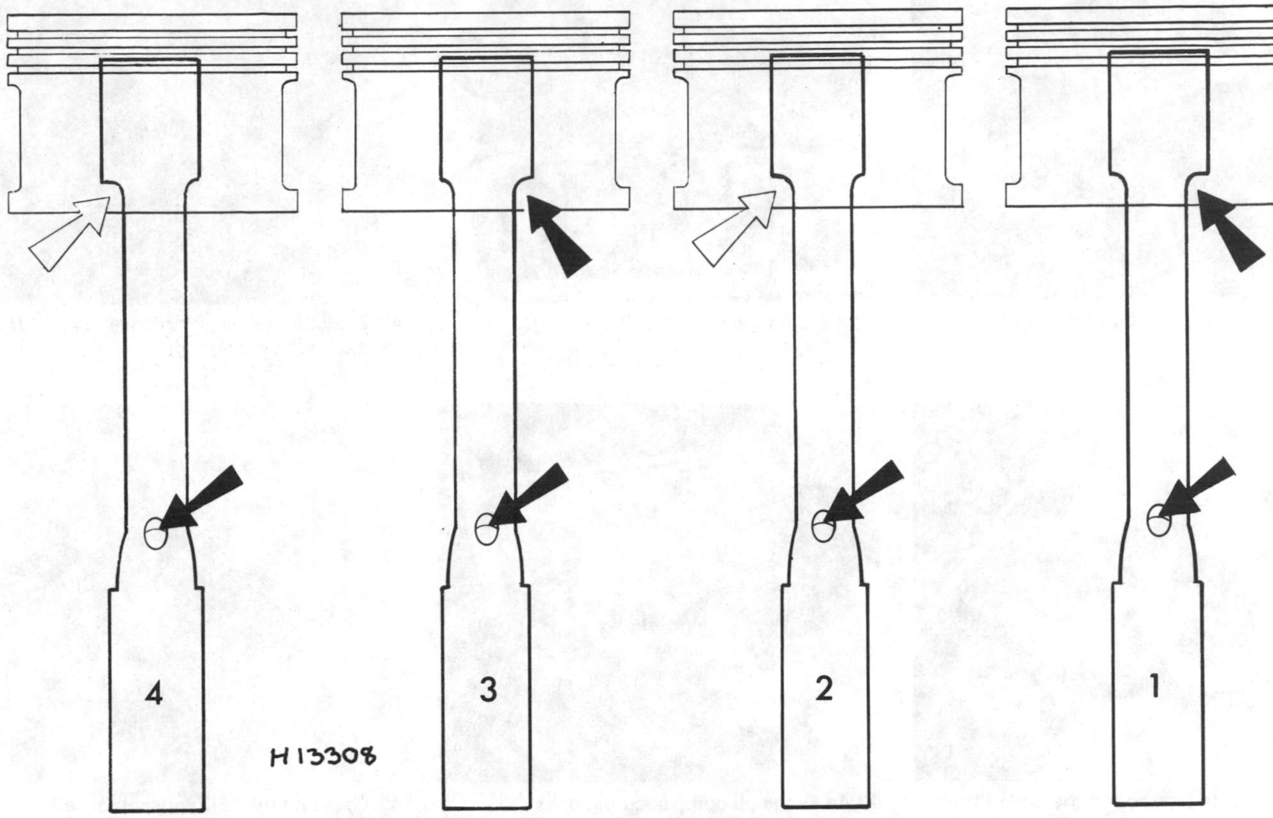

Fig. 1.13 The connecting rod small-end is offset (Sec 40)

41.4 Fitting the connecting rod big-end cap

41.5 Use a torque wrench to tighten the connecting rod big-end bearing cap nuts

42.1 Fitting the oil strainer

clamps, then gently but firmly tap the piston through the piston ring clamp and into the cylinder bore with the shaft of a hammer (photo).

41 Connecting rod to crankshaft – reassembly

1 Wipe the connecting rod half of the big end bearing location and the underside of the shell bearing clean (as for the main bearing shells) and fit the shell bearing in position with its locating tab engaged with the corresponding groove in the connecting rod. Always fit new shells (photo).

2 Generously lubricate the crankpin journals with engine oil and turn the crankshaft so that the crankpin is in the most advantageous position for the connecting rod to be drawn onto it.

3 Fit the bearing shell to the connecting rod cap in the same way as with the connecting rod itself.

4 Generously lubricate the shell bearing and offer up the connecting rod bearing cap to the connecting rod. Fit the connecting rod cap retaining nuts. It wil be observed that these are special twelve sided nuts (photo).

5 Tighten the retaining nuts to a torque wrench setting of 33 lbf ft (4·6 kgf m) (photo).

42 Oil strainer and sump – refitting

1 Refit the oil strainer. Always use a new flange gasket. Fit the two bolts that secure the oil strainer flange to the crankcase and the bolt that secures the support stay (photo).

2 Ensure that all traces of the old gasket have been removed from the mating faces of the sump and crankcase.

3 Apply a jointing compound to the joint faces of the front and rear main bearing caps and fit new cork seals to the caps (photo).

4 Place a new sump gasket on the crankcase. Use a little grease to keep the gasket in position and then fit the oil sump (photo).

5 Refit the sump securing bolts and spring washers; the longer bolt is fitted at the rear right-hand location. Tighten the bolts to the specified torque.

42.3 Use jointing compound when fitting the cork seals

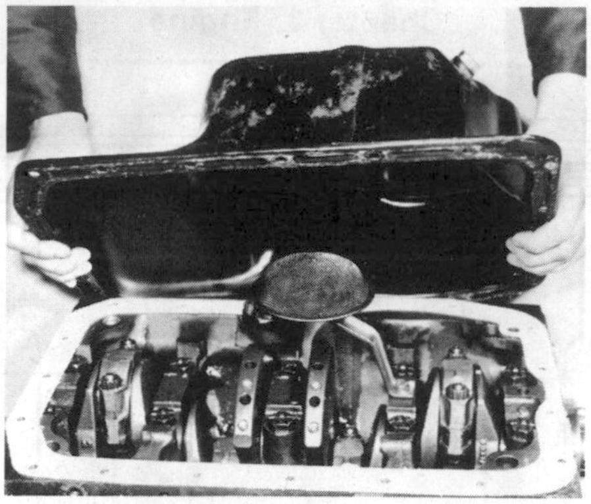

42.4 Fitting the oil sump

43.2 Ensure the oil pump drive key is correctly located in the crankshaft

43.3 Wrap some tape round the end of the crankshaft to prevent it damaging the seal in the oil pump

43.4a Fit the oil pump securing bolts ...

43.4b ... and the LED sensor bracket

44.2 Fit a new oil seal in the engine backplate

44.3 Fitting the engine backplate

44.4 Tightening the backplate securing bolts with a torque wrench

45.1 Screw the oil pressure switch extension into the oil pump housing

46.2 Position a new gasket on the water pump

46.4a Fit the water pump to the block and ...

43 Oil pump, timing pointer and LED sensor timing bracket – refitting

1 Ensure that the joint faces of the crankcase and oil pump housing are clean. Position a new gasket on the oil pump.
2 Fit the oil pump drive key in the crankshaft (photo).
3 Fit a protective sleeve over the end of the crankshaft (photo). Line-up the keyway in the pump with the drive key on the crankshaft and fit the oil pump. Take care not to damage the oil seal.
4 Fit the securing bolts, timing pointer and LED sensor bracket. The long bolt is fitted adjacent to the core plug. Tighten the bolts to the specified torque (photos).
5 Prime the oil pump by removing the bottom plug and injecting oil into the pump.

44 Engine backplate – refitting

1 Lubricate a new oil seal with SAE 90 EP oil and fit the seal in the backplate with the lip of the seal facing to the front of the engine.
2 Ensure that the seal is pressed in square, with the front of the seal flush with the front face of the backplate (photo).
3 Locate the rear oil seal retainer then fit the retainer and backplate securing bolts (photo).
4 Tighten the oil seal retainer bolts to 22 lbf ft (3·0 kgf m) and the backplate bolts to 37 lbf ft (5·1 kgf m) (photo).

45 Oil pressure switch – refitting

1 Screw the oil pressure switch into the extension and then screw the extension into the oil pump housing (photo).
2 Tighten the switch and switch extension to a torque wrench setting of 9 lbf ft (1·2 kgf m).

46 Water pump – refitting

1 Ensure that the mating faces of the water pump and cylinder block are free of old gasket or jointing compound.
2 Smear a little grease on the joint face of the water pump and place a new gasket on the pump (photo).
3 Locate the two bolts that secure the timing belt tensioner in the water pump housing. Use some grease to retain them in the housing while fitting the water pump.
4 Fit the water pump to the cylinder block and secure it in position with the five bolts and the stud. Tighten the stud and five bolts to the specified torque (photos).

47 Crankshaft sprocket, timing disc and pulley – refitting

1 Fit the drive key on the crankshaft (photo).
2 Fit the sprocket flange followed by the sprocket, then the timing disc and the crankshaft pulley (photos).
3 Using a new lockwasher, fit the crankshaft pulley securing bolt.
4 Fit two bolts, temporarily, in the rear flange of the crankshaft and use a lever between them to prevent the crankshaft from turning whilst tightening the crankshaft pulley bolt to a torque wrench setting of 62 lbf ft (8·5 kgf m) (photo).
5 Lock the bolt by bending over two tabs of the lockwasher.

48 Valves – refitting

1 With the valves suitably ground in (see Section 32) and kept in their correct order, start with No 1 cylinder.
2 Lubricate the valve stem with oil and insert the valve into its guide (photo).
3 On inlet valves, fit a new oil seal well lubricated with engine oil

46.4b ... screw in the securing bolts and stud. Make sure the timing bolt tensioner bolts are in position

47.1 Fit the drive key on the crankshaft

47.2a Fit the sprocket flange and sprocket ...

47.2b ... followed by the timing disc and pulley

47.4 Tightening the crankshaft pulley bolt

48.2 Insert the valve in the valve guide

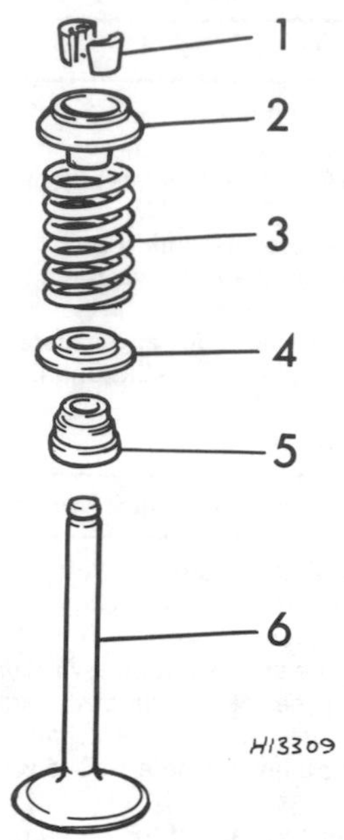

Fig. 1.14 Inlet valve and spring assembly (Sec 48)

1 *Split collet*
2 *Valve spring cup*
3 *Spring*
4 *Valve spring seat*
5 *Oil seal (not fitted on exhaust valves)*
6 *Valve*

48.4a Fit the valve spring seat ...

48.4b ... then valve spring and valve spring cup

48.5 Compress the valve spring and fit the split collets

49.2 Position a new gasket on the cylinder block

49.3 Position the crankshaft at 90° BTDC with the single notch in the timing disc opposite the timing pointer

49.4a Lower the cylinder head onto the block

49.4b Fit the cylinder head bolts ...

49.4c ... and tighten them to the specified torque

49.6 Fitting the valve tappets

49.9 Fit the camshaft cover

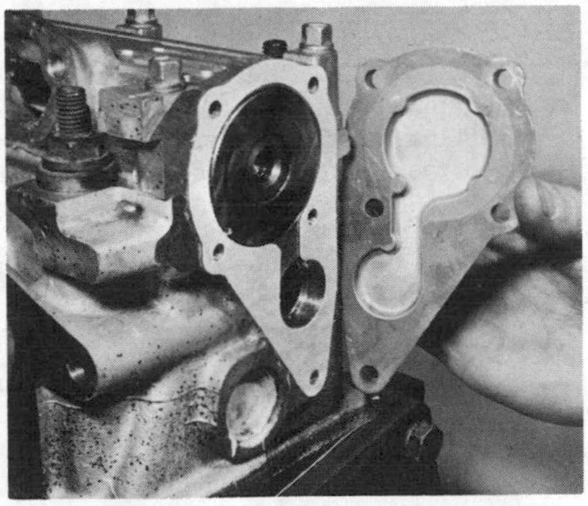

49.10a Use a new gasket when fitting the camshaft rear end cover

49.10b Fit the timing belt cover bracket

50.1 Position the flywheel on the crankshaft flange

50.2 Lock the flywheel with a wedge when tightening the securing bolts

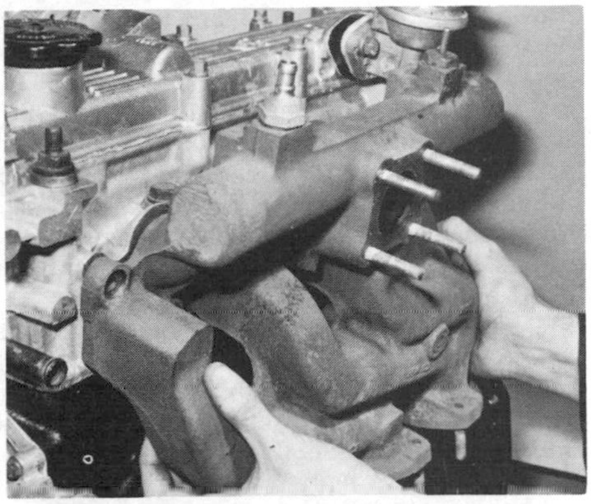

51.1a Fit the inlet/exhaust manifold ...

51.1b ... hot air box ...

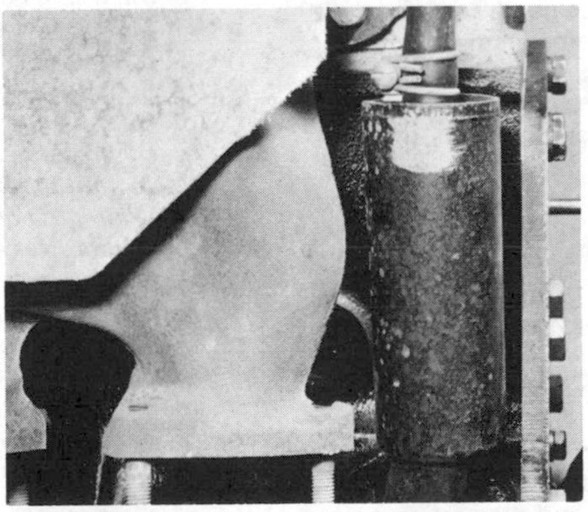

51.1c ... oil separator ...

51.1d ... and heat shield

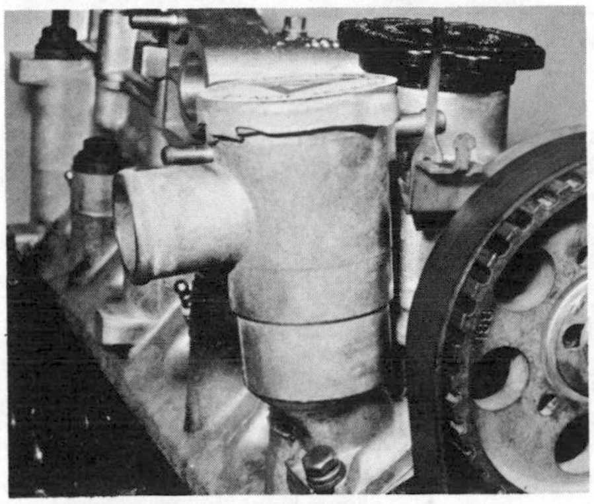

51.3 Fit the thermostat housing

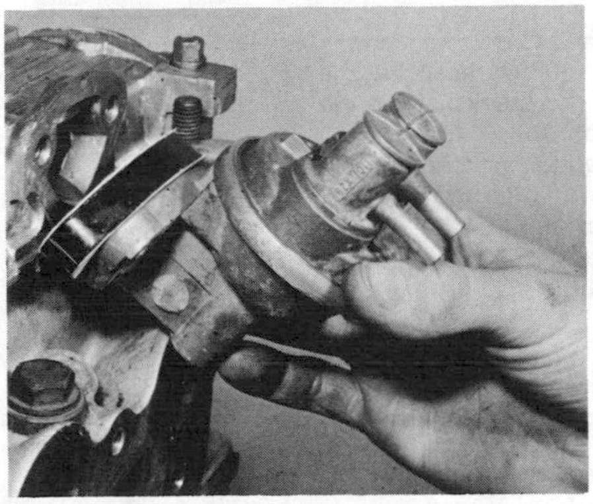

51.5 Fitting the fuel pump

51.7 Left-hand engine mounting

(Fig. 1.14).
4 Fit the valve spring seat, valve spring and valve spring cup over the valve stem (photos).
5 Using a valve spring compressor tool, compress the valve spring until the split collets can be slid into position, then carefully release the valve spring compressor in order not to displace the collets (photo).
6 Refit the other valves in the same way. When they are all fitted, tap the end of each valve stem using a plastic-faced hammer or a hammer with a block of hardwood interposed. This will settle the valve components ready for checking the valve clearances.

49 Cylinder head, camshaft and camshaft cover – refitting

1 Ensure the mating faces of the cylinder block and cylinder head are clean.
2 Position a new cylinder head gasket on the cylinder block, locating it over the dowels in the block (photo). The gasket is pre-coated and must be fitted dry. *Do not use grease or jointing compound on the gasket.*
3 Check that the crankshaft position is 90° BTDC as described in Section 4 paragraph 5 (photo).
4 Lower the cylinder head into position, locating it over the dowels. Smear the bolt threads with engine oil and screw them in finger-tight. Progressively tighten the bolts in the sequence given in Fig. 1.7 to a torque wrench setting of first 35 lbf ft (4·8 kgf m) and then to 60 lbf ft (8·3 kgf m) (photos).
5 Place the valve clearance adjusting shims into their original valve spring cap recesses. If the valves have been ground in more than very lightly, it will usually be necessary to reduce each shim thickness and new shims should be obtained to achieve this or the original ones interchanged (see Section 4).
6 Oil the valve tappets and refit them in their original locations in the cylinder head (photo).
7 Oil the camshaft journals and locate the camshaft on the cylinder head.
8 Adjust the valve clearances as described in Section 4.
9 Ensure that the mating faces of the cylinder head and camshaft cover are clean. Apply a new bead $\frac{1}{16}$ in (1.5 mm) diameter, of RTV silicone sealant then fit the cover immediately and tighten the cover bolts evenly and progressively to the specified torque (photo).
10 Using a new gasket, fit the camshaft rear end cover. Refit the bracket for the timing belt cover (photos).
11 Fit the camshaft front oil seal fully into its housing and oil the outside diameter and the sealing lip. Keep the outside face dry.
12 Fit the camshaft sprocket.
13 Ensure that the crankshaft and camshaft are positioned as described in Section 4, paragraph 5, then fit and tension the timing belt as described in Section 11.
14 If the engine is in the car, reverse the operations described in Section 12, paragraphs 1 to 23.
15 After refitting the cylinder head, run the engine at a fast idle for 15 minutes, then allow the engine to cool. Now slacken each cylinder head bolt, in the specified sequence, a half turn and then retighten to the full torque of 60 lbf ft (8·3 kgf m).

50 Flywheel and clutch – refitting

1 Clean the mating faces of the crankshaft and flywheel. Fit the flywheel, locating it on the dowel in the end of the crankshaft flange and with the alignment marks made at removal (photo).
2 Refit the locking plate and the six securing bolts. Using a suitable wedge, lock the flywheel and tighten the securing bolts in a diagonal and progressive manner to the specified torque (photo).
3 Lock the bolts by bending over the locking plate.
4 Refit the clutch assembly as described in Chapter 5.

51 Final assembly

1 Using a new gasket, refit the inlet/exhaust manifold, the hot air box, the oil separator and the heat shield (photos).
2 Refit the coolant temperature switch.
3 Refer to Chapter 2 and refit the thermostat housing, complete with thermostat (photo).
4 Refit the distributor as described in Chapter 4.
5 Refit the fuel pump as described in Chapter 3 (photo).
6 Refit the alternator as described in Chapter 10.
7 Refit the engine mountings (photo).
8 Refit the oil filter.
9 Refit the oil filler tube (later models).

52 Gearbox – refitting to engine

1 If the engine and gearbox were removed as an assembly, refit the gearbox to the engine.
2 With the engine on a bench or suspended on a hoist, lift up the gearbox and insert the gearbox input shaft into the centre of the clutch so that the input shaft splines pass through the internal splines of the clutch disc. On models fitted with automatic transmission, refer to Chapter 6.
3 If difficulty is experienced in engaging the splines, try turning the gearbox slightly, but on no account allow the weight of the gearbox to rest on the input shaft as it is easily bent.
4 With the gearbox correctly positioned on the engine backplate, secure the gearbox to the engine with the bolts and spring washers.
5 Refit the starter motor.
6 Refit the sump to gearbox bellhousing connecting plate.

53 Engine – refitting in car

1 The refitting of the engine or the engine/gearbox assembly is the reverse of the removal procedure as described in Sections 7 and 6 as appropriate.
2 Ensure that all loose leads, control cables, etc are tucked out of the way. It is easy to trap one as the power unit is hoisted or lowered. This will cause much additional work, and probably expense, after the engine is refitted.
3 Refit or reconnect the following as applicable:

(a) Mounting nuts, bolts and washers
(b) Clutch hydraulic system. Bleed the system as described in Chapter 5
(c) Speedometer cable
(d) Gearchange lever
(e) Oil pressure switch
(f) Coolant temperature switch
(g) Wiring to coil, distributor and alternator
(h) Carburettor and air cleaner
(j) Exhaust manifold to downpipe
(k) Earth strap and starter motor
(l) Radiator, hoses and fan
(m) Heater hoses
(n) Engine breather pipe
(p) Vacuum servo to manifold pipe
(q) Fuel pipes to pump and carburettor

4 Finally check that the drain plugs are fitted, then refill the cooling system as described in Chapter 2. Refill the engine sump to the correct level with the specified oil. Refill the gearbox, if applicable.

54 Initial start-up after major repair or overhaul

1 Ensure that the battery is fully charged.
2 Adjust the throttle stop screw to provide a fast idle speed.
3 Start the engine; keep it running at a fast idle and bring it to normal operating temperature.
4 As the engine warms up there will be odd smells and some smoke from parts getting hot and burning off oil deposits.
5 Check for leaks of water or oil which will be obvious if serious. Check also the exhaust pipe and manifold connections as these do not always find their exact gastight position until the heat and vibration have acted on them. It is almost certain that they will need tightening. This should be done of course, with the engine stopped.
6 Stop the engine and wait a few minutes to see if any lubricant or coolant is leaking out when the engine is stationary.
7 After the engine has run at the fast idle for 15 minutes, allow the engine to cool and then retorque the cylinder head bolts as described in Section 49, paragraph 15.

8 Adjust the engine idle speed as described in Chapter 3.

9 Road test the car to check that the timing is correct and that the engine is giving the necessary smoothness and power. Do not race the engine: if new bearings and/or pistons have been fitted it should be treated as a new engine and run in at a reduced speed for the first 1000 miles (1600 km).

55 Fault diagnosis – engine

Symptom	Reason/s
Engine will not turn over when starter switch is operated	Flat battery Bad battery connections Bad connections at solenoid switch and/or starter motor Starter motor jammed Defective solenoid Starter motor defective
Engine turns over normally but fails to fire and run	No spark at plugs No fuel reaching engine Too much fuel reaching the engine (flooding)
Engine starts but runs unevenly and misfires	Ignition and/or fuel system faults Incorrect valve clearances Burnt out valves
Lack of power	Ignition and/or fuel system faults Incorrect valve clearances Burnt out valves Worn out piston or cylinder bores
Excessive oil consumption	Oil leaks from crankshaft oil seals, camshaft cover, drain plug gasket, or sump joint Worn piston rings or cylinder bores resulting in oil being burnt by engine Smoky exhaust is an indication Worn valve guides and/or defective valve stem seals
Excessive mechanical noise from engine	Wrong valve clearances Worn crankshaft bearings Worn cylinders (piston slap)
Unusual vibration	Misfiring on one or more cylinders Loose mounting bolts

Note: *When investigating starting and uneven running faults do not be tempted into snap diagnosis. Start from the beginning of the check procedure and follow it through. It will take less time in the long run. Poor performance from an engine in terms of power and economy is not normally diagnosed quickly. In any event the ignition and fuel systems must be checked first before assuming any further investigation needs to be made.*

Chapter 2 Cooling system

Contents

Specifications

Type ..... Pressurised system with expansion tank

Thermostat

Type ..... Wax

Opening temperature (standard) ..... 82°C (180°F)

Expansion tank

Cap blow off pressure ..... 15 lbf/in² (1·05 kgf/cm²)

Fan drivebelt

Tension ..... 0·25 in (6 mm) deflection

Water pump

Type ..... Centrifugal

Cooling system

Capacity (with heater) ..... 10 pints (5·7 litres)

Torque wrench settings

	lbf ft	kgf m
Water pump retaining bolts	8	1·1
Pulley and fan to pump flange	9	1·2
Thermostat housing to cylinder head	8	1·1
Cylinder block drain plug	27	3·7

1 General description

The engine cooling water is circulated by a thermo-syphon, water pump assisted system. The coolant is pressurised. This is primarily to prevent premature boiling in adverse conditions and to allow the engine to operate at its most efficient running temperature; this being just under the boiling point of water. The overflow pipe from the radiator is connected to an expansion chamber which makes topping-up virtually unnecessary. The coolant expands when hot, and instead of being forced down an overflow pipe and lost, it flows into the expansion chamber. As the engine cools, the coolant contracts and because of the pressure differential, flows back into the radiator.

The cap on the expansion chamber is set to a pressure of 15 lbf in² (1.05 kgf/cm²) which increases the boiling point of the coolant to 230°F. If the coolant temperature exceeds this figure and the coolant boils, the pressure in the system forces the internal valve of the cap off its seat thus exposing the expansion tank overflow pipe down which the steam from the boiling coolant escapes and so relieves the pressure. It is therefore important to check that the expansion chamber cap is in good condition and that the spring behind the sealing washers has not weakened. Check that the rubber seal has not perished and its seating in the neck is clean to ensure a good seal. A special tool which enables a cap to be pressure tested is available at some garages.

The cooling system comprises the radiator, top and bottom hoses, heater hoses, the impeller water pump (mounted on the front of the engine it carries the fan blades and is driven by a belt from the crankshaft pulley) and the thermostat.

The system functions as follows: Cold coolant from the radiator circulates up the lower radiator hose to the water pump where it is pushed round the water passages in the cylinder block, helping to keep the cylinder bores and pistons cool.

The coolant then travels up into the cylinder head and circulates round the combustion spaces and valve seats absorbing more heat. Then, when the engine is at its normal operating temperature, the coolant travels out of the cylinder head, past the now open thermostat into the upper radiator hose and so into the radiator. The coolant passes along the radiator from one side to the other where it is rapidly cooled by the rush of cold air through the horizontal radiator core. The coolant, now cool, reaches the bottom hose when the cycle is repeated.

When the engine is cold the thermostat (a valve which opens and closes according to the temperature of the coolant) maintains the circulation of the same coolant in the engine and only when the correct minimum operating temperature has been reached, as shown in the Specifications, does the thermostat begin to open thus allowing the coolant to return to the radiator.

2 Cooling system – draining

1 If the engine is cold, remove the pressure relief cap from the expansion tank (photo). If the engine is hot, then turn the cap very slightly to release the pressure in the system. Use a rag over the cap to protect your hands from escaping steam. If the engine is hot and the cap is released suddenly, the drop in pressure can result in the coolant boiling.
2 If the coolant is to be re-used, drain it into a container of at least 10 pints (5.7 litre) capacity.
3 Undo and remove the cylinder block drain plug. This is located on the right-hand side of the block.
4 Slacken the bottom radiator hose clip and pull the hose off the radiator connection.

2.1 Removing the expansion tank cap

5 Remove the coolant filler cap from the thermostat housing and withdraw the thermostat.
6 When the coolant has finished draining out of the cylinder block drain hole, probe the orifice with a short piece of wire to dislodge any particles of rust or sediment which may be causing a blockage thus preventing complete draining.

3 Cooling system – flushing

1 In time, the cooling system will gradually lose its efficiency as the radiator becomes choked with rust, scale deposits from the water, and other sediment. To clean the system out, remove the coolant filler cap from the thermostat housing and lift out the thermostat.
2 Disconnect the bottom radiator hose and remove the cylinder block drain plug, then leave a hose running in the thermostat housing for ten to fifteen minutes.
3 In very bad cases, the radiator should be reverse flushed. This can be done with the radiator in position. Disconnect the top and bottom hoses from the radiator and connect a supply of running water to the bottom hose connection on the radiator.
4 When flushing the cooling system, it is recommended that some polythene sheeting is placed over the engine to prevent water finding its way into the electrical system.

4 Cooling system – filling

1 Fit the cylinder block drain plug and reconnect the bottom hose.
2 Fill the system slowly to ensure that no air-locks develop. Check that the heater control is set at hot, otherwise an air-lock may form in the heater.
3 Use an anti-freeze solution (see Section 12) and keep filling the system until the coolant flows into the expansion tank, then refit the coolant filler cap.
4 Now top-up the expansion tank to the level marked on the tank. Refit the expansion tank cap.
5 Start the engine and run it for half a minute at a fast idle, then switch off.
6 Remove the filler cap and top-up the cooling system through the thermostat housing, then refit the thermostat in the thermostat

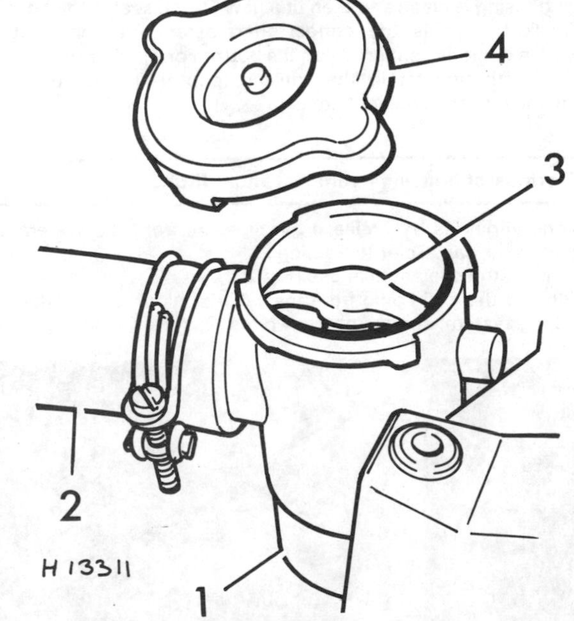

Fig. 2.1 The cooling system is filled through the thermostat housing (Sec 4)

1 Thermostat housing
2 Top hose
3 Thermostat
4 Filler cap

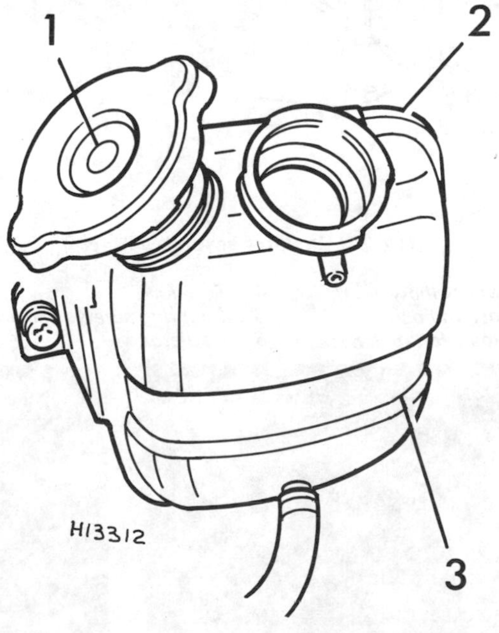

Fig. 2.2 Cooling system expansion tank (Sec 4)

1 Expansion tank cap
2 Expansion tank
3 Coolant level

housing (see Section 7) and refit the filler cap (Fig. 2.1).
7 Run the engine until it has reached its normal operating temperature. Stop the engine and allow it to cool.
8 Top up the expansion tank to the level marked (Fig. 2.2).

5 Radiator – removal and refitting

1 Drain the cooling system as described in Section 2.
2 Slacken the clip that secures the expansion tank hose to the radiator. Carefully ease the hose from the union pipe on the radiator (Fig. 2.3).
3 Slacken the clips that secure the radiator top and bottom hoses to the radiator inlet pipes and carefully ease the two hoses from these pipes (photos).
4 Undo and remove the screws, with spring and plain washers, that secure the two top radiator mounting brackets to the front panel. Lift away these two brackets (photo).
5 The radiator may now be lifted up from its lowest mountings and away from the front of the car.
6 Refitting the radiator is the reverse sequence to removal. Refill the cooling system as described in Section 4. Carefully check to ensure that all hose joints are water tight.

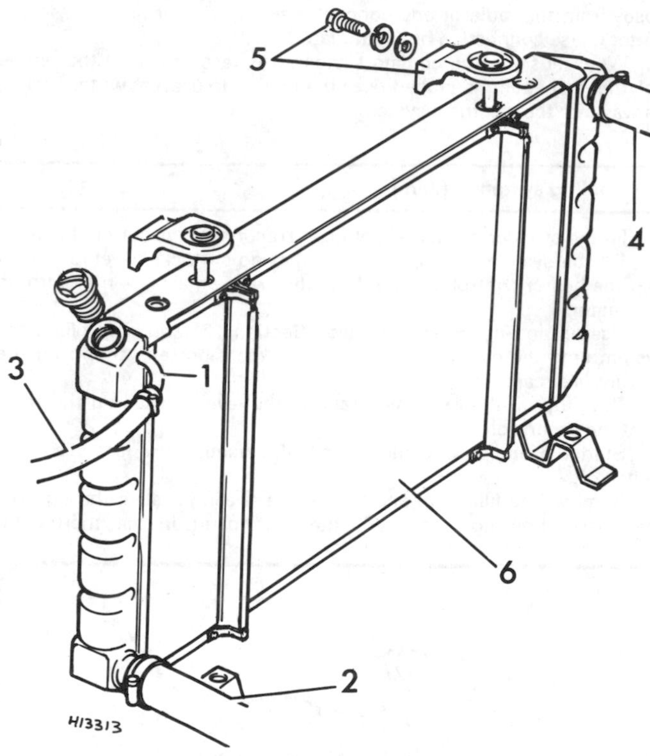

Fig. 2.3 Radiator assembly (Sec 5)

1 Outlet elbow
2 Bottom hose
3 Expansion tank hose
4 Top hose
5 Securing screws and bracket
6 Radiator

6 Radiator – inspection and cleaning

1 With the radiator out of the car, any leaks can be soldered up or temporarily repaired with a suitable proprietary substance. Clean out the inside of the radiator by flushing as described in Section 3.
2 When the radiator is out of the car, it is advantageous to turn it upside down for reverse flushing. Clean the exterior of the radiator by hosing down the radiator matrix with a strong jet of water to clean away road dirt, dead flies etc.
3 Inspect the radiator hoses for cracks, internal and external perishing and damage caused by overtightening of the hose clips. Refit the hoses as necessary. Examine the radiator hose clips and renew them if they are rusted or distorted.

7 Thermostat – removal, testing and refitting

1 If the engine is hot, release the pressure at the expansion tank, refer to Section 2 before removing the filler cap.
2 With the coolant filler cap removed, withdraw the thermostat from the thermostat housing. It may be necessary to use a pair of pliers to pull the thermostat from its seating in the housing (photo).
3 Test the thermostat for correct functioning by suspending it, together with a thermometer on a string, in a container of cold water. Heat the water and note the temperature at which the thermostat begins to open. This should be 82°C (180°F) for a standard thermostat. It is advantageous in winter to fit a thermostat that does not open until 88°C (190°F). Discard the thermostat if it opens too early. Continue heating the water until the thermostat is fully open. Then let it cool down naturally. If the thermostat does not fully open in boiling water, or does not close down as the water cools, then it must be discarded and a new one fitted. If the thermostat is stuck open when cold, this will be apparent when removing it from the housing.
4 When refitting the thermostat, ensure that the inside of the thermostat housing is clean and then fit a new O-ring seal in the housing.
5 Check that it is the correct thermostat. The temperature is stamped in degrees centigrade on the bottom of the thermostat.
6 Fit the thermostat in the housing (photo). Top-up the cooling system and refit the coolant filler cap.

8 Thermostat housing – removal and refitting

1 If the engine is hot, release the pressure from the system at the expansion tank cap. Refer to Section 2.
2 Remove the coolant filler cap (Fig. 2.4).
3 Remove the drain plug from the cylinder block and partially drain the cooling system; approximately 4 pints (2.2 litres) is enough.

5.3a Disconnect the radiator top hose ...

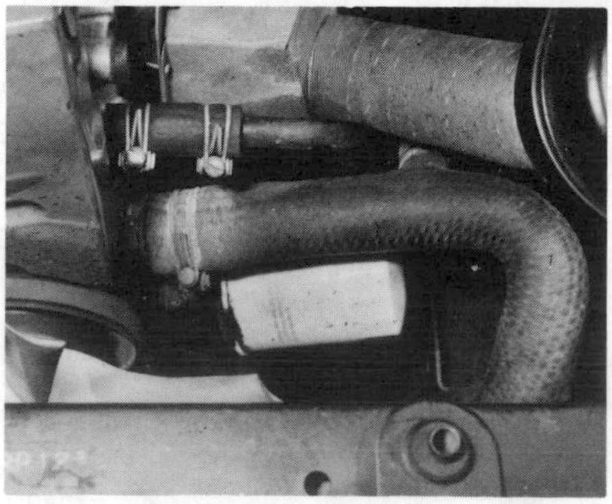

5.3b ... and the bottom hose

5.4 A radiator mounting bracket (arrowed)

7.2 Using pliers to remove the thermostat

7.6 Refitting the thermostat in the housing

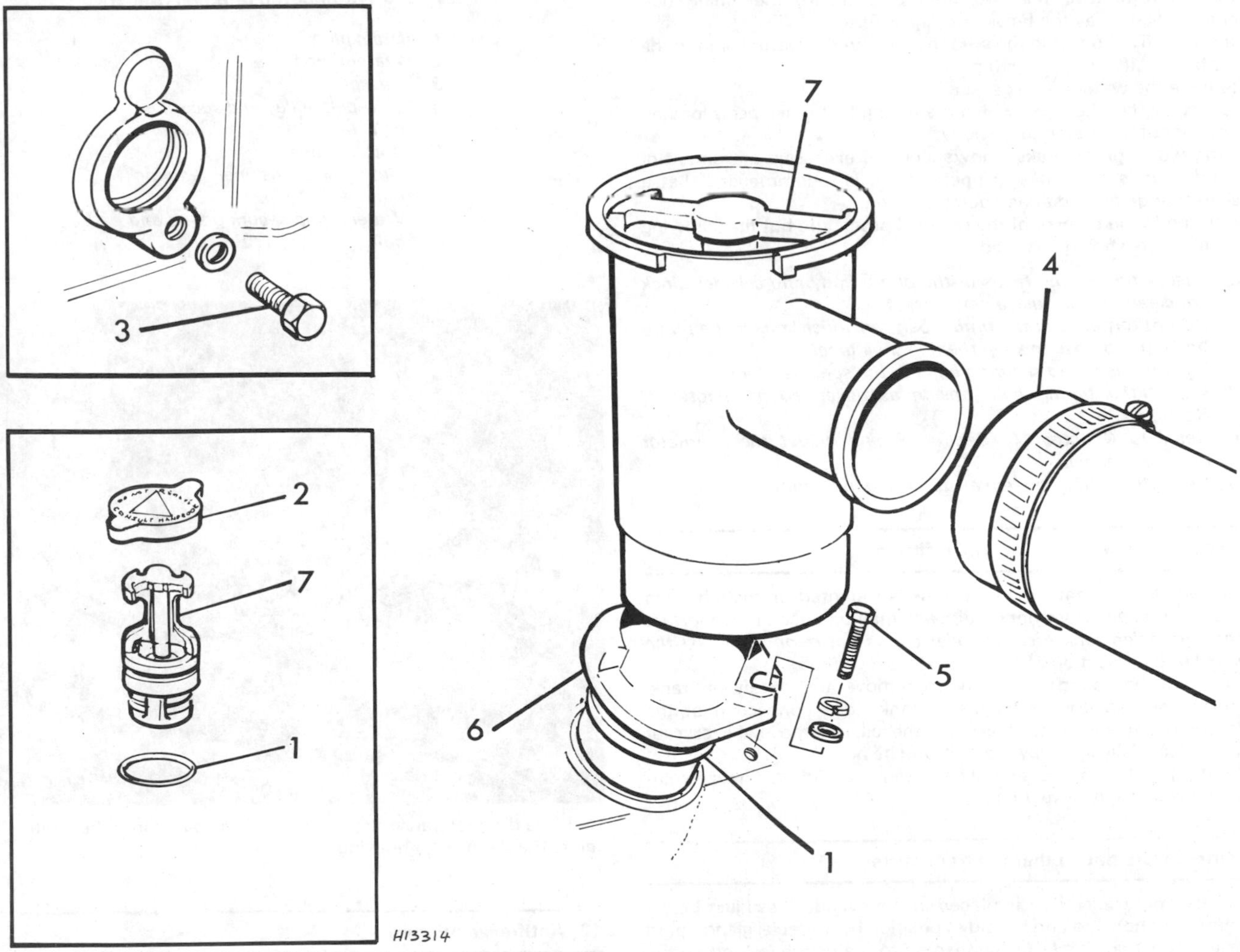

Fig. 2.4 Removing the thermostat housing (Sec 8)

1 O-ring seals
2 Filler cap
3 Drain plug
4 Top hose
5 Housing securing bolt
6 Thermostat housing
7 Thermostat

4 Refit the drain plug.
5 Slacken the securing clip and disconnect the top hose from the thermostat housing.
6 Undo and remove the bolt that secures the housing to the cylinder head and lift away the thermostat housing complete with thermostat. Remove the O-ring seal.
7 Remove the thermostat; see Section 7.
8 Refitting is the reverse of the removal procedure. Ensure that the seating in the cylinder head is clean and fit a new O-ring seal. Tighten the housing securing bolt to the specified torque.
9 Refit the thermostat when refilling the cooling system as described in Section 4.

9 Water pump – removal and refitting

1 Drain the cooling system as described in Section 2.
2 Remove the radiator as described in Section 5.
3 Slacken the alternator pivot and adjusting link bolts and remove the drivebelt.
4 Undo the securing bolts and remove the fan and water pump pulley.
5 Remove the timing belt cover by pulling it off its mounting spigots.
6 Slacken the securing clip and disconnect the bottom hose from the water pump inlet.
7 Undo and remove the two nuts that secure the timing belt tensioner and lift away the tensioner (Fig. 2.5).
8 Undo the five bolts and the stud that secure the water pump to the cylinder block. Lift away the pump.
9 Remove the water pump gasket.
10 Collect the two bolts on which the timing belt tensioner is located from the rear of the water pump body.
11 If the water pump leaks, shows signs of excessive movement of the spindle, or is noisy during operation, it is recommended that a service exchange reconditioned pump is fitted.
12 Refitting is the reverse of the removal procedure but the following additional points should be noted:

(a) Ensure the mating faces of the pump body and cylinder block are clean. Always use a new gasket
(b) Do not forget to fit the timing belt tensioner bolts in the pump body; use grease to keep them in their location
(c) Tighten the pump securing bolts to the specified torque
(d) Adjust the timing belt tension as described in Chapter 1, Section 11
(e) Refer to Chapter 10, Section 9 and adjust the alternator drivebelt tension
(f) Refill the cooling system as described in Section 4

10 Expansion tank – removal and refitting

1 The radiator coolant expansion tank is mounted on the left-hand inner wing panel and does not require any maintenance. *It is important that the expansion tank pressure filler cap is not removed whilst the engine is hot* (see Section 2).
2 Should it be found necessary to remove the expansion tank, disconnect the radiator to expansion tank hose connection at the radiator union, having first slackened the clip. Remove the bracket screws and carefully lift away the tank and its hose.
3 Refitting is the reverse sequence to removal. Add antifreeze solution until it is up to the level mark.

11 Temperature gauge thermal transmitter

1 The thermal transmitter is placed in the cylinder head just below the thermostat housing and is held in position by a special gland nut to ensure a water tight joint. It is connected to the gauge located on the instrument panel by a cable on the main ignition feed circuit and a special bi-metal voltage stabilizer (photo).
2 To remove the thermal transmitter, partially drain the cooling system; usually 4 pints (2.27 litres) is enough. Unscrew the transmitter gland nut from the side of the cylinder head. Withdraw the thermal transmitter. Refitting is the reverse procedure to removal.
3 For information on removing and refitting the gauge refer to Chapter 10.

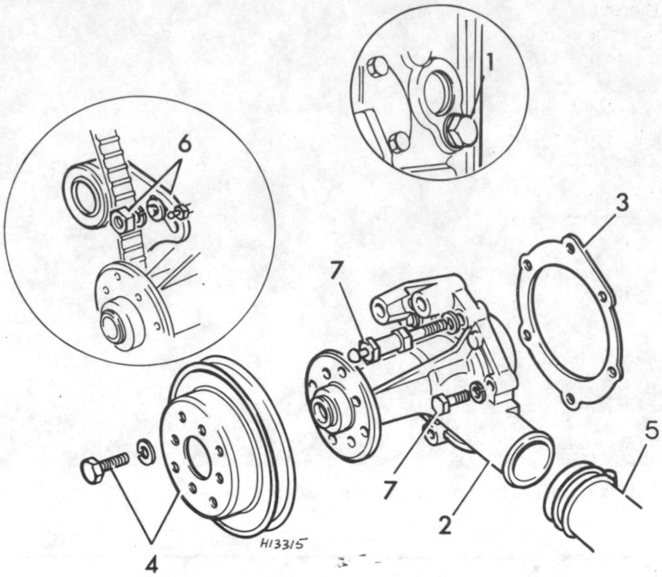

Fig. 2.5 Water pump removal (Sec 9)

1 Drain plug
2 Water pump
3 Gasket
4 Drivebelt pulley and securing bolt
5 Bottom hose
6 Timing belt tensioner securing nuts
7 Water pump securing stud and bolts

11.1 The thermal transmitter (arrowed) is screwed into the head below the thermostat housing

12 Antifreeze mixture

1 Prior to anticipated freezing conditions, it is essential that some antifreeze is added to the cooling system.
2 Any antifreeze which conforms with specification BS 3151 or BS 3152 can be used. Never use an antifreeze with an alcohol base as evaporation is too high.
3 Antifreeze with an anti-corrosion additive can be left in the cooling system for up to two years, but after six months it is advisable to have the specific gravity of the coolant checked at your local garage and

thereafter, every three months.

4 The amounts of antifreeze which should be added to ensure adequate protection down to the temperature given:

Amount of antifreeze	Protection to
33% mixture – 4 pints (2.3 litres)	*-19°C (-2°F)*
50% mixture – $5\frac{1}{2}$ pints (3.1 litres)	*-36°C (-33°F)*

13 Fault diagnosis – cooling system

Symptom	Reason/s
Heat generated in cylinder not being successfully disposed of by radiator	Insufficient water in cooling system Fan belt slipping (Accompanied by a shrieking noise on rapid engine acceleration) Radiator core blocked or radiator grill restricted Bottom water hose collapsed, impeding flow Thermostat not opening properly Ignition advance and retard incorrectly set (Accompanied by loss of power and perhaps misfiring) Carburettor incorrectly adjusted (mixture too weak) Exhaust system partially blocked Oil level in sump too low Blown cylinder head gasket Engine not yet run-in Brakes binding
Too much heat being dispersed by radiator	Thermostat jammed open Incorrect grade of thermostat fitted allowing premature opening of valve Thermostat missing
Leaks in system	Loose clips on water hoses Top or bottom water hoses perished and leaking Radiator core leaking Thermostat housing O-ring leaking Pressure cap spring worn or seal ineffective Blown cylinder head gasket Cylinder wall or head cracked

Chapter 3 Fuel and exhaust systems

Contents

Specifications

Air cleaner

Type	Renewable element with air temperature control valve

Fuel pump

Make and type	SU mechanical AUF 800 or AZX 1800
Pressure (minimum)	6·0 lbf/in² (0·4 bar)

Carburettor

Make and type	SU HIF6
Piston spring colour	Red
Jet size	0·100 in
Needle	BEK
Exhaust CO content	2·5%
Idling speed:	
Manual transmission	750 rpm
Automatic transmission	850 rpm
Fast idle speed	1100 rpm

Fuel tank

Capacity	11·5 gal (52 litre)

Torque wrench settings

	lbf ft	kgf m
Carburettor to manifold	19	2·5
Exhaust downpipe to manifold	21 to 23	2·9 to 3·2
Air cleaner to carburettor	19	2·5

1 General description

The fuel system comprises a fuel tank at the rear of the car, a mechanical pump located on the left-hand side of the camshaft cover and a single horizontally mounted SU carburettor.

A renewable paper element air cleaner is fitted which must be renewed at the recommended mileages. Operation of the individual components is described elsewhere in this Chapter.

2 Fuel pump – general description

The mechanically operated fuel pump is mounted on the left-hand side of the camshaft cover and is operated by a separate lobe on the camshaft.

As the camshaft rotates the rocker lever is actuated, one end of which is connected to the diaphragm operating rod. When the rocker arm is moved by the cam lobe the diaphragm, via a rocker arm, moves downwards causing fuel to be drawn in through the filter, past the inlet valve flap and into the diaphragm chamber. As the cam lobe moves round, the diaphragm moves upwards under the action of the spring, and fuel flows via the large outlet valve to the carburettor float chamber.

When the float chamber has the requisite amount of fuel in it, the needle valve in the top delivery valve line closes and holds the diaphragm down against the action of the diaphragm spring until the needle valve in the float chamber opens to admit more fuel.

3 Fuel pump – testing on engine

Presuming that the fuel lines and unions are in good condition and that there are no leaks anywhere, check the performance of the fuel pump in the following manner. Disconnect the fuel pipe at the carburettor inlet union, and the high tension lead to the coil and, with a suitable container or large rag in position to catch the ejected fuel, turn the engine over. A good spurt of petrol should emerge from the end of the pipe every second revolution.

4 Fuel pump – removal and refitting

1 Remove the fuel inlet and outlet connections from the fuel pump. Plug the ends of the pipes to stop loss of fuel or dirt ingress (Fig. 3.1).
2 Unscrew and remove the two pump mounting flange nuts and washers. Carefully slide the pump off the two studs followed by the insulating block assembly and gasket.
3 Refitting is the reverse sequence to removal. Inspect the gaskets on either side of the insulating block and if damaged, obtain and fit new ones.

5 Fuel pump – testing dry

If the pump is suspect, it may be dry tested by holding a finger over the inlet union and operating the rocker lever through three complete strokes. When the finger is released, a suction noise should be heard. Next, hold a finger over the outlet nozzle and press the rocker arm fully. The pressure generated should hold for a minimum of fifteen seconds.

6 Air cleaner element – renewal

1 Unscrew the wing nut and lift away the cover and element (Fig. 3.2).
2 Clean out the body of the air cleaner and fit a new element. Refit the cover.
3 Ensure that the sealing ring between the air cleaner and carburettor adaptor is not damaged or perished.
4 Refit the air cleaner to the adaptor and secure it with the fibre washer and wing nut.
5 To check the operation of the air temperature control valve, rotate and disconnect the air intake ducts from the control valve. Depress the valve plate and check that when released, it returns to its original position; if not, it must be renewed. Refit the air intake ducts (Fig. 3.3).

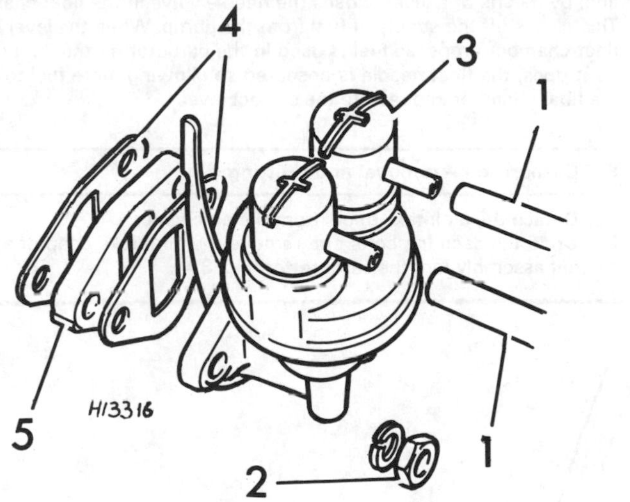

Fig. 3.1 Fuel pump removal (Sec 4)

1 *Fuel pipes*
2 *Nut and washer*
3 *Fuel pump*
4 *Gaskets (typical)*
5 *Insulator block*

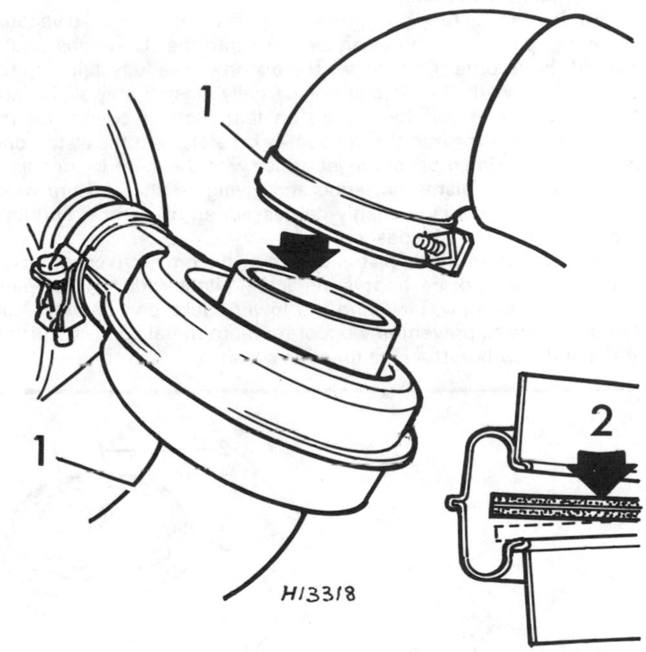

Fig. 3.3 Checking the air temperature control valve (Sec 6)

1 *Intake ducts*
2 *Valve plate*

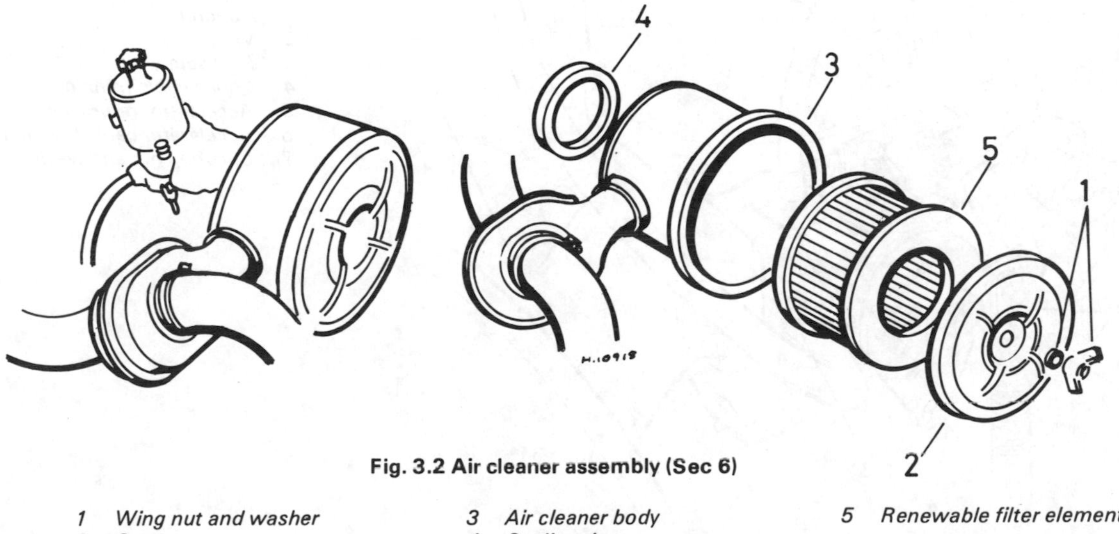

Fig. 3.2 Air cleaner assembly (Sec 6)

1 *Wing nut and washer*
2 *Cover*
3 *Air cleaner body*
4 *Sealing ring*
5 *Renewable filter element*

7 Carburettor – description

The variable choke SU carburettor as shown in Fig. 3.5 is a relatively simple instrument, and is basically the same irrespective of its size and type. It differs from most other carburettors in that instead of having a number of various sized fixed jets for different conditions, only one variable jet is fitted to deal with all possible conditions.

Air passing rapidly through the carburettor draws petrol from the jet so forming the petrol/air mixture. The amount of petrol drawn from the jet depends on the position of the tapered carburettor needle, which moves up and down the jet orifice according to the engine load and throttle opening, thus effectively altering the size of jet so that exactly the right amount of fuel is metered for the prevailing conditions.

The position of the tapered needle in the jet is determined by engine vacuum. The shank of the needle is held at its top end in a piston which slides up and down the dashpot in response to the degree of manifold vacuum.

With the throttle fully open, the full effect of inlet manifold vacuum is felt by the piston which has an air bleed into the choke tube on the outside of the throttle. This causes the piston to rise fully, bringing the needle with it. With the accelerator partially closed, only slight inlet manifold vacuum is felt by the piston (although of course, on the engine side of the throttle the vacuum is greater), and the piston only rises a little, blocking most of the jet orifice with the metering needle.

To prevent the piston fluttering and giving a richer mixture when the accelerator pedal is suddenly depressed, an oil damper and light spring are fitted inside the dashpot.

The only portion of the piston assembly to come into contact with the piston chamber or dashpot is the actual piston rod. All other parts of the piston assembly, including the lower choke portion, have sufficient clearance to prevent any direct metal to metal contact which is essential if the carburettor is to function correctly.

The jet is held in place by a horizontal arm. This is made of a bi-metallic material, so will vary the jet height to give compensation for temperature changes. These would otherwise give mixture variation due to fuel viscosity changes. This jet mounting arm is connected through a pivot to a lever. The lever is moved by a screw in the side of the carburettor body to adjust the mixture. The screw head may be hidden under a seal.

The rich mixture needed for cold starting is provided by a special jet. This has a progressive control to allow partial enrichment and is worked by turning a cam lever on the carburettor side opposite to that having the mixture control screw. When the cam lever is moved to enrich the mixture, the cam will push up the fast idle screw to open the throttle. The valve that controls this cold start mixture is a hollow inner core that is rotated within a cylindrical sleeve to bring a hole in it in-line with one in the sleeve.

An emulsion bypass passage runs from the jet bridge to the throttle. At small throttle openings, unevaporated fuel droplets will be drawn along this passage and will be mixed with the faster travelling air. To match this passage there is a slot cut out of the base of the piston.

The correct level of the petrol in the carburettor is determined by the level of the float chamber. When the level is correct the float rises and, by means of a lever, closes the needle valve in the float chamber. This closes off the supply of fuel from the pump. When the level in the float chamber drops, as fuel is used in the carburettor, the float drops. As it does, the float needle is unseated so allowing more fuel to enter the float chamber and restore the correct level.

8 Carburettor – removal and refitting

1 Detach the air intake ducts from the air cleaner.

2 Undo the securing bolts and remove the air intake adaptor and air cleaner assembly fom the carburettor (Fig. 3.4).

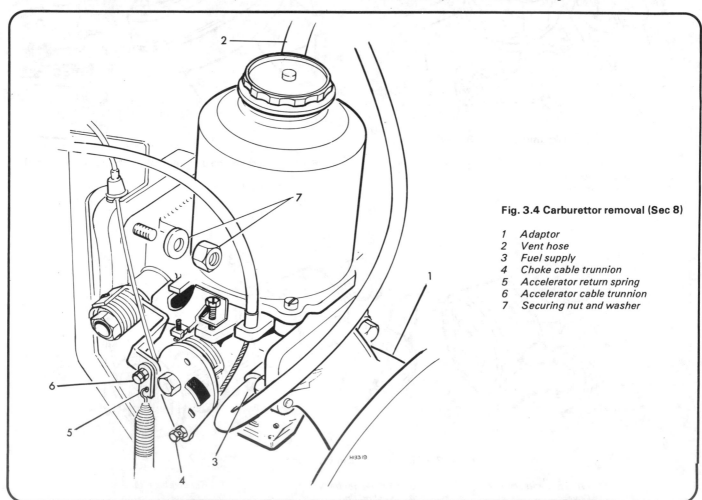

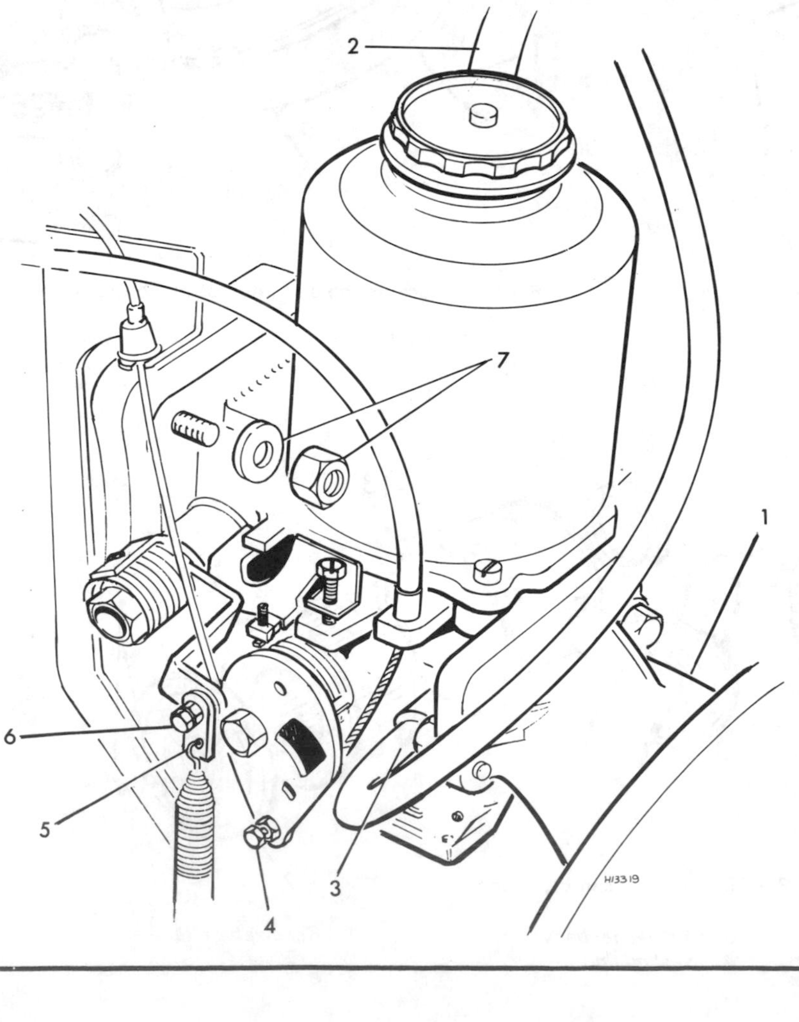

Fig. 3.4 Carburettor removal (Sec 8)

1 *Adaptor*
2 *Vent hose*
3 *Fuel supply*
4 *Choke cable trunnion*
5 *Accelerator return spring*
6 *Accelerator cable trunnion*
7 *Securing nut and washer*

3 Disconnect the fuel pipe from the carburettor. Plug the end to prevent the ingress of dirt.
4 Slacken the clip and ease off the engine breather pipe from the union on the carburettor body.
5 Disconnect both the accelerator and choke cables from the carburettor.
6 Unhook the accelerator return spring.
7 On automatic transmission models, disconnect the downshift cable from the carburettor linkage.
8 Undo and remove the four nuts and washers that secure the carburettor to the manifold studs. Withdraw the heater tube support bracket from the two bottom studs and lift away the carburettor.
9 Refitting is the reverse of the removal procedure. Always fit new gaskets to the inlet manifold flange and insulator block. Refer to Section 12 and adjust the choke control cable. Refer to Section 11 for details of the accelerator cable adjustment.

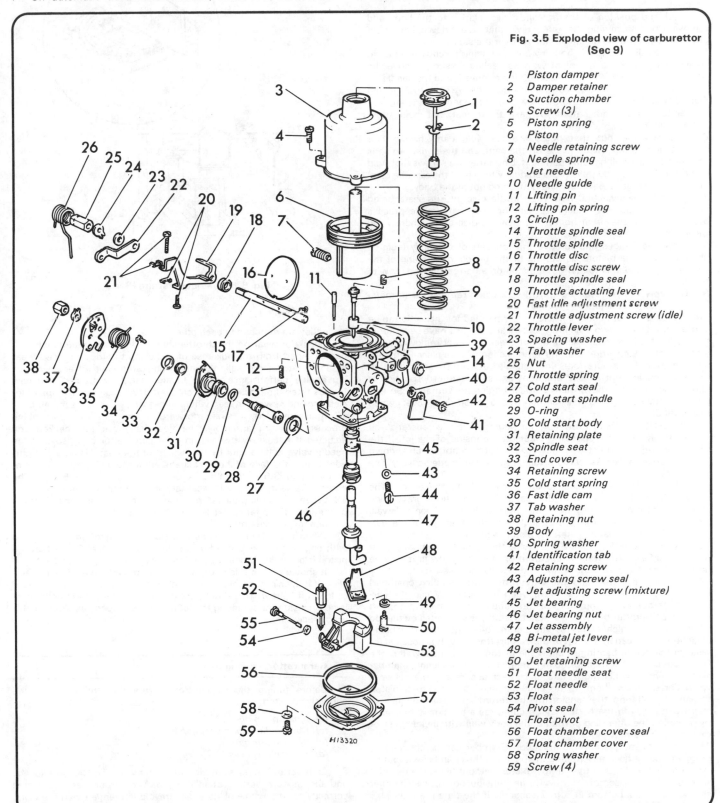

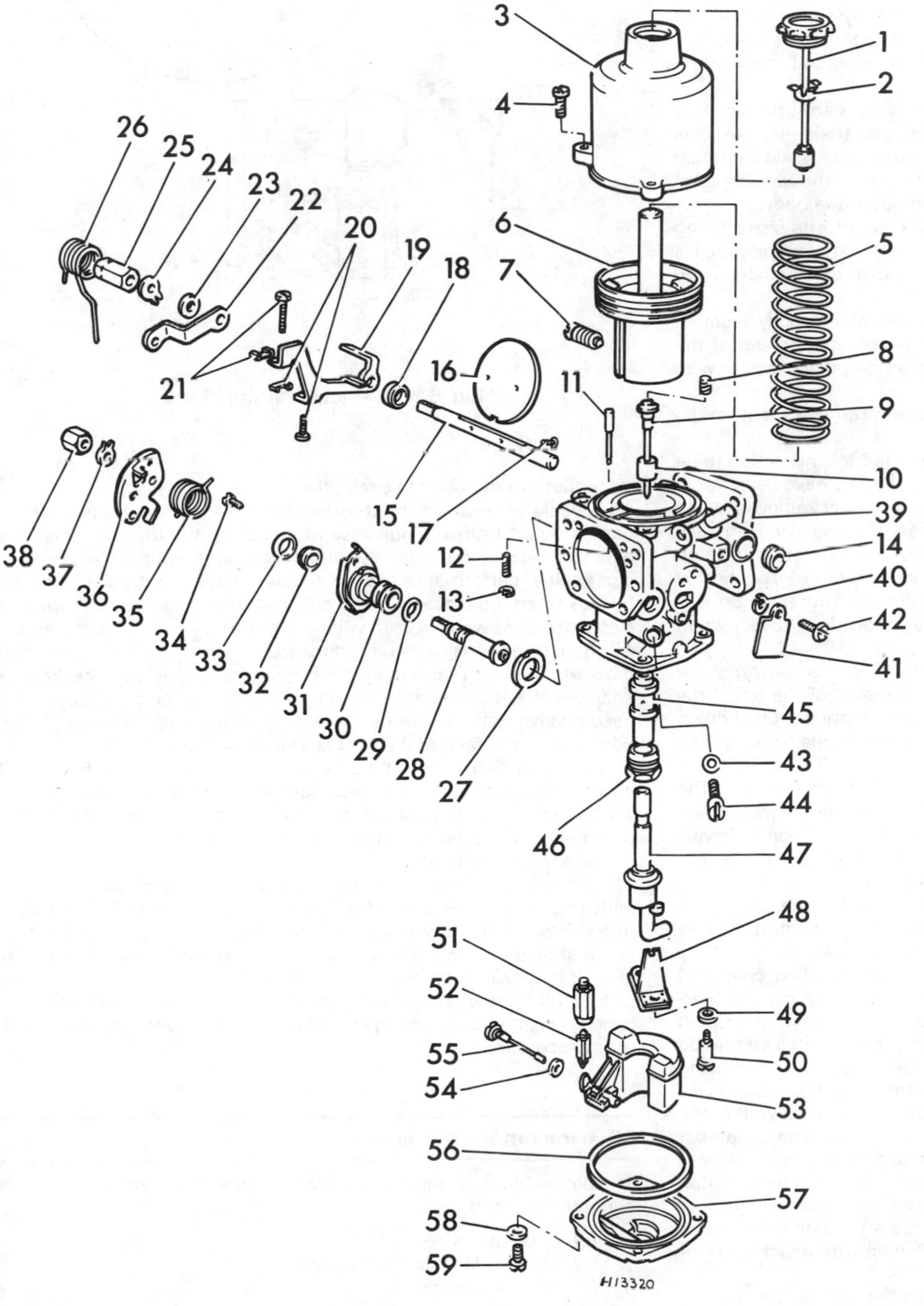

Fig. 3.5 Exploded view of carburettor (Sec 9)

1 *Piston damper*
2 *Damper retainer*
3 *Suction chamber*
4 *Screw (3)*
5 *Piston spring*
6 *Piston*
7 *Needle retaining screw*
8 *Needle spring*
9 *Jet needle*
10 *Needle guide*
11 *Lifting pin*
12 *Lifting pin spring*
13 *Circlip*
14 *Throttle spindle seal*
15 *Throttle spindle*
16 *Throttle disc*
17 *Throttle disc screw*
18 *Throttle spindle seal*
19 *Throttle actuating lever*
20 *Fast idle adjustment screw*
21 *Throttle adjustment screw (idle)*
22 *Throttle lever*
23 *Spacing washer*
24 *Tab washer*
25 *Nut*
26 *Throttle spring*
27 *Cold start seal*
28 *Cold start spindle*
29 *O-ring*
30 *Cold start body*
31 *Retaining plate*
32 *Spindle seat*
33 *End cover*
34 *Retaining screw*
35 *Cold start spring*
36 *Fast idle cam*
37 *Tab washer*
38 *Retaining nut*
39 *Body*
40 *Spring washer*
41 *Identification tab*
42 *Retaining screw*
43 *Adjusting screw seal*
44 *Jet adjusting screw (mixture)*
45 *Jet bearing*
46 *Jet bearing nut*
47 *Jet assembly*
48 *Bi-metal jet lever*
49 *Jet spring*
50 *Jet retaining screw*
51 *Float needle seat*
52 *Float needle*
53 *Float*
54 *Pivot seal*
55 *Float pivot*
56 *Float chamber cover seal*
57 *Float chamber cover*
58 *Spring washer*
59 *Screw (4)*

9 Carburettor – dismantling, inspection and reassembly

1 Assuming that you have the carburettor on the workbench, start by cleaning the exterior thoroughly well with paraffin or a degreasing solvent, using a stiff brush where necessary.
2 Undo the cap at the top of the carburettor and withdraw it complete with the small damper piston and retainer. Empty the oil from the dashpot.
3 Mark the position of the bottom cover relative to the body and remove it by unscrewing the four screws that hold it down. Empty out any fuel still in the fuel chamber and recover the seal.
4 The float is held to the body by a pivot having a screw head on it. Unscrew and remove the pivot with its sealing washer, remove the float, unscrew the needle valve socket and remove it and the needle.
5 Dismantle the various control linkages, being sure by studying Figs. 3.5 and 3.7 that you know how they fit together. It is an easy matter to sort this out before you take them apart but much more difficult when they are dismantled.
6 Unscrew the nut that holds the fast idle cam, having first straightened its tab washer; take off the cam, and the spring which is contained in a small housing behind it. Undo the two screws that hold down this housing and pull on the spindle which held the fast idle cam. The whole cold start assembly will now come out of the body.
7 Undo the screws that hold the throttle disc into its shaft, being careful not to put too much pressure on the shaft in the process (support it with the other hand). Remove the disc and withdraw the throttle shaft.
8 Mark the flanges and remove the top part of the body (suction chamber) and the piston. Be careful of the needle on the end of the piston. A good idea is to stand the piston on a narrow-necked jar with the needle hanging inside it.
9 Unscrew the jet retaining (pivot) screw and remove the bi-metal assembly holding the jet.
10 The carburettor is now sufficiently dismantled for inspection to be carried out. One or two adjusting screws and the like have been left in the body, but it is recommended that these are only removed when you are actually ensuring that the various channels are clear. Generally speaking, the SU carburettor is very reliable but even so it may develop faults which are not readily apparent unless a careful inspection is carried out, yet may nevertheless affect engine performance. So it is well worthwhile giving the carburettor a good look over when dismantled.
11 Inspect the carburettor needle for ridging. If this is apparent, you will probably find corresponding wear on the inside of the jet. If the needle is ridged, it must be renewed. Do not attempt to rub it down with abrasive paper as carburettor needles are made to very fine tolerances.
12 When fitting the needle, locate it carefully in the piston. The shoulder should be flush with the piston face and the engraved line should point directly away from the channel in the piston sidewall. Note that this makes the needle incline in the direction of the carburettor air cleaner flange when the piston is fitted.
13 Inspect the jet for wear. Wear inside the jet will accompany wear on the needle. If any wear is apparent on the jet, renew it. It may be unhooked from the bi-metal spring and this may be used again.
14 Inspect the piston and the carburettor body (suction chamber) carefully for signs that these have been in contact. When the carburettor is operating, the main piston should not come into contact with the carburettor body. The whole assembly is supported by the rod of the piston which slides in the centre bearings, this rod being attached to the cap in the top of the carburettor body. It is possible for wear in the centre bearing to allow the piston to touch the wall of the body. Check for this by assembling the piston in the suction chamber and spinning it whilst horizontal. If contact occurs and the cause is worn parts, renew them. In no circumstances try to correct piston sticking by altering the tension of the return spring, although very slight contact with the body may be cured (as a temporary measure) by polishing the offending portion of the body wall with metal polish or extremely fine emery cloth.
15 The fit of the piston in the suction chamber can be checked by plugging the air hole, assembling the piston in the chamber without its return spring and fitting the damper piston without filling the dashpot with oil. If the assembly is now turned upside down, the chamber should fall to the bottom in 5 to 7 seconds. If the time is appreciably less than this, the piston and suction chamber should both be renewed since they are matched to each other.
16 Check for wear on the throttle shaft and bushes through which it passes. Apart from the nuisance of a sticking throttle, excessive wear here can cause air leaks in the induction system thus adversely affecting engine performance. Worn bushes can be extracted and new bushes fitted if necessary. The cold start device can be dismantled for cleaning and new parts used where necessary, when reassembling.
17 Reassembly is a straightforward reversal of the dismantling process. During reassembly, the float level can be checked and adjusted if necessary by inverting the carburettor body so that the needle valve is held shut by the weight of the float. Using a straight-edge across the face of the float chamber measure the gap at the point arrowed (Fig. 3.6). It should be 0.04 ± 0.02 in (1.0 ± 0.5 mm). The arm can be bent carefully, if necessary, to obtain the dimensions.
18 When assembling the jet, position the adjusting screw so that the upper edge of the jet comes level with the bridge. This gives the initial position for jet adjustment.
19 When the carburettor is assembled, the dashpot should be filled with engine oil as described in Routine Maintenance. Check that the piston is operating properly by lifting it with the lifting pin and letting it fall. It should hit the bridge of the carburettor with an audible metallic click. If it does not, perhaps the needle is fouling the jet (it is supposed to touch it lightly). This should not occur with careful assembly; there is no provision for centering the jet but if it is properly assembled, this is not necessary.

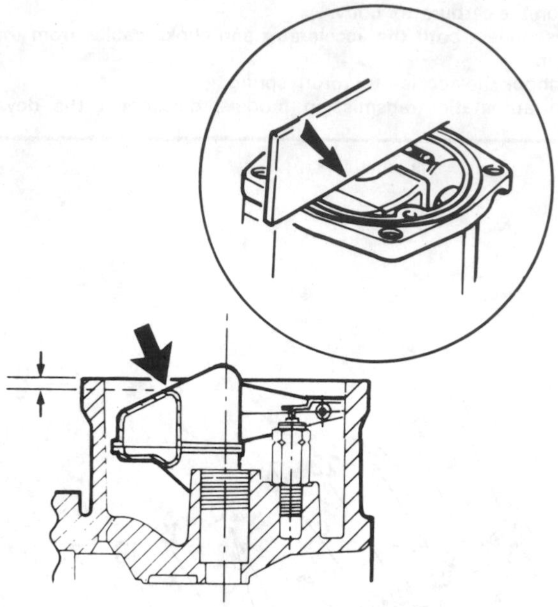

Fig. 3.6 Float level setting (Sec 9)

10 Carburettor – tuning

1 Before tuning the carburettor, ensure that the following are correctly adjusted:

(a) Ignition timing
(b) Contact breaker points gap
(c) Spark plugs gap
(d) Valve clearances

2 Carburettor tuning is limited to setting the idle and fast idle speeds and the mixture setting at idle speed. The carburettor is adjusted correctly for the whole of its engine revolution range when the idling mixture strength is correct.

3 If the carburettor has been dismantled, the initial setting is to turn the mixture adjusting screw until the top of the jet is flush with the bridge of the main body.
4 If the idle throttle setting has been lost, unscrew the throttle stop screw till the throttle is completely shut. Then screw it in again one complete turn. Screw in the fast idle screw until it is close to, but not touching, the cam on the choke mechanism.
5 Start and warm-up the engine.
6 Once the engine is warmed up, the fast idle screw should be adjusted to give 1100 rpm with the fast idle cam arrow aligned (Fig. 3.7). Push in the choke control and adjust the throttle stop screw to give the correct idle speed as given in the Specifications section.
7 Now adjust the mixture. The most accurate setting can be achieved using a gas analyser. If using an exhaust gas analyser, connect it to the engine in accordance with the manufacturer's instructions.
8 To weaken the mixture, screw the screw out (anti-clockwise). As the screw is coupled through the bi-metal there may be some lag in the movement of the jet. Tap the carburettor body to encourage it to find its new position. Also note where the mixture seems best when screwing in one direction; count the $\frac{1}{4}$ turns of the screw, going on past the correct position, and then coming back again, still counting the distance the screw has turned to try and note the correct position. This is half-way between the two positions which make the engine slow down because of mixture being too weak or too rich.
9 The correct setting for the jet can be found from a combination of engine speed, which should be as fast as possible, and listening to the exhaust note. The exhaust note should be smooth. If it is haphazardly irregular accompanied by a burping noise, with the engine still running fairly fast, a weak mixture is indicated. This can be confirmed by lifting the piston about 0.1 in (3 mm), either with the lifting device on the side of the carburettor or a very fine screwdriver; the engine should

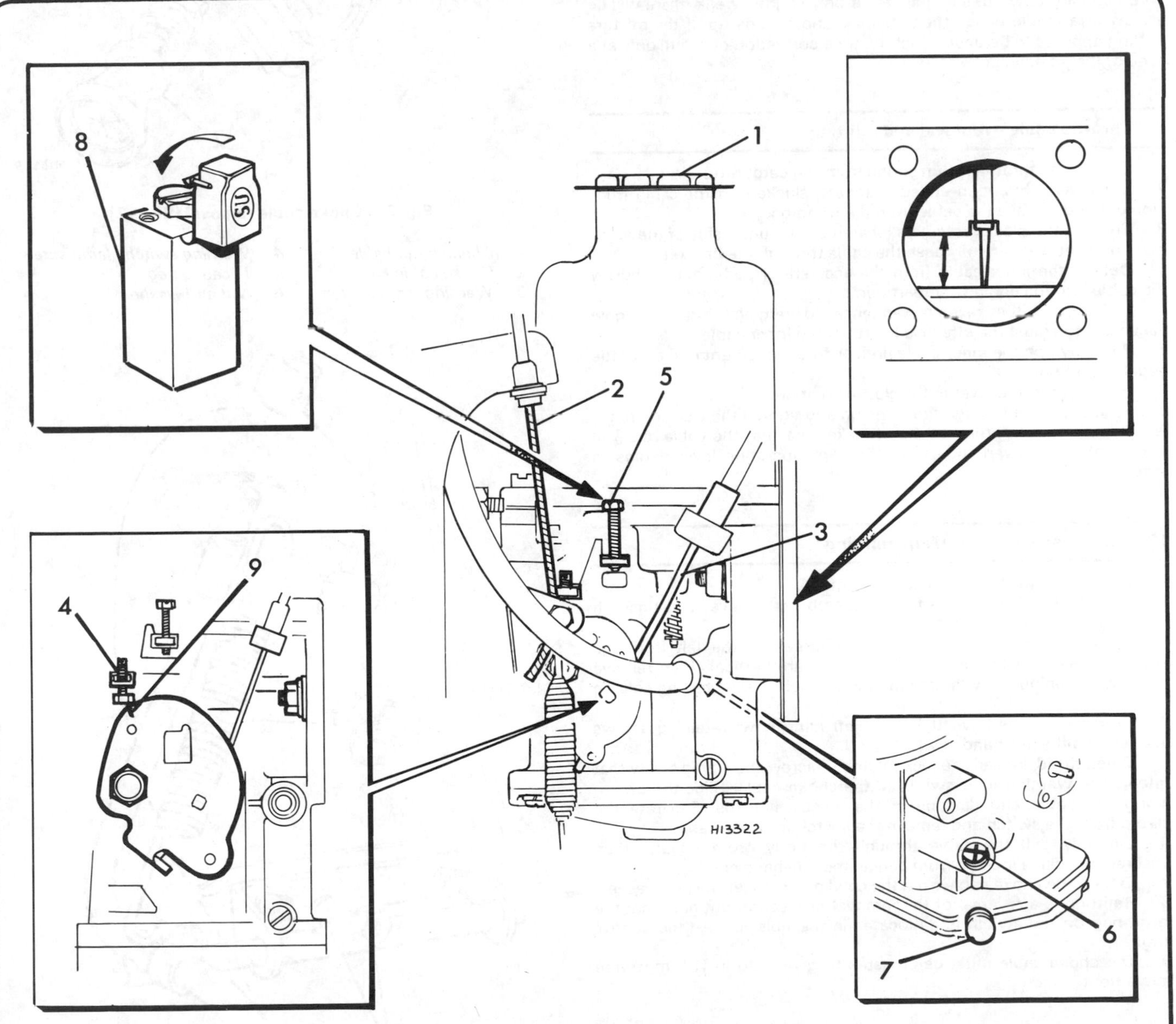

Fig. 3.7 Carburettor adjustments (Secs 10, 11 and 12)

1 Piston damper
2 Throttle cable
3 Choke cable
4 Fast idling adjustment screw
5 Idling speed adjustment screw
6 Mixture adjustment screw
7 Plug
8 Idling screw cover
9 Fast idle cam and arrow

immediately slow down and will be very liable to stall. If the mixture is too rich the idle speed will tend to be low, accompanied by a rhythmic sound from the exhaust. Lifting the piston about 0.1 in (3 mm) will result in the engine speeding up.

10 Moving the jet a $\frac{1}{4}$ turn of the adjusting screw from the correct setting should give an indication of weak or rich mixture. It is best to err in the direction of a weak mixture.

11 Finally, having got the mixture correct, recheck that the idle speed is correct.

12 If at times the engine does not seem to respond to adjusting of the mixture correctly, blip the throttle a few times. This will clear the petrol that will collect in the inlet manifold and burn the soot off the spark plugs.

13 On the road it might be found that the mixture has been set a trace rich or weak. Richness is apparent by the car idling well when cool, but when hot becoming lumpy and, after a few seconds, slower and more and more uneven. A weak mixture is indicated by a liability to stall when coming down to idling speed. If only slightly weak, there will be a slow erratic idle which then steadies and speeds up. If the mixture setting appears to be unsatisfactory, try a correction of $\frac{1}{4}$ turn only at a time of the jet adjusting screw.

11 Throttle cable – removal and refitting

1 Detach the throttle return spring from the carburettor.

2 Using two thin open-ended spanners, slacken off the cable trunnion nut and remove the cable from the trunnion.

3 Press in the plastic retainers located on the underside of the abutment bracket and carefully ease the cable through the bracket.

4 Detach the inner cable from the accelerator pedal and withdraw the cable into the engine compartment.

5 Refitting is the reverse sequence to removal but it is now necessary to adjust the effective length of the inner cable.

6 Pull down on the inner cable until all free movement of the throttle pedal is eliminated.

7 Hold the throttle lever in the closed position.

8 Feed the cable through the trunnion and tighten the trunnion nut.

9 Depress the throttle pedal and make sure that the cable has $\frac{1}{16}$ in (1.6 mm) free movement before the cam operating lever begins to move.

12 Choke cable – removal and refitting

1 Disconnect the battery negative terminal.

2 Detach the choke cable from the carburettor lever and unclip it from the throttle cable (Fig. 3.8).

3 Move the steering wheel ninety degrees to the left from the straight-ahead position so that the spokes are vertical. Then remove the cowl retaining screw and withdraw the right-hand steering column cowl.

4 Unscrew and remove the three left-hand cowl retaining screws and remove the left-hand cowl.

5 When fitted, loosen the locknut and unscrew the clamp screw that retains the warning light switch to the choke cable. Slide the switch along the cable and disconnect the electrical leads. Separate the clamp from the switch and remove the switch from the cable.

6 Carefully pull the cable through the body grommet and then unscrew the cable locknut from behind the left-hand cowl.

7 Extract the choke cable from the cowl and recover the lockwasher.

8 Refitting is a reversal of the removal procedure, but note that the large peg on the switch body locates in the hole nearest the control knob.

9 The choke cable must be adjusted to give 0.06 in (1.5 mm) free movement.

13 Throttle pedal – removal and refitting

1 Disconnect the throttle return spring and retaining clip (Fig. 3.9).

2 Detach the throttle cable from the end of the pedal.

3 Undo and remove the two nuts and spring washers that secure the pedal bracket to the bulkhead panel.

4 Lift the pedal assembly from the mounting studs.

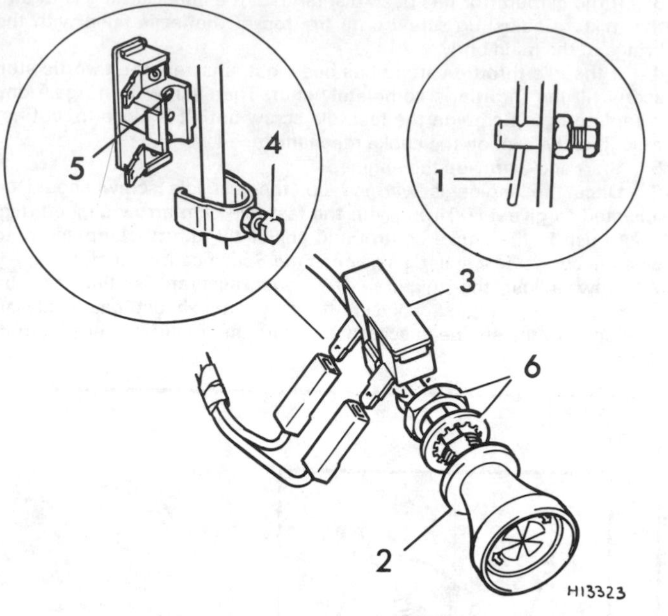

Fig. 3.8 Choke cable removal (Sec 12)

1 Choke inner cable
2 Choke cable knob
3 Warning lamp switch
4 Warning switch clamp screw
5 Locating peg
6 Nut and washer

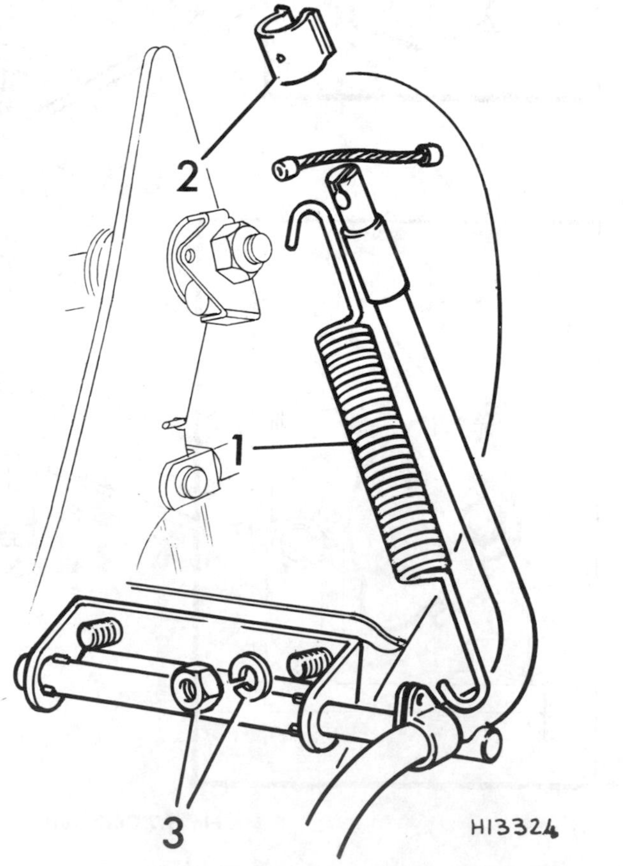

Fig. 3.9 Throttle pedal components (Sec 13)

1 Return spring
2 Retaining clip
3 Bracket retaining nut and washer

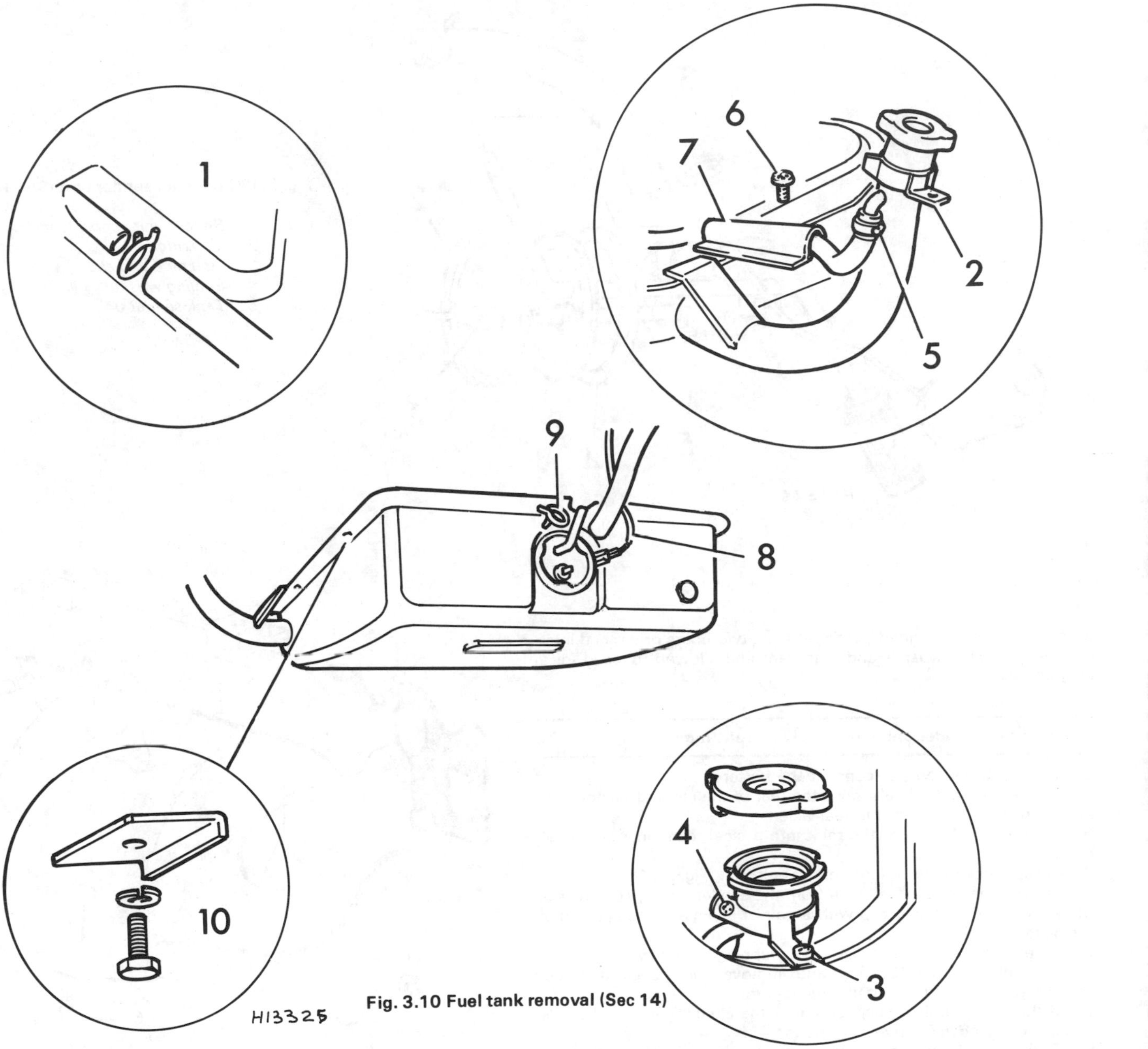

Fig. 3.10 Fuel tank removal (Sec 14)

1 Fuel pipe to tank connection
2 Filler neck clamp
3 Clamp screw
4 Clamp screw
5 Vent pipe
6 Self-tapping screw
7 Cover
8 Sender unit supply lead
9 Hose clip
10 Support plate

5 Refitting the throttle pedal is the reverse of the removal procedure.

14 Fuel tank – removal and refitting

1 For safety reasons, disconnect the battery.
2 Chock the front wheels, raise the rear of the car and support it on axle-stands located under the rear axle.
3 Syphon the fuel from the tank into a clean container of suitable capacity.
4 Unscrew and remove the retaining screw. Detach the filler neck clamp from the body then slacken the filler neck clamp screw. Turn the clamp clear of the body (Fig. 3.10).
5 Unscrew and remove the retaining screws and withdraw the pipe protective cover into the boot (saloon models only).
6 Disconnect the vent pipe from the filler neck.
7 Detach the cable terminal from the fuel tank sender unit.
8 Using a pair of pliers, compress the retaining clip that secures the fuel hose to the tank and ease off the hose.
9 Undo and remove the screws, spring washers and support plates that secure the fuel tank. Lower the tank from the body.
10 Refitting the fuel tank is the reverse of the removal procedure.

15 Fuel tank – cleaning

1 With time, it is likely that sediment will collect in the bottom of the fuel tank. Condensation, resulting in rust and other impurities, will usually be found in the fuel tank of any car more than three or four years old.
2 When the tank is removed, it should be vigorously flushed out and turned upside down. If facilities are available at the local garage, the tank may be steam cleaned and the exterior repainted with a lead based paint.
3 Never weld or bring a naked light close to an empty fuel tank until

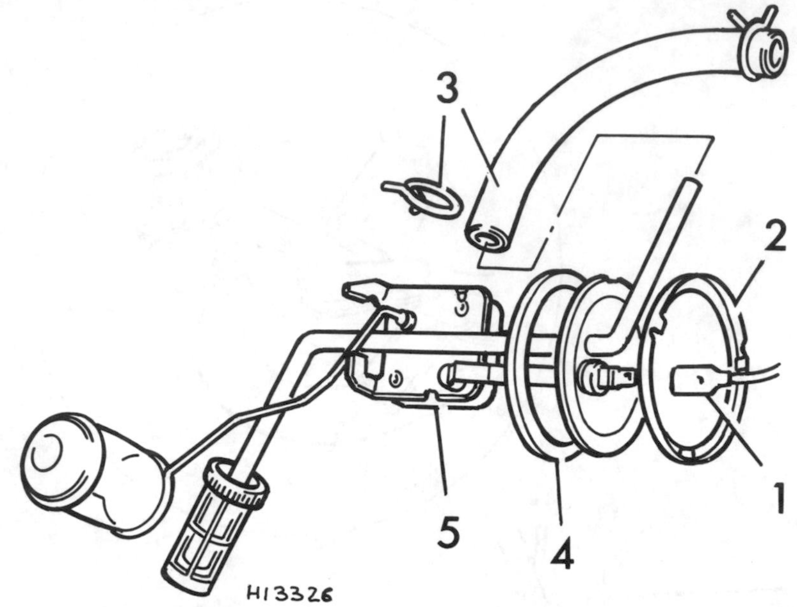

Fig. 3.11 Fuel tank sender unit (Sec 16)

1 Sender unit supply lead
2 Lockring
3 Fuel pipe and clip
4 Sealing washer
5 Tank sender unit

it has been steam cleaned out for at least two hours or washed internally with boiling washer and detergent and allowed to stand for at least three hours.

16 Fuel tank sender unit – removal and refitting

1 For safety reasons, disconnect the battery.
2 Chock the front wheels, raise the rear of the car and support it on axle-stands placed under the rear axle.
3 Syphon the fuel from the tank into a clean container of suitable capacity.
4 Disconnect the fuel gauge sender unit lead (Fig. 3.11).
5 Detach the flexible fuel hose from the sender unit by compressing the ears of the retaining clip with a pair of pliers and then pulling off the hose.
6 Release the sender unit lockring by tapping it round with a suitable punch. Unscrew the lockring and remove the gauge tank unit assembly. Take care not to bend the float rod.
7 If the sender unit is suspect, check the circuit, gauge and sender unit as described in Chapter 10, Section 48.
8 Refitting is the reverse of the removal procedure. Always fit a new sealing washer between the fuel gauge tank unit and the tank.

17 Exhaust system – general

The exhaust system consists of a cast iron manifold, a front pipe and muffler and a rear pipe and resonator.

The system is attached to the floor pan by two brackets and flexible support straps.

At regular intervals the system should be checked for corrosion, joint leakage, the condition and security of the flexible mountings and the tightness of the joints.

18 Combined inlet and exhaust manifold – removal and refitting

1 Jack up the front of the car and support it on axle stands or blocks.
2 From underneath the car, undo and remove the six nuts and washers securing the exhaust front pipe to the manifold studs.
3 Slacken the retaining clip and detach the front pipe from the gearbox steady bracket. Slide the front pipe down off the manifold studs and allow it to rest on blocks.

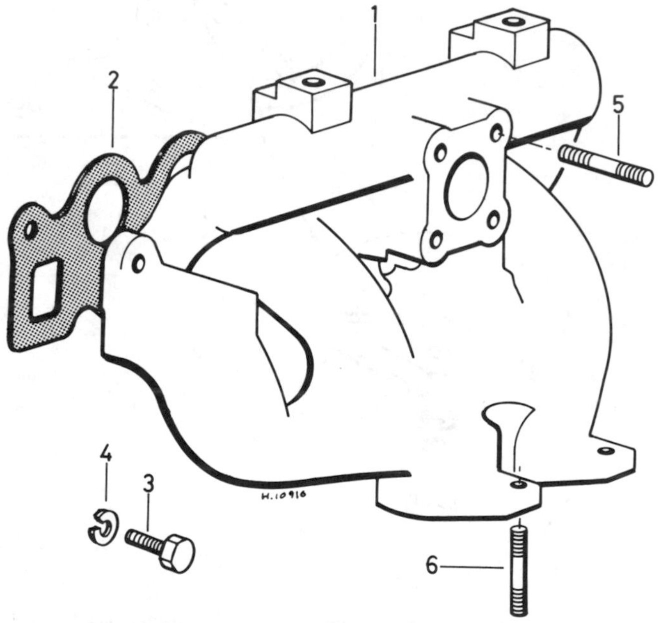

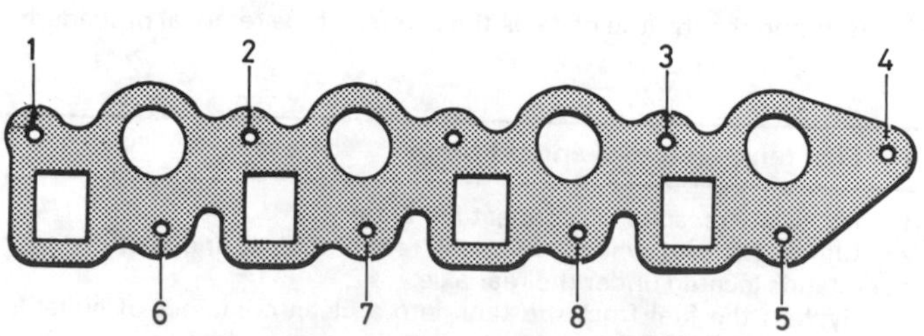

Fig. 3.12 Inlet and exhaust manifold assembly (Sec 18)

1 Manifold assembly
2 Gasket
3 Retaining bolt
4 Washer
5 Carburettor retaining stud
6 Front pipe retaining stud
Inset: Bolt tightening sequence

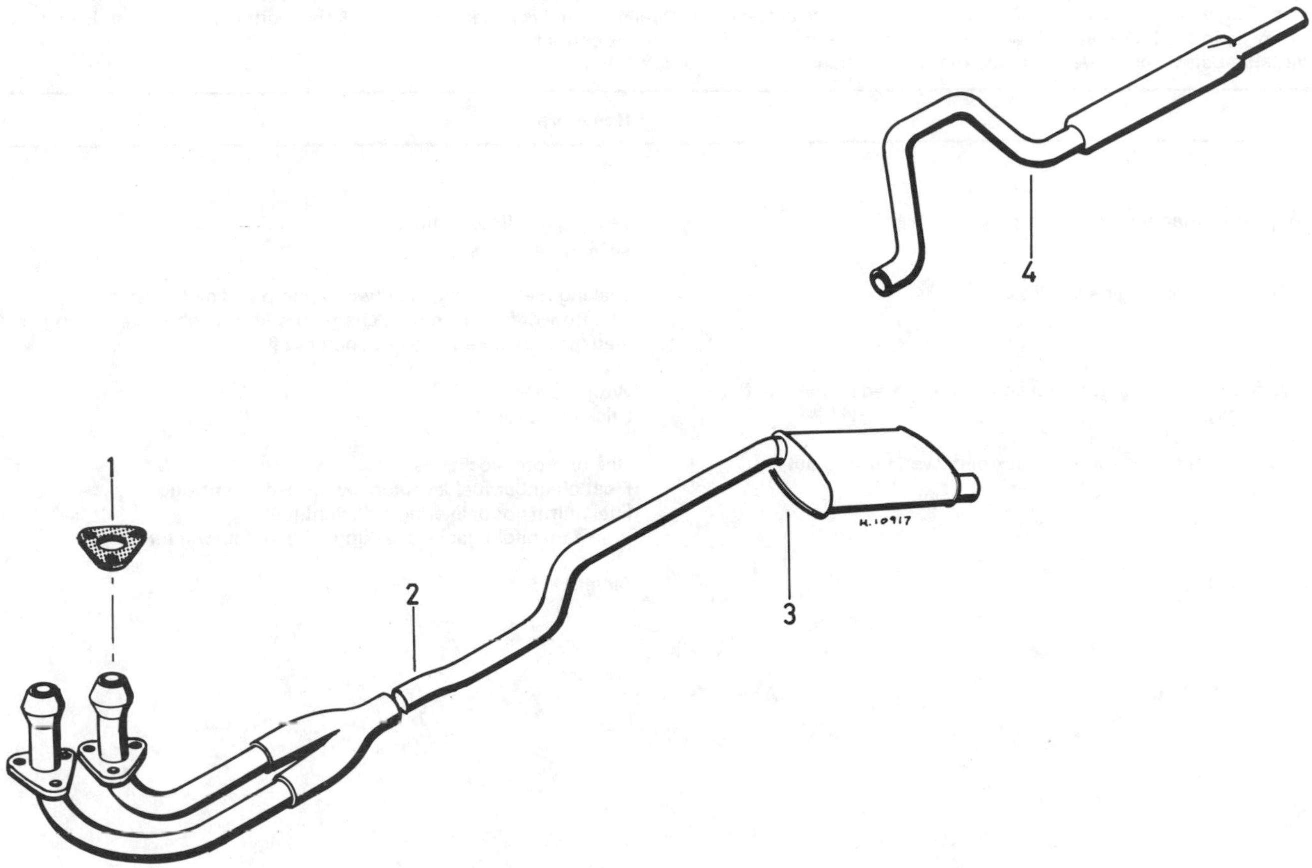

Fig. 3.13 Layout of exhaust system (Sec 19)

1 Gasket
2 Front pipe assembly
3 Silencer
4 Rear pipe assembly

4 Refer to Chapter 3 and remove the air cleaner assembly and the carburettor.
5 Detach the distributor vacuum advance hose from the union on the inlet manifold.
6 Remove the bolt securing the brake servo vacuum pipe banjo union to the inlet manifold. Recover the two copper washers.
7 Undo and remove the bolts securing the combined inlet and exhaust manifold to the cylinder head and lift away the manifold. Recover the gasket.
8 Before refitting the manifold, clean away all traces of old gasket from the mating faces of the cylinder head and manifold.
9 Make up two locating pegs by cutting the heads off two suitable bolts, then cut a screwdriver slot in the end of each one.
10 Refer to Fig. 3.12 and fit the locating pegs to holes 4 and 7 in the cylinder head.
11 Place a new gasket over the pegs then refit the manifold.
12 Refit the retaining bolts to the unoccupied holes and tighten bolt 5 finger-tight.
13 Remove the locating pegs, refit the remaining two bolts, then tighten all the bolts to the specified torque in the order shown.
14 The remainder of refitting is a reverse of the removal sequence.

19 Exhaust system – removal and refitting

1 Chock the wheels on the right-hand side of the car, jack up the left-hand side and support it on axle-stands.
2 Using a piece of wire or rope, support the front pipe.
3 Undo the six nuts that secure the twin flanges of the exhaust pipe to the manifold.
4 Detach the left-hand rear shock absorber from its mounting plate and push it inwards.
5 Remove the bolts that secure the flexible support straps.
6 Lift the exhaust assembly clear of the rear axle and withdraw it from under the car.
7 The front and rear pipes can be separated by undoing the securing clamp.
8 Refitting the exhaust system is the reverse of the removal procedures. Always use new seals between the exhaust pipe and manifold flanges. Do not tighten any of the clamps until the complete system is fitted, and then tighten them ensuring that the system is not under stress.

20 Fault diagnosis – fuel and exhaust systems

Unsatisfactory engine performance and excessive fuel consumption are not necessarily the fault of the fuel system or carburettor. In fact they more commonly occur as a result of ignition faults. Before acting on the fuel system it is necessary to check the ignition system first. Even though a fault may lie in the fuel system it will be difficult to trace unless the ignition is correct.

The table below therefore, assumes that the ignition system is in order:

Symptom	Reason/s
Smell of petrol when engine is stopped	Leaking fuel lines or unions Leaking fuel tank
Smell of petrol when engine is idling	Leaking fuel line unions between pump and carburettor Overflow of fuel from float chamber due to wrong level setting or ineffective needle valve or punctured float
Excessive fuel consumption for reasons not covered by leaks or float chamber faults	Worn needle Sticking needle
Difficult starting, uneven running, lack of power, cutting out	One or more blockages Float chamber fuel level too low or needle sticking Fuel pump not delivering sufficient fuel Intake manifold gaskets leaking, or manifold fractured
Fast idle; erratic	Air leak

Chapter 4 Ignition system

Contents

Specifications

Spark plugs

Type	Champion BN 9Y	
Electrode gap	0·035 in (0·90 mm)	
Firing order	1 – 3 – 4 – 2	
Ignition coil		
Type	Lucas HA 12 or Lucas 16 C6	
Primary resistance at 20°C (68°F):		
Lucas HA 12	3·1 to 3·5 ohms	
Lucas 16 C6	1·3 to 1·45 ohms	
Ballast resistance:		
Lucas 16 C6	1·3 to 1·5 ohms	
Distributor		
Type	Lucas 48D4	
Direction of rotation	Anti-clockwise	
Dwell angle	57° ± 5°	
Contact breaker gap	0·014 to 0·016 in (0·36 to 0·40 mm)	
Condenser capacity	0·18 to 0·24 mfd	
Static ignition timing	8° BTDC	
Dynamic timing at 1500 rpm	14° BTDC (vacuum pipe disconnected)	
Vacuum advance:		
Starts	5 in (127 mm) Hg	
Maximum	20° (crankshaft) at 13 in (330) mm) Hg	
Torque wrench settings	**lbf ft**	**kgf m**
Distributor flange securing nuts	18	2·5
Spark plugs	7	1·0

1 General description

In order that the engine may run correctly, it is necessary for an electrical spark to ignite the fuel/air mixture in the combustion chamber at exactly the right moment in relation to engine speed and load. The ignition system is based on supplying low tension voltage from the battery to the ignition coil, where it is converted into high tension voltage. The high tension voltage is powerful enough to jump the spark plug gap in the cylinders many times a second under high compression pressure, providing that the ignition system is in good working order and that all adjustments are correct.

The ignition system comprises two individual circuits known as the low tension and high tension circuits.

The low tension circuit (sometimes known as the primary circuit) comprises the battery, the lead to the ignition switch, the lead to the low tension or primary coil windings (terminal SW, via a ballast resistor on some models) and the lead from the low tension coil windings (terminal CB) to the contact breaker points and condenser in the distributor.

The high tension (secondary circuit) comprises the high tension or secondary coil windings, the heavily insulated ignition lead from the centre of the coil to the centre of the distributor cap, the rotor arm, the spark plug leads and the spark plugs.

The complete ignition system operation is as follows: Low tension voltage from the battery is changed within the ignition coil to high tension voltage by the opening and closing of the contact breaker points in the low tension circuit. High tension voltage is then fed via the carbon broush in the centre of the distributor cap to the rotor arm of the distributor. The rotor arm revolves inside the distributor cap and each time it comes into line with one of the four metal segments in the cap, these being connected to the spark plug leads, the opening and closing of the contact breaker points causes the high tension voltage to build up, jump the gap from the rotor arm to the appropriate metal segment and so, via the spark plug lead, to the spark plug where it finally jumps the gap between the two spark plug electrodes, one being connected to the earth system.

The ignition timing is advanced and retarded automatically to ensure the spark occurs at just the right instant for the particular load at the prevailing engine speed.

The ignition advance is controlled by a mechanical and vacuum operated system.The mechanical governor mechanism comprises two lead weights which move out under centrifugal force from the central distributor shaft as the engine speed rises. As they move outwards they rotate the cams relative to the distributor shaft, and so advance the spark. The weights are held in position by two springs, and it is the tension of the springs which is largely responsible for correct spark advancement.

The vacuum control comprises a diaphragm, one side of which is connected via a small bore tube to the inlet manifold, and the other side to the contact breaker plate. Depression in the induction manifold and carburettor, which varies with engine speed and throttle opening, causes the diaphragm to move, so moving the contact breaker plate and advancing or retarding the spark. A fine degree of control is achieved by a spring in the vacuum assembly.

2 Contact breaker points – adjustment

1 To adjust the contact breaker points so that the correct gap is obtained, first undo the two screws that secure the distributor cap to the distributor body, and lift away the cap (photo). Clean the inside and outside of the cap with a dry cloth. It is unlikely that the four segments will be badly burned or scored, but if they are, the cap must be renewed. If only a small deposit is on the segments it may be scraped away using a small screwdriver.

2 Push in the carbon bush located in the top of the cap several times to ensure that it moves freely. The bush should protrude at least $\frac{1}{4}$ in (6·35 mm).

3 Gently prise the contact breaker points open to examine the condition of their faces. If they are rough, pitted or dirty it will be necessary to remove them to enable new points to be fitted.

4 Presuming the points are satisfactory, or that they have been cleaned or renewed, measure the gap between the points by turning the engine over until the contact breaker arm is on the peak of one of the four cam lobes. A 0·014 to 0·016 in (0·36 to 0·40 mm) feeler gauge should now just fit between the points (photo).

5 If the gap varies from this amount, slacken the contact plate securing screw and adjust the contact gap by inserting a screwdriver in the notched hole at the end of the plate, turning clockwise to decrease, and anti-clockwise to increase the gap. Tighten the securing screw and recheck the gap (Fig. 4.1).

6 Refit the rotor arm and distributor cap. Tighten the distributor cap securing screws.

2.1 Distributor with cap and rotor removed

3 Contact breaker points – removal and refitting

1 If the contact breaker points are burned, pitted or badly worn, they must be renewed.

2 Undo the distributor cap securing screws and lift off the cap.

3 Pull the rotor arm off the cam spindle.

4 Undo the screw that secures the contact set to the moving plate, then lift up the contact set. Press the contact set spring and release the terminal plate from the spring.

5 Lift away the contact set assembly.

6 To refit the contact set, connect the terminal plate with the black lead uppermost, into the end of the contact spring.

7 Locate the contact set securing screw, spring and plain washer in the slot in the contact breaker adjustable plate.

2.4 Measuring the points gap with a feeler gauge

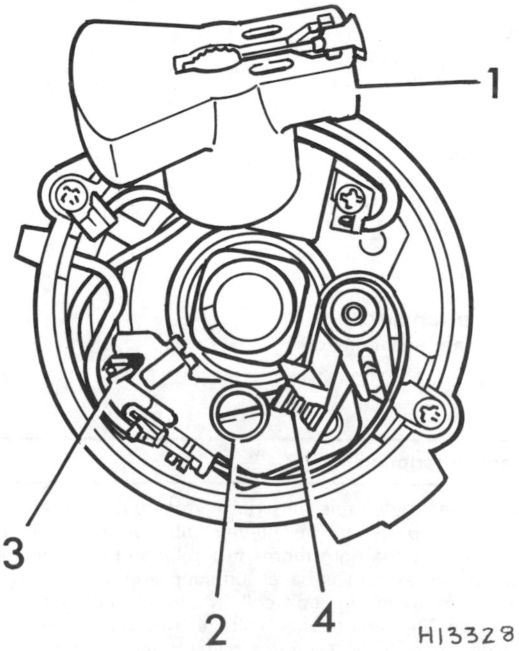

Fig. 4.1 Contact breaker points gap adjustment (Sec 2)

1 Rotor arm
2 Contact set securing screw
3 Adjustment notch
4 Contact points

8 Locate the base plate peg in the fork of the contact set, then press the pivot post into the plate and tighten the securing screw.
9 Check and adjust the contact breaker points gap as described in Section 2.
10 Refit the rotor arm and distributor cap.
11 Check the points gap after 500 miles (800 km)

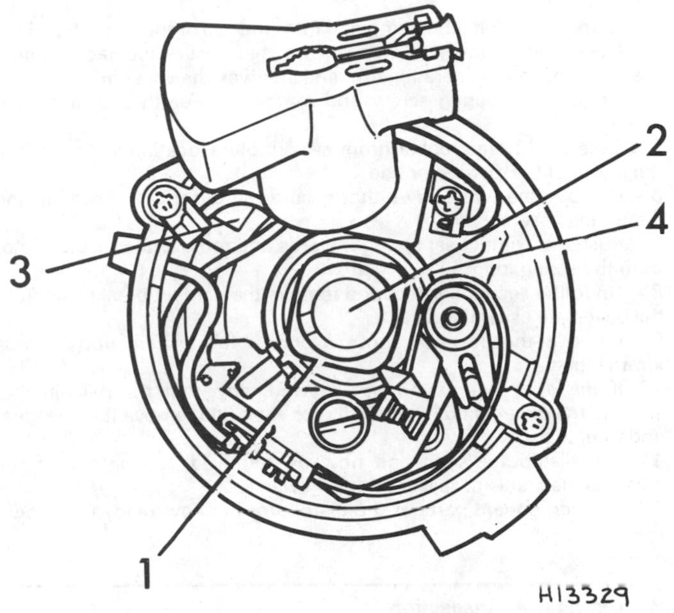

Fig. 4.2 Distributor lubrication points (Sec 4)

1 Spindle cam
2 Oil pad
3 Advance mechanism oiling point
4 Moving plate bearing groove

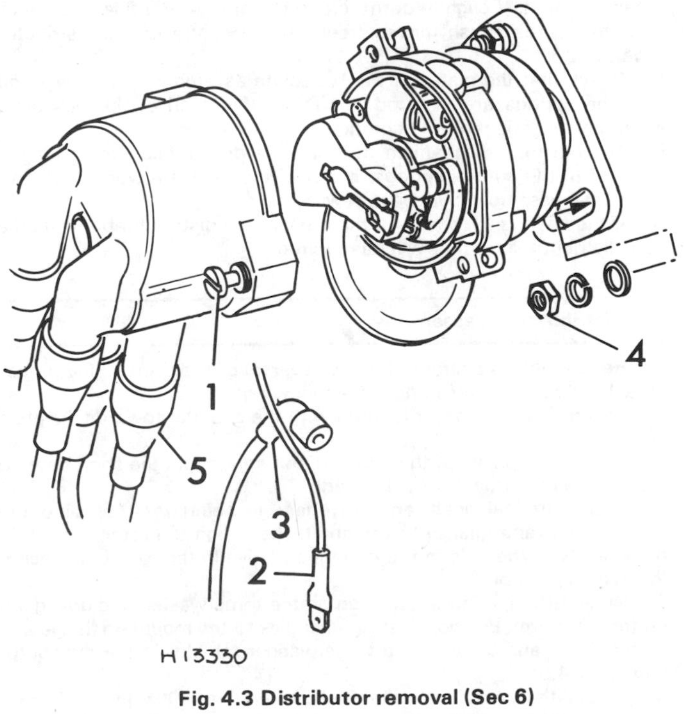

Fig. 4.3 Distributor removal (Sec 6)

1 Distributor cap securing screw
2 LT lead
3 Vacuum pipe
4 Distributor securing nut
5 HT leads

4 Distributor – lubrication

1 It is important that the distributor is lubricated every 6000 miles (10 000 km) as follows:

(a) *Smear the cam lightly with grease*
(b) *Put a few drops of oil on the pad in the top of the cam spindle (Fig. 4.2).*
(c) *Lubricate the advance mechanism by adding one or two drops of oil through the gap in the baseplate*

2 Every 24 000 miles (40 000 km), lubricate the moving plate bearing groove with one drop of oil.
3 Great care should be taken not to use too much lubricant, as any excess that might find its way onto the contact breaker points could cause burning and misfiring. Clean off any surplus lubricant and make sure that the breaker points are dry.

5 Condenser – removal and refitting

1 The purpose of the condenser (capacitor) is to ensure that when the contact breaker points open there is no sparking across them which would waste voltage and cause wear.
2 The condenser is fitted in parallel with the contact breaker points. If it develops a short circuit it will cause ignition failure, as the points will be prevented from interrupting the low tension circuit.
3 If the engine becomes very difficult to start, or begins to miss after several miles running, and the breaker points show signs of excessive burning, then the condition of the condenser must be suspect. A further test can be made by separating the points by hand with the ignition switched on. If this is accompanied by a flash it is indicative that the condenser has failed.
4 Without special test equipment, the only sure way to diagnose condenser trouble is to replace a suspected unit with a new one and note if there is any improvement. They are not expensive.
5 To remove the condenser from the distributor, remove the distributor cap and the rotor arm. Undo the screw that secures the earth lead and the condenser and lift away the condenser after disconnecting the LT lead.
6 Refitting is the reverse of the removal procedure.

6 Distributor – removal and refitting

1 Undo the securing screws and lift off the distributor cap (Fig. 4.3).
2 Disconnect the LT lead at the connector.
3 Detach the vacuum pipe from the distributor vacuum advance unit.
4 Rotate the crankshaft until the 8° BTDC notch on the timing disc lines up with the timing pointer, with No 1 cylinder on compression stroke (both valves closed). The rotor arm will be pointing to approximately the 3 o'clock position.
5 Undo and remove the two nuts that secure the distributor to the camshaft cover (photo).
6 Lift the distributor away from the camshaft cover.
7 Remove the O-ring seal from the distributor body.
8 When refitting the distributor, always fit a new O-ring seal on the distributor body.
9 Check that the contact breaker points gap is correctly set, refer to Section 2.
10 Ensure that the crankshaft is still at 8° BTDC on the compression stroke and then rotate the crankshaft to 90° BTDC to line up the notch in the timing disc with the timing pointer. Check that the dimple on the camshaft gear is opposite the pointer on the camshaft cover (photo). Now rotate the crankshaft clockwise to 8° BTDC.
11 Position the rotor arm so that it is pointing towards the 1 o'clock position.
12 Hold the distributor so that the vacuum unit is 45° below horizontal and to the left, then fit the distributor into position. As the gear on the distributor engages with the gear on the camshaft, note that the rotor arm turns to approximately the 3 o'clock position (photo).
13 Fit the distributor securing nuts finger-tight. Now turn the distributor slowly to the point where the contact breaker points are just beginning to open, then tighten the two securing nuts.
14 Refit the distributor cap and reconnect the LT lead.

6.5 Remove the distributor securing nuts

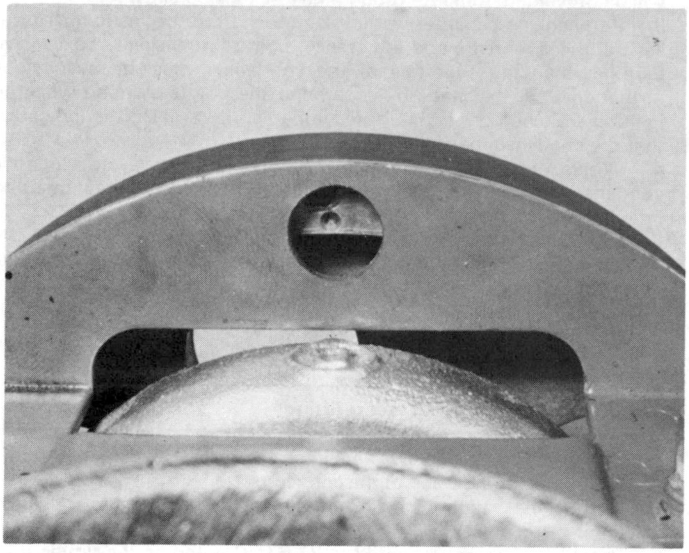

6.10 Check that the camshaft gear dimple is opposite the pointer

6.12 Fitting the distributor

15 Run the engine and check the timing with a timing light, refer to Section 10.
16 Reconnect the vacuum pipe to the vacuum advance unit.

7 Distributor – dismantling

1 With the distributor removed from the engine and on the bench, remove the distributor cap securing screws and lift off the cap. Pull off the rotor arm.
2 Remove the felt lubricating pad from the top of the shaft (Fig. 4.4).
3 Undo and remove the two screws that secure the vacuum unit, then disengage the operating arm and lift away the vacuum unit.
4 Undo the securing screw and remove the earth lead and condenser.
5 Ease the LT lead rubber grommet out of its location and push it to the inside of the distributor body.
6 Remove the two screws that secure the baseplate assembly and lift out the assembly.
7 Push the contact set spring inwards and remove the LT connector from the spring loop.
8 Undo the securing screw and remove the contact breaker set from the baseplate assembly.
9 Remove the advance control springs, taking care not to overstretch them.
10 If the drivegear has to be renewed, drive out the roll pin that secures the drivegear on the distributor shaft and remove the drivegear and thrustwasher.
11 The distributor shaft can now be removed complete with the advance mechanism, spacer and plain washer.
12 The component parts of the distributor are now ready for inspection.

8 Distributor – inspection

1 Thoroughly wash all mechanical parts in paraffin and wipe dry using a clean non-fluffy rag.
2 Check the contact breaker points as described in Section 3. Check the distributor cap for signs of tracking, indicated by a thin black line between the segments. Renew the cap if evident.
3 If the metal portion of the rotor arm is badly burned or loose, renew the arm. If slightly burnt, clean the arm with a fine file. Check that the carbon brush moves freely in the centre of the distributor cover.
4 Check that the plates of the baseplate assembly move freely and that the springs are in good condition. If defective, the baseplate assembly must be renewed complete.
5 The advance mechanism must not be dismantled except for the removal of the control springs. If any of the pads are worn, the complete shaft assembly must be renewed.
6 If the shaft is a loose fit in the bush in the distributor body, then a new distributor assembly will be required.

9 Distributor – reassembly

1 Reassembly is a straightfoward reversal of the dismantling procedure. In addition however, note the following:
2 Ensure that the spacer is fitted with the chamfer towards the plain washer.
3 Grease the pivots of the weights and spring and the shaft bearing area with a little Rocol MP (Molypad).
4 If the original shaft and drivegear is being refitted, slide the thrustwasher and gear onto the shaft, then align the rotor keyway in the shaft with the hole in the drivegear. Drive in the roll pin to secure the gear in position.
5 When fitting a new shaft, slide on the thrustwasher and drivegear. Fit the rotor arm. Position it at right-angles to the mounting flange and with its rear end towards the thrustwasher tab slot in the distributor body (Fig. 4.5).
6 To set the endfloat of the shaft, insert an 0·005 in (0·13 mm) feeler gauge between the thrustwasher and drivegear.
7 Align the hole in the drivegear with the edge of the thrustwasher tab slot and, holding the shaft and gear tightly together, drill a $\frac{1}{8}$ in (3·2 mm) hole through the shaft.

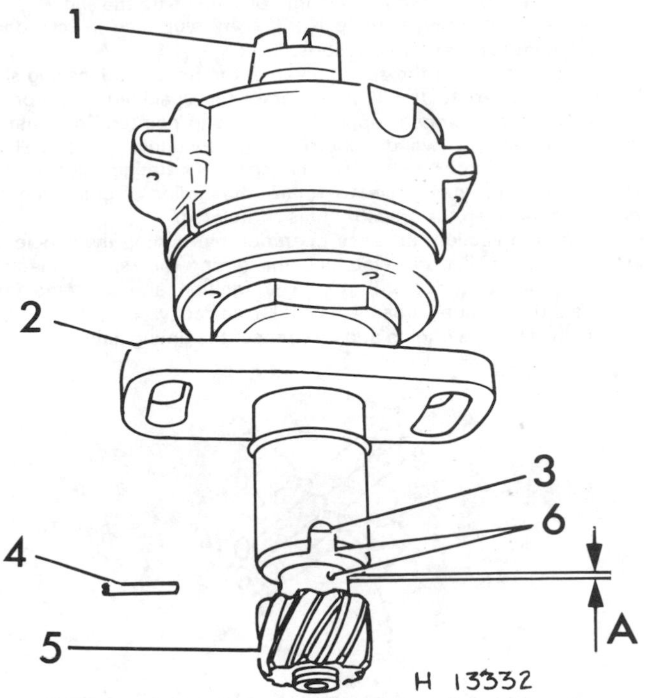

Fig. 4.5 Fitting the drivegear on a new shaft (Sec 9)

1 Rotor	*5 Drivegear*
2 Distributor flange	*6 Hole aligned with edge of slot*
3 Thrustwasher tab slot	*Dimension A - 0.005 in (0.13 mm)*
4 Roll pin	

8 Remove the feeler gauge and drive in the roll pin to secure the drivegear in position.
9 Proceed with the remainder of the reassembly.
10 Lubricate the distributor as described in Section 4
11 Adjust the contact breaker points gap as described in Section 2.

10 Ignition – timing

1 To make an accurate check of the ignition timing, it is necessary to use an LED (light emitting diode) optical sensor or a stroboscopic timing light, whereby the timing is checked with the engine running at a specified speed. For the DIY owner, the stroboscopic timing light is the obvious method as the LED checking equipment is expensive. The bracket for the LED optical sensor is mounted on the front of the engine to the left of the timing disc. The procedure for checking the ignition using a timing light is given in the following paragraphs:
2 Paint the specified timing mark on the timing disc and the tip of the timing pointer with quick drying white paint (or white chalk) to highlight these points.

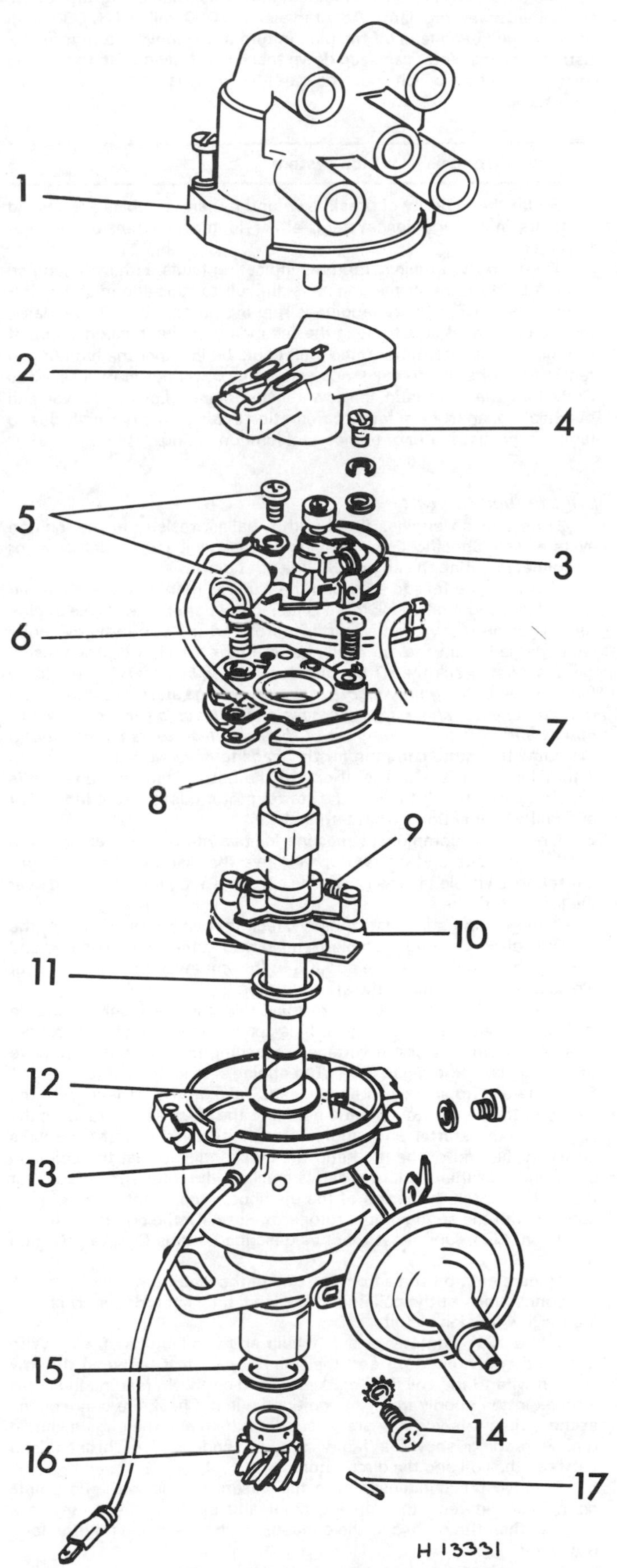

Fig. 4.4 Exploded view of the distributor assembly (Secs 7 and 9)

1 Distributor cap	*10 Shaft assembly*
2 Rotor arm	*11 Spacer*
3 Contact set	*12 Plain washer*
4 Contact set securing screw	*13 Grommet*
5 Condenser and securing screw	*14 Vacuum unit and securing screw*
6 Baseplate securing screw	*15 Thrustwasher*
7 Baseplate assembly	*16 Drivegear*
8 Felt oil pad	*17 Roll pin*
9 Control springs	

3 Disconnect the distributor vacuum pipe and plug the end.
4 Connect the timing light to No 1 spark plug lead in accordance with the manufacturer's instructions.
5 Run the engine at the speed given in the Specifications and shine the timing light onto the timing marks; the specified notch on the timing disc should appear opposite the timing pointer. To adjust the timing, slacken the two distributor flange securing nuts and slowly rotate the distributor to obtain the specified timing; clockwise to advance and anti-clockwise to retard the timing. Tighten the two securing nuts when the adjustment is completed.
6 Check the vacuum advance operation by running the engine at a fast idle then whilst still watching the timing marks, reconnect the vacuum pipe to the distributor. The ignition should immediately advance if the vacuum capsule is working correctly.
7 Finally, stop the engine and disconnect the timing light.

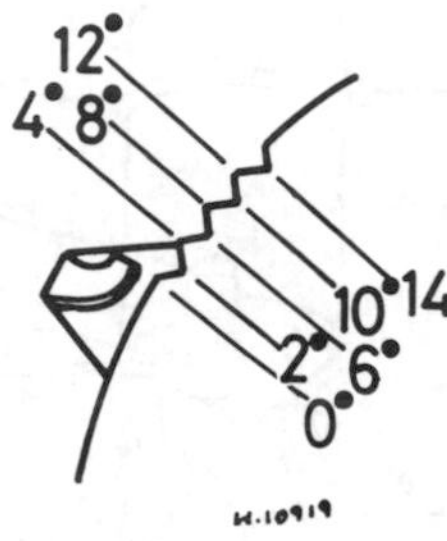

Fig. 4.6 Timing disc and pointer, showing timing marks (Sec 10)

11 Spark plugs and HT leads

1 The correct functioning of the spark plugs is vital for the proper running and efficient operation of the engine.
2 At intervals of 6000 miles (10 000 km) the plugs should be removed, examined, cleaned, and if worn excessively, renewed. The condition of the spark plug can also tell much about the general condition of the engine.
3 If the insulator nose of the spark plug is clean and white with no deposits, this is indicative of a weak mixture or too hot a plug (a hot plug transfers heat away from the electrode slowly – a cold plug transfers heat away quickly).
4 If the insulator nose is covered with hard black looking deposits, then this is indicative that the mixture is too rich. Should the plug be black and oily, then it is likely that the engine is fairly worn as well as the mixture being too rich.
5 If the insulator nose is covered with light tan to greyish brown deposits, then the mixture is correct and it is likely that the engine is in good condition.
6 If there are any traces of long brown tapering stains on the outside of the white portion of the plug, then the plug will have to be renewed as this showns that there is a faulty joint between the plug body and the insulator and compression is being allowed to leak away.
7 Plugs should be cleaned by a sand blasting machine, which will free them from carbon more than by cleaning by hand. The machine will also test the condition of the plugs under compression. Any plug that fails to spark at the recommended pressure should be renewed.
8 The spark plug gap is of considerable importance as, if it is too large or too small the size of the spark and its efficiency will be seriously impaired. The spark plug gap should be set to 0·035 in (0·90 mm).
9 To set the gap, measure it, with a feeler gauge and then bend open, or close, the outer plug electrode until the correct gap is achieved. The centre electrode should never be bent as this may crack the insulation and cause plug failure, if nothing worse.
10 When fitting the plugs, ensure that both the seatings and threads in the cylinder head and plugs are thoroughly cleaned. Screw in the plugs finger-tight then, using a 16 mm plug spanner, gently tighten them a further $\frac{1}{16}$ of a turn. **Caution:** *Do not overtighten the plugs as the conical face of the seatings can lock together making it very difficult (if not impossible) to remove them.* Note that torque settings are given in the Specifications.
11 The plug leads require no maintenance other than being kept clean and wiped over regularly. At intervals of 6000 miles (10 000 km) however, pull each lead off the plug in turn and remove them from the distributor cap. Water can seep down these joints giving rise to a white corrosive deposit which must be carefully removed from the end of each cable.

12 Fault diagnosis – ignition system

By far the majority of breakdown and running troubles are caused by faults in the ignition system, either in the low tension or high tension circuits.

There are two main symptoms indicating faults. Either the engine will not start or fire, or the engine is difficult to start and misfires,. If it is a regular misfire (ie the engine is running on only 2 or 3 cylinders), the fault is almost sure to be in the secondary or high tension circuit. If the misfiring is intermittent, the fault could be in either the high or low tension circuits. If the car stops suddenly or will not start at all, it is likely that the fault is in the low tension circuit. Loss of power and overheating, apart from faulty carburation settings, are normally due to faults in the distributor or to incorrect ignition timing.

Engine fails to start

Note that on engines fitted with a ballast resistor in the ignition system (see Specification), should the resistor become defective or disconnected, then the engine will fire but not run.
1 If the engine fails to start and the car was running normally when it was last used, first check there is fuel in the petrol tank. If the engine turns over normally on the starter motor and the battery is evidently well charged, then the fault may be in either the high or low tension circuits. First check the HT circuit. **Note:** *If the battery is known to be fully charged, the ignition light comes on and the starter motor fails to turn the engine, check the tightness of the leads on the battery terminals and also the secureness of the earth lead to its connection to the body.* It is quite common for the leads to have worked loose, even if they look and feel secure. If one of the battery terminal posts gets very hot when trying to work the starter motor this is a sure indication of a faulty connection to that terminal.
2 One of the commonest reasons for bad starting is wet or damp spark plug leads and distributor. Remove the distributor cap. If condensation is visible internally, dry the cap with a rag and also wipe over the leads. Refit the cap.
3 If the engine still fails to start, check that voltage is reaching the plugs by disconnecting each plug lead in turn at the spark plug end and holding the end of the cable about $\frac{3}{16}$ in (5 mm) away from the cylinder block. Spin the engine on the starter motor.
4 Sparking between the end of the cable and the block should be fairly strong with a strong regular blue spark. Hold the lead with rubber to avoid electric shocks. If voltage is reaching the plugs, then remove them and clean and regap them. The engine should now start.
5 If there is no spark at the plug leads, take off the HT lead from the centre of the distributor cap and hold it to the block as before. Spin the engine on the starter once more. A rapid succession of blue sparks between the end of the lead and the block indicate that the coil is in order and that the distributor cap is cracked, the rotor arm is faulty, or the carbon brush in the top of the distributor cap is not making good contact with the spring on the rotor arm. Possibly, the points are in bad condition. Clean and reset them as described in this Chapter, Section 3.
6 If there are no sparks from the end of the lead from the coil, check the connections at the coil end of the lead. If it is in order, start checking the low tension circuit.
7 Use a 12v voltmeter or a 12v bulb and two lengths of wire. With the ignition switched on and the points open, test between the low tension wire to the coil (it is marked +) and earth. No reading indicates a break in the supply from the ignition switch. Check the connections at the switch to see if any are loose. Refit them and the engine should run. A reading shows a faulty coil or condenser, or broken lead between the coil and the distributor.
8 Remove the condenser from the baseplate and with the points open, test between the moving point and earth. If there now is a reading, then the fault is in the condenser. Fit a new one and the fault is cleared.
9 With no reading from the moving point to earth, take a reading

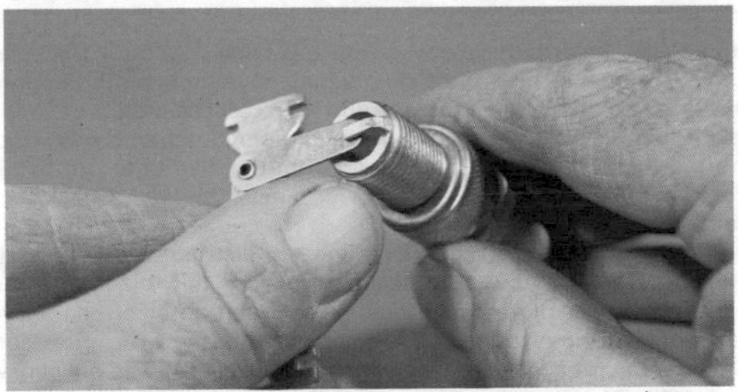

Measuring plug gap. A feeler gauge of the correct size (see ignition system specifications) should have a slight 'drag' when slid between the electrodes. Adjust gap if necessary

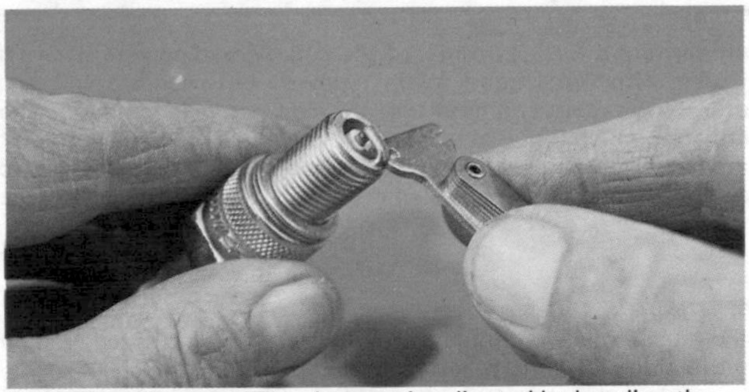

Adjusting plug gap. The plug gap is adjusted by bending the earth electrode inwards, or outwards, as necessary until the correct clearance is obtained. Note the use of the correct tool

Normal. Grey-brown deposits, lightly coated core nose. Gap increasing by around 0.001 in (0.025 mm) per 1000 miles (1600 km). Plugs ideally suited to engine, and engine in good condition

Carbon fouling. Dry, black, sooty deposits. Will cause weak spark and eventually misfire. Fault: over-rich fuel mixture. Check: carburettor mixture settings, float level and jet sizes; choke operation and cleanliness of air filter. Plugs can be re-used after cleaning

Oil fouling. Wet, oily deposits. Will cause weak spark and eventually misfire. Fault: worn bores/piston rings or valve guides; sometimes occurs (temporarily) during running-in period. Plugs can be re-used after thorough cleaning

Overheating. Electrodes have glazed appearance, core nose very white – few deposits. Fault: plug overheating. Check: plug value, ignition timing, fuel octane rating (too low) and fuel mixture (too weak). Discard plugs and cure fault immediately

Electrode damage. Electrodes burned away; core nose has burned, glazed appearance. Fault: pre-ignition. Check: as for 'Overheating' but may be more severe. Discard plugs and remedy fault before piston or valve damage occurs

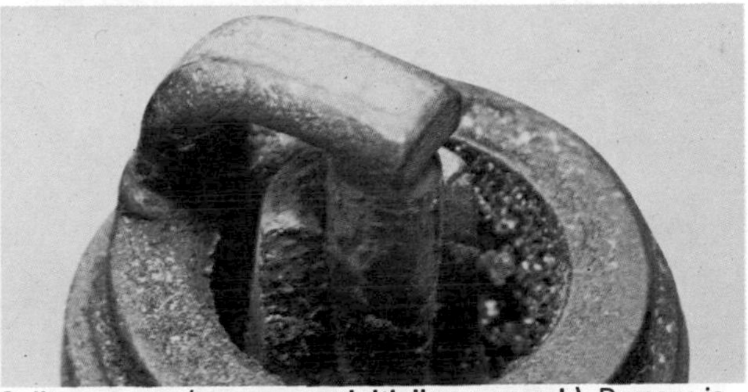

Split core nose (may appear initially as a crack). Damage is self-evident, but cracks will only show after cleaning. Fault: pre-ignition or wrong gap-setting technique. Check: ignition timing, cooling system, fuel octane rating (too low) and fuel mixture (too weak). Discard plugs, rectify fault immediately

between earth and the – terminal of the coil. A reading here shows a broken wire which will need to be renewed between the coil and distributor. No reading confirms that the coil has failed and must be renewed, after which the engine will run once more. Remember to refit the condenser to the baseplate. For these tests it is sufficient to separate the points with a piece of dry paper whilst testing with the points open.

Engine misfires

10 If the engine misfires regularly, run it at a fast idling speed. Pull off each of the plug caps in turn and listen to the note of the engine. Hold the plug cap in a dry cloth or with a rubber glove as additional protection against a shock from the HT supply.

11 No difference in engine running will be noticed when the lead from the defective circuit is removed. Removing the lead from one of the good cylinders will accentuate the misfire.

12 Remove the plug lead from the end of the defective plug and hold it about $\frac{3}{16}$ in (5 mm) away from the block. Restart the engine. If sparking is fairly strong and regular the fault must lie in the spark plug.

13 The plug may be loose, the insulation may be cracked, or the points may have burnt away giving too wide a gap for the spark to jump. Worse still, one of the points may have broken off. Either renew the plug or clean it, reset the gap and then test the plug.

14 If there is no spark at the end of the plug lead, or if it is weak and intermittent, check the ignition lead from the distributor to the plug. If the insulation is cracked or perished, renew the lead. Check the connections at the distributor cap.

15 If there is still no spark, examine the distributor cap carefully for tracking. This can be recognised by a very thin black line running between two or more electrodes, or between an electrode and some other part of the distributor. These lines are paths which now conduct electricity across the cap thus letting it run to earth. The only answer is a new distributor cap.

16 Apart from the ignition timing being incorrect, other causes of misfiring have already been dealt with under the Section dealing with the failure of the engine to start.

17 If the ignition timing is too far retarded, it should be noted that the engine will tend to overheat and there will be quite a noticeable drop in power. If the engine is overheating and power is down, and the ignition is correct, then the carburettor should be checked, as it is likely that this is where the fault lies. See Chapter 3 for details.

Chapter 5 Clutch

Contents

Specifications

Make	Borg and Beck or Laycock	
Type	Diaphragm spring	
Driveplate diameter	8·5 in (216 mm)	
Number of damper springs	6	
Master cylinder bore	0·5 in (12·70 mm)	
Slave cylinder bore	0·875 in (22·22 mm)	
Torque wrench settings	**lbf ft**	**kgf m**
Clutch to flywheel bolts	17	2·3

1 General description

The Marina 1700 models are fitted with an 8.5 in diameter diaphragm spring clutch operated hydraulically by a master and slave cylinder.

The clutch comprises a steel cover which is bolted and dowelled to the rear face of the flywheel and contains the pressure plate and clutch disc or driven plate.

The pressure plate, diaphragm spring and release plate are all attached to the clutch assembly cover.

The clutch disc is free to slide along the splined first motion shaft and is held in position between the flywheel and pressure plate by the pressure of the diaphragm spring.

Friction lining material is riveted to the clutch disc which has a spring cushioned hub to absorb transmission shocks and to help ensure a smooth take off.

The clutch is acutated hydraulically. The pendant clutch pedal is connected to the clutch master cylinder and hydraulic fluid reservoir by a short pushrod. The master cylinder and hydraulic reservoir are mounted on the engine side of the bulkhead in front of the driver.

Depressing the clutch pedal moves the piston in the master cylinder forwards so forcing hydraulic fluid through the clutch hydraulic pipe to the slave cylinder.

The piston in the slave cylinder moves forward on the entry of the fluid and actuates the clutch assembly arm by means of a short pushrod. The opposite end of the release arm is forked and is located behind the release bearing.

As this pivoted clutch release arm moves backwards, it bears against the release bearing pushing it forwards to bear against the release plate, so moving the centre of the diaphragm spring inwards. The spring is sandwiched between two annular rings which act as fulcrum points. The centre of the spring is pushed out, so moving the pressure plate backwards and dis-engaging the pressure plate from the clutch disc.

When the clutch pedal is released, the diaphragm spring forces the pressure plate into contact with the high friction linings on the clutch disc and at the same time pushes the clutch disc a fraction of an inch forwards on its splines so engaging the clutch disc with the flywheel. The clutch disc is now firmly sandwiched between the pressure plate and the flywheel so the drive is taken up.

As the friction linings on the clutch disc wear, the pressure plate automatically moves closer to the disc to compensate. There is therefore no need to periodically adjust the clutch.

2 Clutch hydraulic system – bleeding

1 Gather together a clean jam jar, a length of rubber tubing which fits tightly over the bleed nipple on the slave cylinder, a tin of hydraulic brake fluid, and someone to help.

2 Check that the master cylinder is full. If it is not, fill it and cover the bottom 2 in of the jar with hydraulic fluid.

3 Remove the rubber dust cap from the bleed nipple (if fitted) on the slave cylinder, and with a suitable spanner open the bleed nipple approximately three quarters of a turn.

4 Place one end of the tube securely over the nipple and insert the other end in the jam jar so that the tube orifice is below the level of the fluid (Fig. 5.1).

5 The assistant should now depress the pedal and hold it down at the end of its stroke. Close the bleed screw and allow the pedal to return to its normal position.

6 Continue this series of operations until clear hydraulic fluid, without any traces of air bubbles, emerges from the end of the tubing.

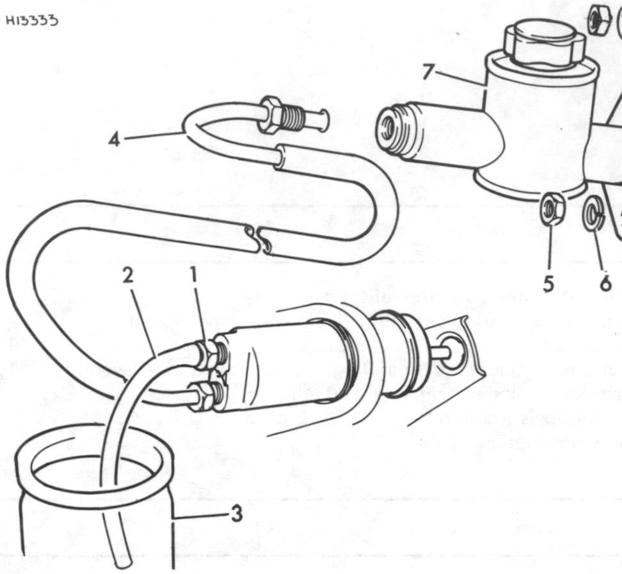

Fig. 5.1 Bleeding the clutch hydraulic system (Sec 2)

1	*Bleed nipple*	*5*	*Nut*
2	*Bleed tube*	*6*	*Washer*
3	*Glass jar*	*7*	*Master cylinder*
4	*Flexible pipe*		

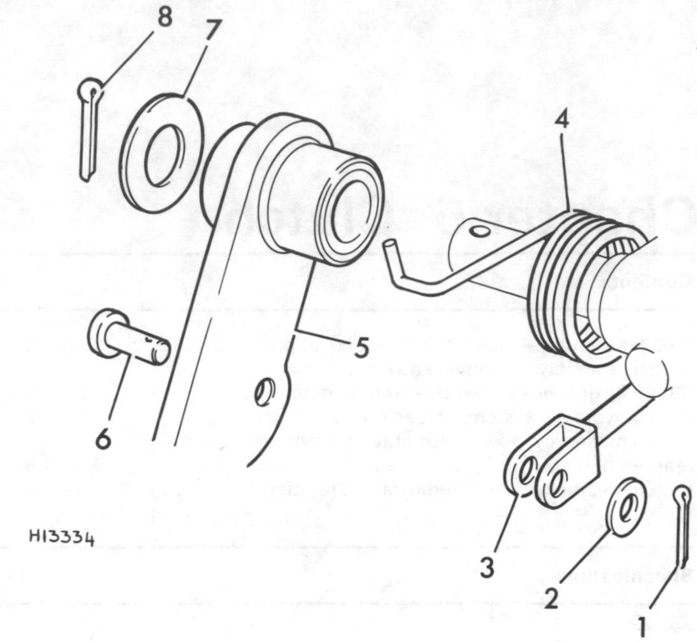

Fig. 5.2 Clutch pedal mounting (Sec 3)

1	*Split-pin*	*5*	*Clutch pedal*
2	*Plain washer*	*6*	*Clevis pin*
3	*Pushrod*	*7*	*Plain washer*
4	*Spring*	*8*	*Split-pin*

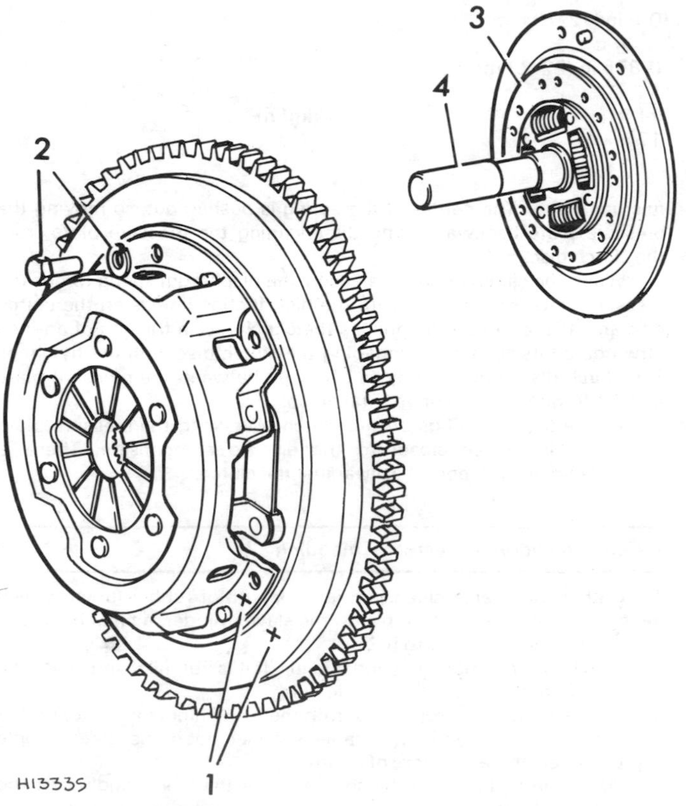

Fig. 5.3 Clutch assembly mounted on flywheel (Sec 4)

1	*Alignment marks*	*3*	*Clutch drive plate*
2	*Securing bolt and spring washer*	*4*	*Aligning tool*

Make sure that the reservoir is checked frequently to ensure that the hydraulic fluid does not drop too far thus letting air into the system.

7 When no more air bubbles appear, tighten the bleed nipple on the downstroke.

8 Refit the rubber dust cap (if fitted) over the bleed nipple.

3 Clutch pedal – removal and refitting

1 Disconnect and withdraw the supply tube from the face level vent.

2 Straighten the legs and extract the split-pin that retains the master cylinder operating rod yoke to pedal clevis pin. Lift away the plain washer and withdraw the clevis pin (Fig. 5.2).

3 Straighten the legs and extract the split-pin from the clutch pedal end of the pedal pivot shaft. Lift away the plain washer.

4 Carefully release the pedal return spring from the pedal and slide the pedal from the end of the shaft.

5 Inspect the pedal bush for signs of wear which if evident, will mean either the old bush being drifted out and a new one fitted, or a new pedal assembly obtained.

6 Refitting is the reverse sequence to removal. Lubricate the pedal bush and shaft and also the spring coils to prevent squeaking.

4 Clutch assembly – removal and refitting

1 Remove the gearbox as described in Chapter 6, Section 2.

2 With a scriber or file, mark the relative position of the clutch cover and flywheel to ensure the correct refitting of the original parts if they are to be re-used (Fig. 5.3).

3 Remove the clutch assembly by unscrewing the six bolts that hold the cover to the rear face of the flywheel. Unscrew the bolts diagonally half a turn at a time to prevent distortion of the cover flange, also to prevent an accident caused by the cover flange binding on the dowels and suddenly flying off.

4 With the bolts and spring washers removed, lift the clutch assembly off the locating dowels. The driven plate or clutch disc will fall out at this stage, as it is not attached to either the clutch cover

assembly or the flywheel. Carefully make a note of which way round it is fitted.

5 It is important that no oil or grease gets on the clutch disc friction linings or the pressure plate and flywheel faces. It is advisable to handle the parts with clean hands and to wipe down the pressure plate and flywheel faces with a clean dry rag before inspection or refitting commences.

6 To refit the clutch, place the clutch disc against the flywheel with the clutch spring housing facing outwards away from the flywheel. On no account should the clutch disc be refitted the wrong way round as it will be found quite impossible to operate the clutch with the friction disc incorrectly fitted.

7 Refit the clutch cover assembly loosely on the dowels. Refit the six bolts and spring washers and tighten them finger-tight so that the clutch disc is gripped but can still be moved.

8 The clutch disc must now be centralised so that when the engine and gearbox are mated, the gearbox input shaft splines will pass through the splines in the centre of the hub.

9 Centralisation can be carried out quite easily by inserting a round bar or long screwdriver through the hole in the centre of the clutch, so that the end of the bar rests in the small hole in the end of the crankshaft containing the input shaft bearing bush. Moving the bar sideways or up and down will move the clutch disc in whichever direction is necessary to achieve centralisation.

10 Centralisation is easily judged by removing the bar and viewing the driven plate hub in relation to the hole in the centre of the diaphragm spring. When the hub appears exactly in the centre of the diaphragm spring hole all is correct. Alternatively, if an old input shaft can be borrowed this will eliminate all the guesswork as it will fit the bush and centre of the clutch hub assembly exactly, obviating the need for visual alignment.

11 Tighten the clutch bolts firmly in a diagonal sequence to ensure that the cover plate is pulled down evenly and without distortion of the flange.

12 Refit the gearbox, refer to Chapter 6. Bleed the slave cylinder if the pipe was disconnected and check the clutch for correct operation.

5 Clutch assembly – inspection and renovation

1 In the normal course of events, clutch dismantling and reassembly is the term used for simply fitting a new clutch pressure plate and friction disc. Under no circumstances should the diaphragm spring clutch unit be dismantled. If a fault develops in the pressure plate assembly, an exchange replacement unit must be fitted.

2 If a new clutch disc is being fitted, it is false economy not to renew the release bearing at the same time. This will preclude having to renew it at a later date when wear on the clutch linings is very small.

3 Examine the clutch disc linings for wear or loose rivets and the disc for rim distortion, cracks and worn splines.

4 It is always best to renew the clutch driven plate as an assembly to preclude further trouble, but if it is wished to merely renew the linings, the rivets should be drilled out and not knocked out with a centre punch. The manufacturers do not advise that the linings only are renewed and personal experience dictates that it is far more satisfactory to renew the driven plate complete than to try to economise by fitting only new friction linings.

5 Check the machined faces of the flywheel and the pressure plate. If either is badly grooved, it should be machined until smooth, or replaced with a new item. If the pressure plate is cracked or split it must be renewed.

6 Examine the hub splines for wear and also make sure that the centre hub is not loose.

6 Clutch flexible hose – removal and refitting

1 Wipe the slave cylinder end of the translucent hose to prevent dirt ingress. Obtain a clean and dry glass jam jar and have it ready to catch the hydraulic fluid during the next operation.

2 Carefully unscrew the union at the slave cylinder end and drain the fluid into the jar.

3 Unscrew the union at the master cylinder end.

4 Where applicable, remove the screw which secures the hose clip to the body, then remove the hose.

5 Refitting is the reverse of the removal procedure but ensure that the screw threads are clean.

6 It will be necessary to bleed the clutch hydraulic system as described in Section 2 of this Chapter.

7 Clutch master cylinder – removal and refitting

1 Drain the fluid from the clutch master reservoir by attaching a rubber tube to the slave cylinder bleed nipple. Undo the nipple by approximately three quarters of a turn and then pump the fluid out into a suitable container by means of operating the clutch pedal. Note that the pedal must be held in against the floor at the completion of each stroke and the bleed nipple tightened before the pedal is allowed to return. When the pedal has returned to its normal position, loosen the bleed nipple and repeat the process until the clutch master cylinder is empty.

2 Place a rag under the master cylinder to catch any hydraulic fluid that may be spilt. Unscrew the union nut from the end of the metal pipe where it enters the clutch master cylinder and gently pull the pipe clear.

3 Straighten the legs of the split-pin and extract it from the operating fork clevis pin on the pedal.

4 Unscrew and remove the two nuts and spring washers that secure the master cylinder and lift it away. Take care not to allow any hydraulic fluid to come into contact with the paintwork as it acts as a solvent.

5 Refitting the master cylinder is the reverse sequence to removal. Bleed the system as described in Section 2 of this Chapter.

8 Clutch master cylinder – dismantling, examination and reassembly

1 Ease back the rubber dust cover from the pushrod end.

2 Using a pair of circlip pliers, release the circlip that retains the pushrod assembly. Lift away the pushrod complete with rubber boot and plain washer (Fig. 5.4).

3 By shaking hard, the piston with its seal, dished washer, second seal, and spring retainer may be removed from the cylinder bore.

4 Lift away the long spring, noting which way round it is fitted.

5 If they prove stubborn, carefully use a foot pump air jet on the hydraulic pipe connection and this should move the internal parts, but do take care as they will fly out. We recommend placing a pad over the pushrod end to catch the parts.

6 Carefully ease the secondary cup seal from the piston, noting which way round it is fitted.

7 Thoroughly clean the parts in brake fluid or methylated spirits. After drying the items, inspect the seals for signs of distortion, swelling, splitting or hardening, although it is recommended new rubber parts are always fitted after dismantling as a matter of course.

8 Inspect the bore and piston for signs of deep scoring marks which if evident, will mean that a new cylinder should be fitted. Make sure the bypass ports are clear by poking gently with a piece of thin wire.

9 As the parts are refitted to the cylinder bore, make sure that they are thoroughly wetted with clean hydraulic fluid.

10 Refit the secondary cup seal onto the piston making sure it is the correct way round.

11 Insert the spring with its retainer into the master cylinder bore.

12 Next refit the main cup seal with its flat end facing towards the open end of the bore.

13 Replace the dished washer and carefully insert the piston into the bore. The small end of the piston should be towards the dished washer. Make sure that the lip of the seal does not roll over as it enters the bore.

14 Smear a little rubber grease onto the ball end of the pushrod and refit the pushrod assembly. Slide down the plain washer and secure it in position with the circlip.

15 Pack the rubber dust cover with rubber grease and place it over the end of the master cylinder.

9 Clutch slave cylinder – removal and refitting

1 Wipe the top of the master cylinder reservoir and unscrew the cap. Place a piece of polythene sheet over the top of the reservoir and refit

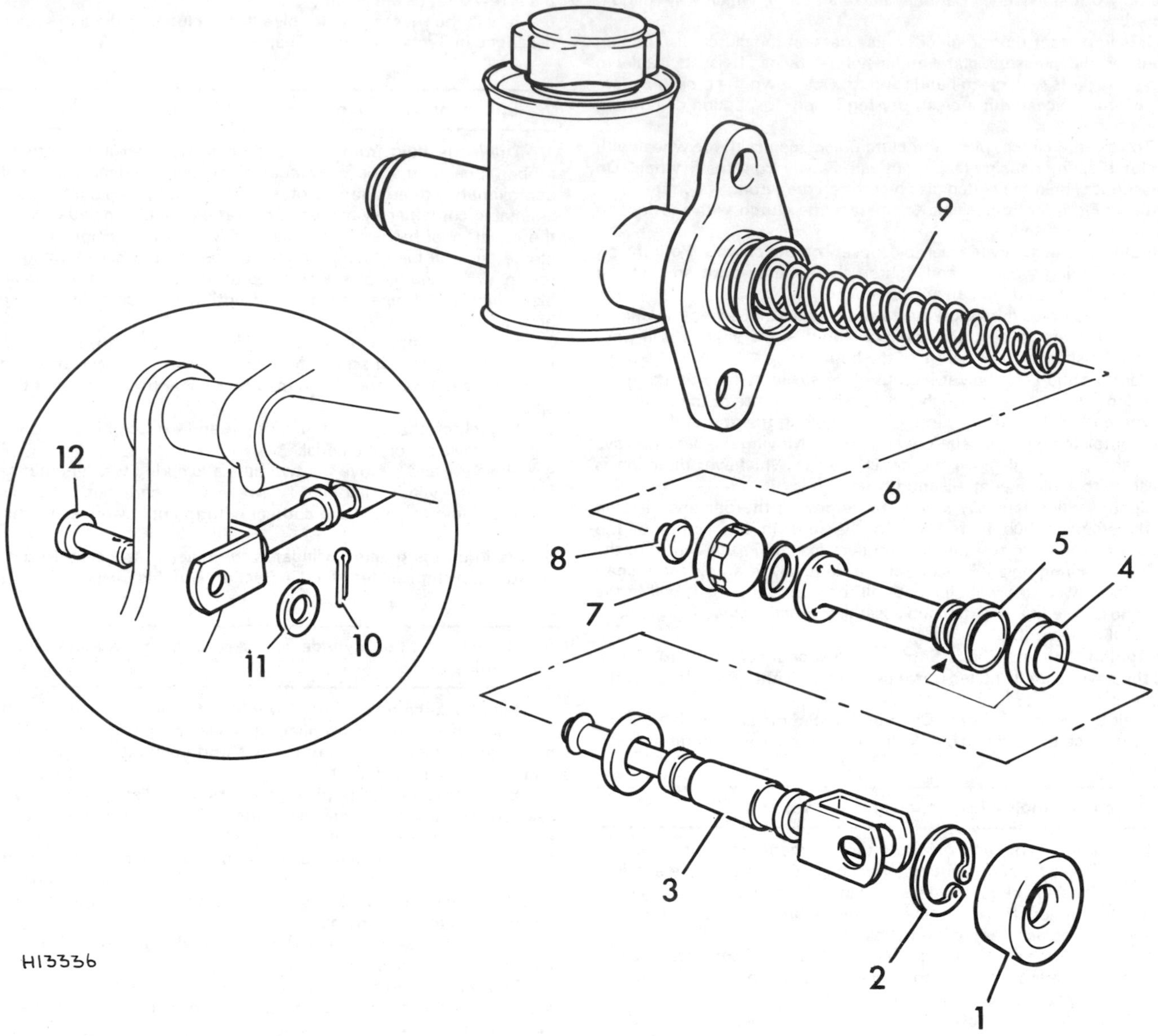

Fig. 5.4 Clutch master cylinder components (Sec 8)

1 *Rubber dust cover*
2 *Circlip*
3 *Pushrod*
4 *Seal*
5 *Piston*
6 *Dished washer*
7 *Seal*
8 *Spring retainer*
9 *Spring*

Inset: Pushrod attachments
10 *Split-pin*
11 *Plain washer*
12 *Clevis pin*

the cap. This will stop hydraulic fluid syphoning out during subsequent operations.

2 Wipe the area around the hydraulic pipe on the slave cylinder and disconnect the metal pipe from the slave cylinder.

3 Turn the slave cylinder until the flat on the shoulder faces the clutch housing.

4 Using a piece of metal bar or a large screwdriver, carefully draw the clutch release lever rearwards until it is possible to lift out the slave cylinder. It will be found helpful to push the operating rod into the slave cylinder to clear the operating lever, then remove the rod and slave cylinder.

5 Refitting the slave cylinder is the reverse sequence to removal. It is very important that the bleed nipple is uppermost otherwise it will be impossible to bleed all air from the system.

6 Bleed the clutch hydraulic system as described in Section 2.

10 Clutch slave cylinder – dismantling, examination and reassembly

1 Clean the outside of the slave cylinder before dismantling.

2 Pull off the rubber dust cover and by shaking hard, the piston, seal, filler and spring should come out of the cylinder bore (Fig. 5.5).

3 If they prove stubborn, carefully use a foot pump air jet on the hydraulic hose connection and this should remove the internal parts, but do take care as they will fly out. We recommend placing a pad over the dust cover end to catch the parts.

4 Wash all internal parts with either brake fluid or methylated spirits and dry using a non-fluffy rag.

5 Inspect the bore and piston for signs of deep scoring which if evident, means a new cylinder should be fitted.

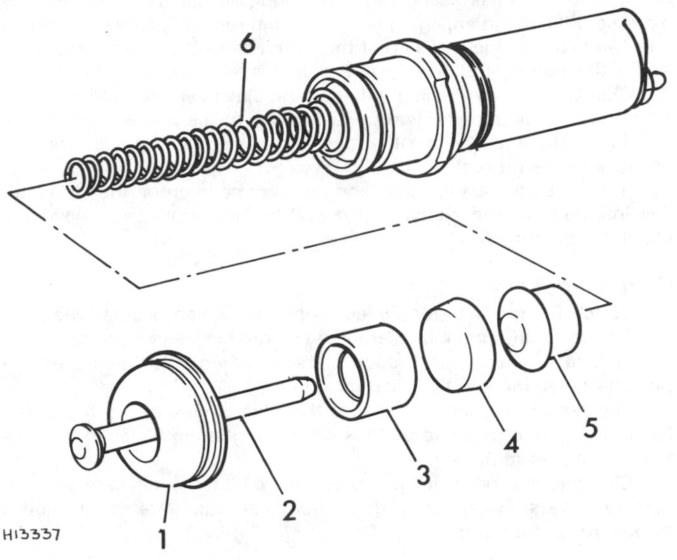

Fig. 5.5 Clutch slave cylinder components (Sec 10)

1 Rubber dust cover
2 Pushrod
3 Piston
4 Seal
5 Filler
6 Spring

6 Carefully examine the rubber components for signs of swelling, distortion, splitting, hardening or other wear, although it is recommended new rubber parts are always fitted after dismantling.
7 All parts should be reassembled wetted with clean hydraulic fluid.
8 Refit the spring, large end first into the cylinder bore.
9 Refit the cup filler into the bore.
10 Fit a new cup seal and fit the piston, making sure that both are fitted the correct way round.
11 Apply a little rubber grease to both ends of the pushrod and also pack the dust cover.
12 Fit the dust cover over the end of the slave cylinder, engaging the lips over the groove in the body.
13 Fit the pushrod to the slave cylinder by pushing through the hole in the dust cover.

11 Clutch release bearing assembly – removal, inspection and refitting

1 To gain access, it is necessary to remove the gearbox as described in Chapter 6.
2 Detach the operating lever from the release bearing and slide off the bearing assembly (Fig. 5.6).
3 If the bearing is worn or showing signs of overheating, it may be removed using a large bench vice or a press and suitable packing.
4 When refitting a new bearing, always apply the load to the inner race.
5 Fit the release bearing onto the gearbox first motion shaft front end cover.
6 Engage the pivots of the operating lever into the groove in the release bearing and at the same time engage the lever retaining spring clip with the fulcrum pin in the gearbox housing.
7 Press the operating lever fully into position.
8 Refitting the gearbox is now the reverse sequence to removal.

12 Fault diagnosis – clutch

There are four main faults to which the clutch and release mechanism are prone. They may occur by themselves, or in conjunction with any of the other faults. They are clutch squeal, slip, spin and judder.

Clutch squeal

1 If on taking up the drive or when changing gear the clutch squeals, this is indicative of a badly worn clutch release bearing.

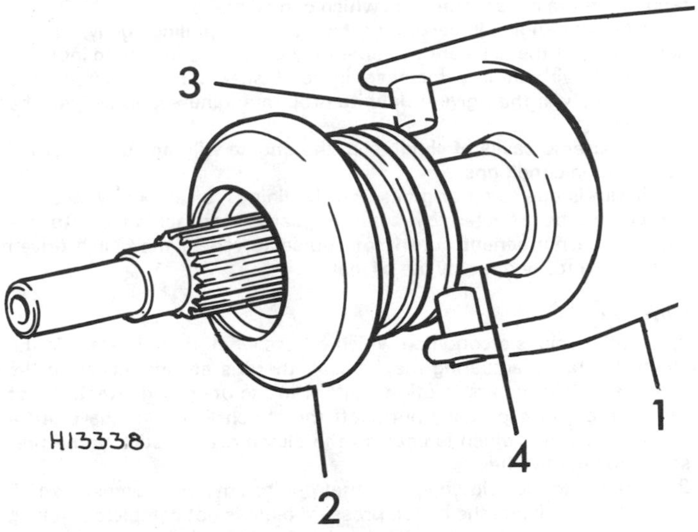

Fig. 5.6 Clutch release mechanism (Sec 11)

1 Operating lever
2 Release bearing
3 Operating lever dowels
4 First motion shaft front end cover

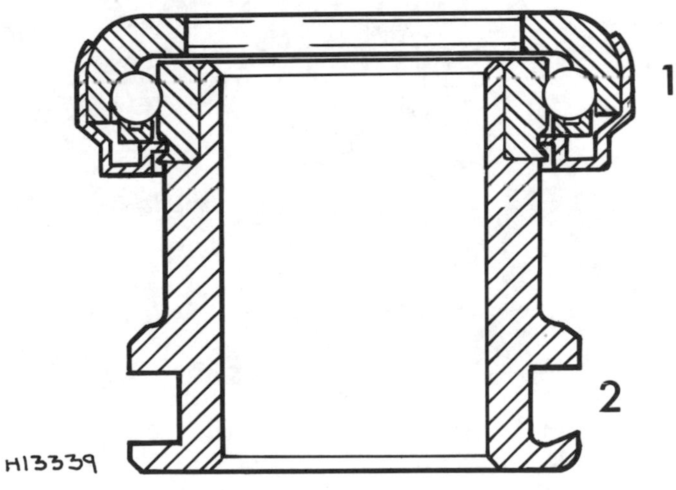

Fig. 5.7 Cross-section through clutch release bearing (Sec 11)

1 Bearing assembly
2 Bearing carrier

2 As well as regular wear due to normal use, wear of the clutch release bearing is much accentuated if the clutch is ridden or held down for long periods in gear with the engine running. To minimise wear of this component, the car should always be taken out of gear at traffic lights and for similar hold ups.
3 The clutch release bearing is not an expensive item but difficult to get at.

Clutch slip

1 Clutch slip is a self evident condition which occurs when the clutch friction plate is badly worn, oil or grease have got onto the flywheel or pressure plate faces, or the pressure plate itself is faulty.
2 The reason for clutch slip is that due to one of the faults above, there is either insufficient pressure from the pressure plate, or insufficient friction from the friction plate to ensure solid drive.
3 If small amounts of oil get onto the clutch, they will be burnt off under the heat of the clutch engagement and in the process, gradually darken the linings. Excessive oil on the clutch will burn off leaving a carbon deposit which can cause quite bad slip, or fierceness, spin and judder.

4 If clutch slip is suspected, and confirmation of this condition is required, there are several tests which can be made.
5 With the engine in second or third gear and pulling lightly, sudden depression of the accelerator pedal may cause the engine to increase its speed without any increase in road speed. Easing off on the accelerator will then give a definite drop in engine speed without the car slowing.
6 In extreme cases of clutch slip, the engine will race under normal acceleration conditions.
7 If slip is due to oil or grease on the linings, a temporary cure can sometimes be effected by squirting carbon tetrachloride into the clutch. The permanent cure is of course to renew the clutch driven plate and trace and rectify the oil leak.

Clutch spin

1 Clutch spin is a condition which occurs when there is a leak in the clutch hydraulic actuating mechanism, there is an obstruction in the clutch either in the first motion shaft or in the operating lever itself, or the oil may have partially burnt off the clutch lining and have left a resinous deposit which is causing the clutch disc to stick to the pressure plate or flywheel.
2 The reason for clutch spin is that due to any, or a combination of, the faults just listed, the clutch pressure plate is not completely freeing from the centre plate even with the clutch pedal fully depressed.
3 If clutch spin is suspected, the condition can be confirmed by extreme difficulty in engaging first gear from rest, difficulty in changing gear, and very sudden take up of the clutch drive at the fully depressed end of the clutch pedal travel as the clutch is released.
4 Check the clutch master cylinder and slave cylinder and the connecting hydraulic pipe for leaks. Fluid in one of the rubber dust covers fitted over the end of either the master or slave cylinder is a sure sign of a leaking piston seal.
5 If these points are checked and found to be in order, then the fault lies internally in the clutch and it will be necessary to remove the clutch for examination.

Clutch judder

1 Clutch judder is a self evident condition which occurs when the gearbox or engine mountings are loose or too flexible, when there is oil in the face of the clutch friction plate, or when the clutch pressure plate has been incorrectly adjusted.
2 The reason for clutch judder is that due to one of the faults just listed, the clutch pressure plate is not freeing smoothly from the friction disc and is snatching.
3 Clutch judder normally occurs when the clutch pedal is released in first or reverse gears and the whole car shudders as it moves backwards or forwards.

Chapter 6 Manual gearbox and automatic transmission

Contents

Specifications

Manual gearbox

Type	4 forward speeds, 1 reverse. Synchromesh fitted to all forward speeds
Oil capacity	1·5 pints (0·85 litres)
Gearbox ratios	
Fourth (top)	1·000 : 1
Third	1·307 : 1
Second	1·916 : 1
First	3·111 : 1
Reverse	3·422 : 1
Overall ratios	
Fourth (top)	3·636 : 1
Third	4·751 : 1
Second	7·003 : 1
First	11·313 : 1
Reverse	12·444 : 1
Road speed per 1000 rpm in top gear	18·1 mph (29·2 kph)
Tolerances	
2nd and 3rd gear endfloat on bushes	0·002 to 0·006 in (0·050 to 0·152 mm)
Endfloat of bushes on shaft	0·004 to 0·006 in (0·101 to 0·152 mm)
Thrustwasher sizes available:	
Colour code:	
Plain	0·152 to 0·154 in (3·860 to 3·911 mm)
Green	0·156 to 0·158 in (3·962 to 4·013 mm)
Blue	0·161 to 0·163 in (4·089 to 4·140 mm)
Orange	0·165 to 0·167 in (4·191 to 4·241 mm)
Yellow	0·169 to 0·171 in (4·293 to 4·343 mm)
Laygear needle roller retaining rings fitted depth:	
Inner	0·840 to 0·850 in (21·336 to 21·590 mm)
Outer	0·010 to 0·015 in (0·254 to 0·381 mm)
Centre bearing to circlip endfloat	0·000 to 0·002 in (0·000 to 0·050 mm)
Thrustwasher sizes available:	
Colour code:	
Plain	0·119 to 0·121 in (3·022 to 3·073 mm)
Green	0·122 to 0·124 in (3·123 to 3·173 mm)
Blue	0·125 to 0·127 in (3·198 to 3·248 mm)
Orange	0·128 to 0·130 in (3·273 to 3·323 mm)
Reverse idler gear bush fitted depth	Flush to 0·010 in (0·254 mm) below gear face

Torque wrench settings	**lbf ft**	**kgf m**
Flywheel housing retaining bolts	28 to 30	3·9 to 4·1
Rear extension to gearbox bolts	18 to 20	2·4 to 2·7
Drive flange nuts	90 to 100	12.4 to 13.8

Automatic transmission

Type	Borg Warner model 65	
Fluid capacity	11·5 pints (6·54 litres)	
Gear ratios		
First	2·39 : 1	
Second	1·45 : 1	
Third	1·00 : 1	
Reverse	2·09 : 1	
Shift speeds		
(D selected with accelerator in kickdown position)		
Upshift:		
1 – 2	34 to 42 mph (57 to 68 km/h)	
2 – 3	60 to 70 mph (97 to 113 km/h)	
Downshift:		
3 – 2	56 to 64 mph (90 to 102 km/h)	
2 – 1	23 to 36 mph (37 to 58 km/h)	
Torque wrench settings	**lbf ft**	**kgf m**
Driveflange nut	55 to 60	7·6 to 8·3
Driveplate to crankshaft	50	6·9
Converter to driveplate bolts	25 to 30	3·4 to 4·1
Oil pan to gearbox bolts	7	1·0
Drain plug	11	1·5
Starter inhibitor switch clamp bolt	4 to 6	0·5 to 0·8
Converter to driveplate bolts	28	3·9

PART A: MANUAL TRANSMISSION

1 General description

The manual gearbox fitted contains four forward and one reverse gear. Synchromesh is fitted to all four forward gears.

The gearchange lever is mounted on the extension housing and operates the selector mechanism in the gearbox by a long shaft. When the gearchange lever is moved sideways, the shaft is rotated so that the pins in the gearbox end of the shaft locate in the appropriate selector fork. Forward or rearward movement of the gearchange lever moves the selector fork, which in turn moves the synchromesh unit outer sleeve until the gear is firmly engaged. When reverse gear is selected, a pin on the selector shaft engages with a lever and this in turn moves the reverse idler gear into mesh with the laygear reverse gear and mainshaft. The direction of rotation of the mainshaft is thereby reversed.

The gearbox input shaft is splined and it is onto these splines that the clutch driven plate is located. The gearbox end of the input shaft is in constant mesh with the laygear cluster, and the gears formed on the laygear are in constant mesh with the gears on the mainshaft with the exception of the reverse gear. The gears on the mainshaft are able to rotate freely which means that when the neutral position is selected the mainshaft does not rotate.

When the gearchange lever moves the synchromesh unit outer sleeve via the selector fork, the synchromesh cup first moves and friction caused by the conical surfaces meeting takes up initial rotational movement until the mainshaft and gear are both rotating at the same speed. This condition achieved, the sleeve is able to slide over the dog teeth of the selected gear and thereby gives a firm drive. The synchromesh unit inner hub is splined to the mainshaft and because the outer sleeve is splined to the inner hub, engine torque is passed to the mainshaft and propeller shaft.

2 Gearbox – removal and refitting

1 The gearbox can be removed in unit with the engine as described in Chapter 1. An alternative method is to separate the gearbox bellhousing from the engine end plate, lower the gearbox and remove from under the car, leaving the engine in position. Use this method if only clutch and/or gearbox repairs are to be made.

2 Disconnect the battery, raise the car and put on axle-stands if a ramp is not available. The higher the car is off the ground the easier it will be to work underneath.

3 Undo the gearbox drain plug and drain the oil into a clean container. When all oil has drained out refit the drain plug.

4 Undo and remove the six nuts and washers that secure the exhaust pipes to the manifold.

5 Wipe the top of the clutch master cylinder and unscrew the cap. Place a piece of thin polythene sheet over the filler neck and refit the cap. This is to stop the clutch hydraulic fluid syphoning out during subsequent operations.

6 Undo the union nut that secures the clutch hydraulic pipe to the end of the slave cylinder. Unscrew the hydraulic pipe clip securing screw and tie back the hydraulic pipe.

7 With a scriber or file mark the gearbox and propeller shaft driveflanges to ensure correct refitting. Then undo and remove the four locknuts and bolts that secure the gearbox and propeller shaft driveflange. Using string or wire, tie the propeller shaft to the torsion bar.

8 Undo and remove the bolt and spring washer that secure the speedometer drive cable retaining clip on the side of the gearbox extension. Lift away the clip and carefully withdraw the speedometer cable.

9 Make a note of the electric cable connections to the starter motor, detach the cables and undo and remove the two bolts and spring washers that secure the starter motor. Carefully lift away the starter motor.

10 Using a hoist or jack, support the weight of the engine and gearbox. If a jack is being used, place it under the rear of the sump with a piece of wood between jack and sump.

11 Undo and remove the two bolts, plain and spring washers that secure the sump connecting plate to the underside of the gearbox bellhousing.

12 Undo and remove the two bolts, flat and spring washers that secure the gearbox rear crossmember to the underside of the bodyframe.

13 Undo and remove the nut and spring washer that secure the rear crossmember to the rear mounting.

14 Place the gear lever in the neutral position.

15 Lower the gearbox and engine assembly until the gear lever retaining cover can be reached from beneath the car, then press down the cover and turn it anti-clockwise to release its bayonet fixing. This will release the gear lever to allow the gearbox to be lowered.

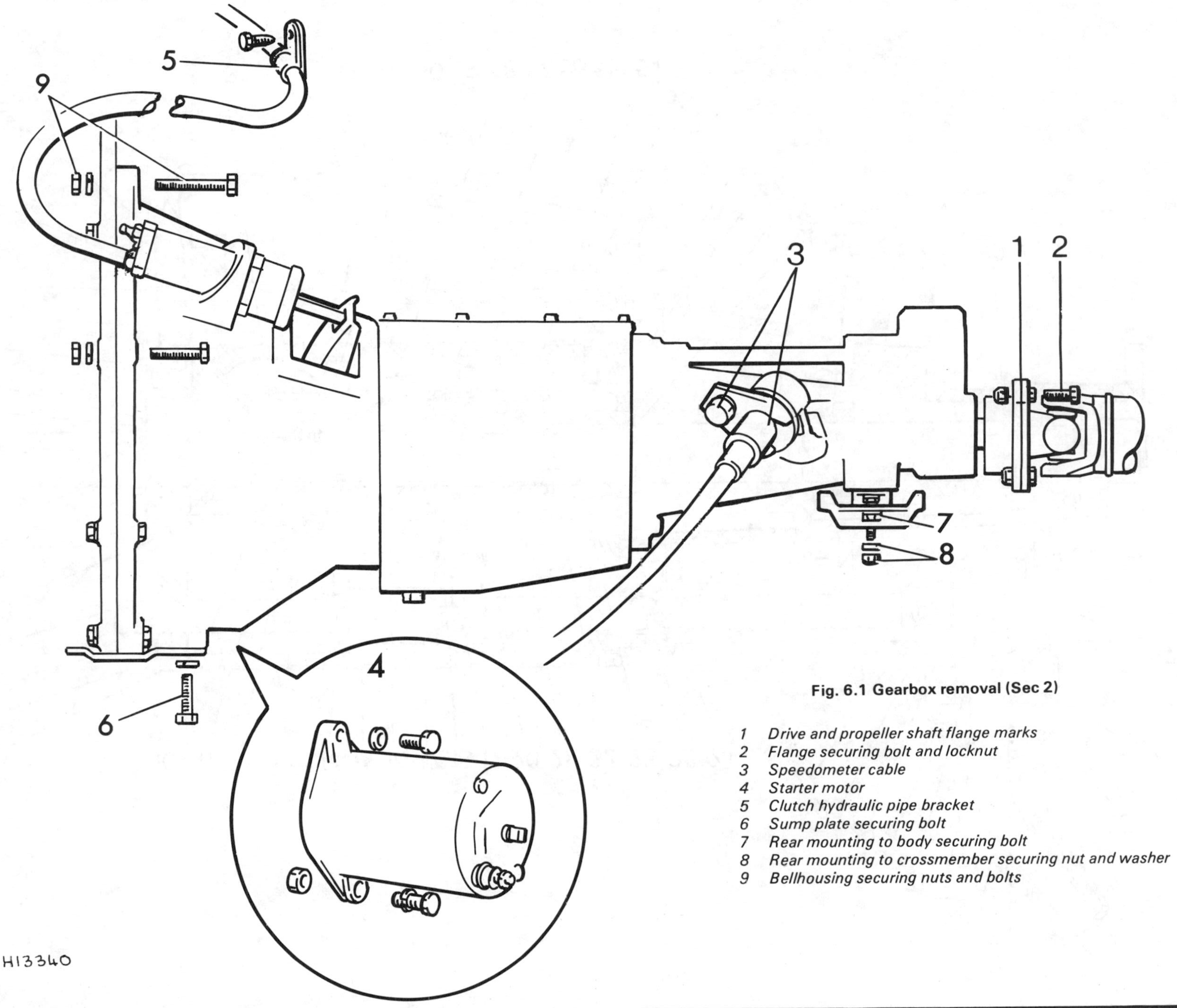

Fig. 6.1 Gearbox removal (Sec 2)

1 *Drive and propeller shaft flange marks*
2 *Flange securing bolt and locknut*
3 *Speedometer cable*
4 *Starter motor*
5 *Clutch hydraulic pipe bracket*
6 *Sump plate securing bolt*
7 *Rear mounting to body securing bolt*
8 *Rear mounting to crossmember securing nut and washer*
9 *Bellhousing securing nuts and bolts*

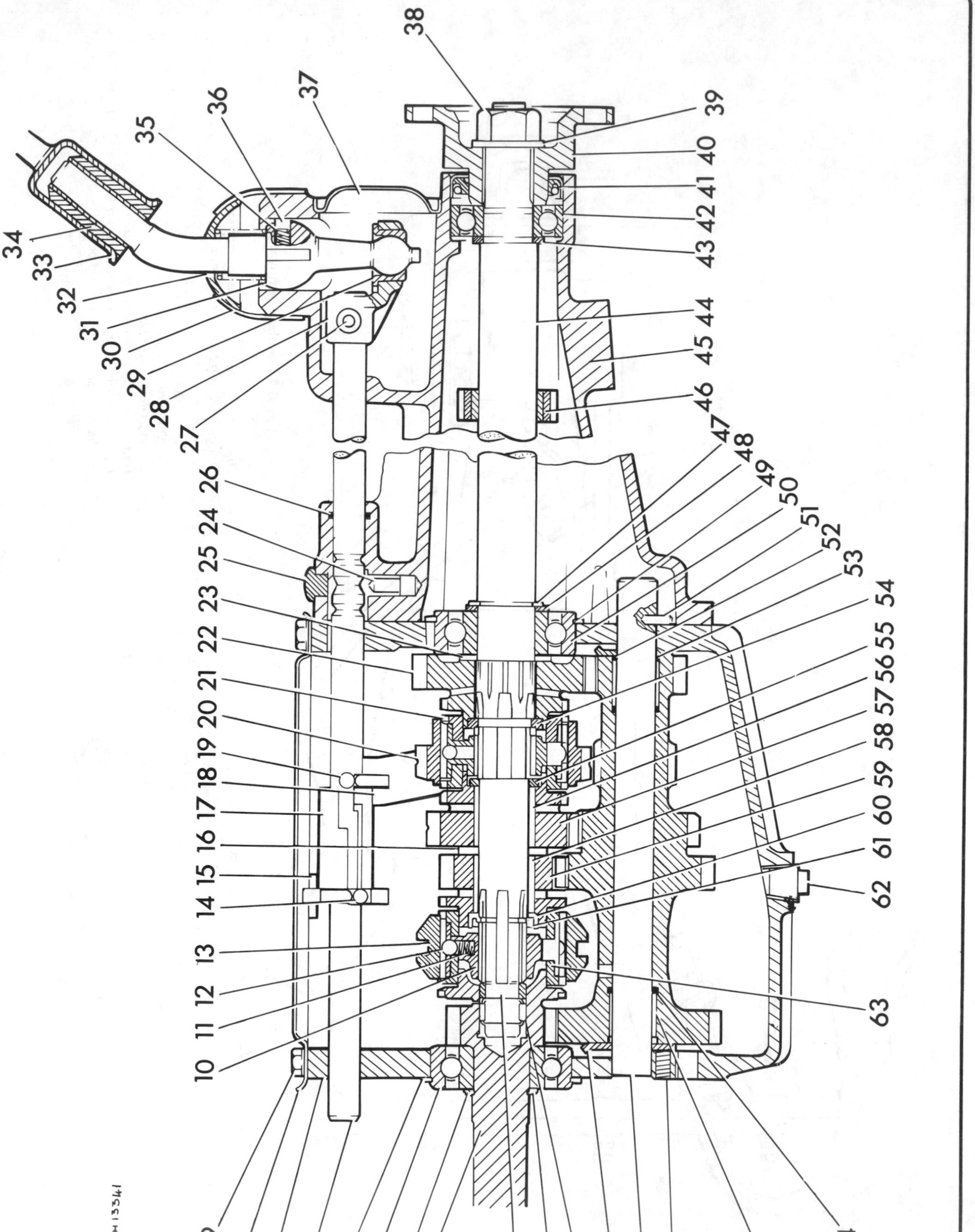
1
2
3
4
5
6
7
8
9
10
11
12
13
14
15
16
17
18
19
20
21
22
23
24
25
26
27
28
29
30
31
32
33
34
35
36
37
38
39
40
41
42
43
44
45
46
47
48
49
50
51
52
53
54
55
56
57
58
59
60
61
62
63
64
65
66
67
68
69
70
H13541

Fig. 6.2 Cross-sectional view of manual gearbox (Sec 1)

1 1st motion shaft
2 Circlip
3 Front ball-bearing
4 Snap ring
5 Gear selector shaft
6 Gearbox case
7 Top cover
8 Spacer
9 Top cover bolt
10 3rd and 4th speed synchromesh hub
11 Spring
12 Ball
13 3rd and 4th speed operating sleeve
14 Selector shaft pin
15 Interlock spool plate
16 Selective washer
17 Interlock spool
18 Reverse operating lever
19 Selector shaft roll pin
20 Mainshaft reverse gear and 1st/2nd operating sleeve
21 Synchromesh cup
22 1st speed gear
23 Thrustwasher
24 Detent plunger
25 Plug
26 O-ring
27 Yoke pin
28 Gear lever yoke
29 Seat
30 Retaining cap
31 Lower gear lever
32 Seal
33 Upper gear lever
34 Bush
35 Anti-rattle spring
36 Anti-rattle plunger
37 End cover
38 Self-locking nut
39 Flange washer
40 Flange and stone-guard assembly
41 Seal
42 End ball-bearing
43 Thrustwasher
44 Mainshaft
45 Gearbox rear extension
46 Speedometer wheel
47 Circlip
48 Selective washer
49 Snap ring
50 Centre ball-bearing
51 Layshaft dowel
52 Retaining ring
53 Rear thrust washer
54 Split collar
55 Thrustwasher
56 Gear bush
57 3rd speed gear
58 Gear bush
59 2nd speed gear
60 Thrustwasher
61 Circlip
62 Drain plug
63 Synchromesh cup
64 Retaining ring
65 Needle roller bearing
66 Laygear preload springs
67 Layshaft
68 Thrustwasher
69 Needle roller bearing
70 Circlip backing washer

16 Undo and remove the seven bolts and spring washers that secure the flywheel housing to the mounting plate.
17 Disconnect the power unit braided earthing strap.
18 Disconnect the leads from the reverse lamp switch.
19 Make sure the weight of the gearbox is not allowed to be taken solely on the first motion shaft as it bends easily.
20 Lower the rear of the engine until there is sufficient clearance between the top of the bellhousing and underside of the body. Carefully draw the gearbox rearwards. Lift away the gearbox from the underside of the car.
21 Refitting the gearbox is the reverse sequence to removal. Do not forget to refill the gearbox if the oil has been previously drained. It will be necessary to bleed the clutch hydraulic system as described in Chapter 5.

3 Gearbox – dismantling

1 Before commencing work, clean the exterior of the gearbox thoroughly using a solvent such as paraffin. After the solvent has been applied and allowed to stand for a time, a vigorous jet of water will wash off the solvent together with all oil and dirt. Finally wipe down the exterior of the unit with a dry non-fluffy rag.
2 Note that all numbers in brackets contained in the following text refer to Fig. 6.3 unless stated otherwise.
3 Detach the operating lever from the release bearing and slide off the bearing assembly.
4 Undo and remove the five bolts that secure the clutch bellhousing to the gearbox casing. Note that the lowermost bolt has a plain copper washer, whereas the remaining bolts have spring washers.
5 Lift away the bellhousing. Collect the three laygear preload springs (95) and the paper gasket from the front of the gearbox casing.
6 Undo and remove the nine bolts and spring washers that secure the top cover to the main casing.
7 Lift off the top cover and remove the paper gasket.
8 Note which way up the interlock spool plate is fitted and lift it from the top of the main casing.
9 Undo and remove the one bolt and spring washer that secure the reverse lift plate (14) to the rear extension. Lift away the lift plate.
10 Using a screwdriver, carefully remove the rear extension end cover (12).
11 With a mole wrench, hold the driveflange (86) and using a socket wrench, undo and remove the locking nut (88) and plain washer (87).
12 Tap the driveflange (86) from the end of the mainshaft (83).
13 Lift out the speedometer drive pinion and housing assembly (24 to 27) from the rear extension.
14 Make a special note of the location of the selector shaft pegs and interlock spool (16) so that there will be no mistakes on reassembly.
15 Using a suitable diameter parallel pin punch carefully remove the roll pin (17) from the bellhousing end of the selector shaft (20).
16 Undo and remove the eight bolts and spring washers that secure the rear extension (11) to the gearbox casing (1).
17 Draw the rear extension rearwards whilst at the same time feeding the interlock spool (16) from the selector shaft (20).
18 With the rear extension (1) and selector shaft (20) away from the gearbox casing, lift out the interlock spool (16).
19 Recover the paper gasket (7) from the rear face of the gearbox casing (1).
20 If oil was leaking from the end of the rear extension, or the bearing (85) requires renewal, the oil seal must be removed and discarded. It must never be refitted but always renewed. Ease it out with a screwdriver noting which way round the lip is fitted.
21 To remove the bearing, obtain a long metal drift and tap it out working from inside the rear extension. Note which way round the bearing is fitted as indicated by the lettering.
22 Slide the washer from over the end of the mainshaft.
23 Make a special note of the location of the speedometer drivegear (77) on the mainshaft, if necessary by taking a measurement.
24 Using a tapered but blunt drift, drive the speedometer drivegear from the mainshaft. **Caution**: *Beware, because it is very tight and can break.*
25 Using a suitable diameter drift, tap out the selector fork shaft (45) towards the front of the gearbox casing.
26 Note the location of the two forward gear selector forks (43, 44) and lift these from the synchromesh sleeve (54, 68).
27 Using a suitable diameter drift, tap out the layshaft (93) working

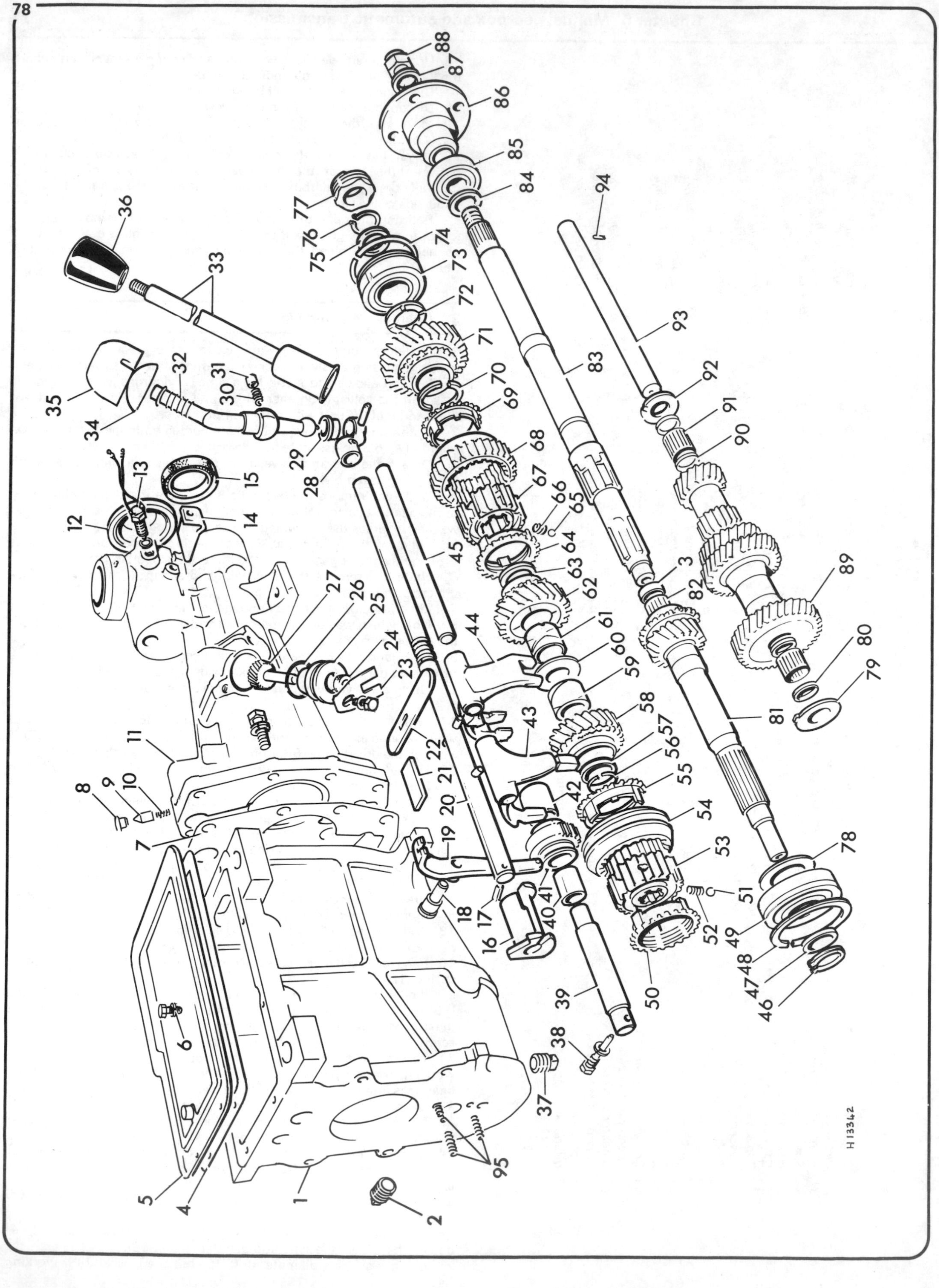
1
2
3
4
5
6
7
8
9
10
11
12
13
14
15
16
17
18
19
20
21
22
23
24
25
26
27
28
29
30
31
32
33
34
35
36
37
38
39
40
41
42
43
44
45
46
47
48
49
50
51
52
53
54
55
56
57
58
59
60
61
62
63
64
65
66
67
68
69
70
71
72
73
74
75
76
77
78
79
80
81
82
83
84
85
86
87
88
89
90
91
92
93
94
95
H 13342

Fig. 6.3 Exploded view of manual gearbox (Sec 3)

1 Gearbox case
2 Oil filler/level plug
3 Spacer
4 Gaskets
5 Top cover
6 Top cover bolt
7 Gasket
8 Plug
9 Detent plunger
10 Detent spring
11 Rear extension
12 End cover
13 Reverse light switch
14 Reverse lift plate
15 Oil seal
16 Interlock spool
17 Selector shaft roll pin
18 Reverse operating lever pin
19 Reverse operating lever
20 Gear selector shaft
21 Magnet
22 Interlock spool plate
23 Retaining clip
24 Seal
25 Housing
26 O-ring
27 Speedometer pinion
28 Gear lever yoke
29 Seat
30 Spring
31 Anti-rattle plunger
32 Lower gear lever
33 Upper gear lever
34 Dust cover washer
35 Retaining cup
36 Knob
37 Drain plug
38 Reverse idler spindle locating screw
39 Reverse idler spindle
40 Reverse idler gear bush
41 Reverse idler gear
42 Reverse idler distance-piece
43 3rd/4th speed selector forks
44 1st/2nd speed selector forks
45 Selector fork shaft
46 Circlip
47 Backing washer
48 Snap ring
49 Ball-bearing
50 Synchromesh hub
51 Ball
52 Spring
53 3rd/4th speed synchromesh hub
54 3rd/4th speed operating sleeve
55 Synchromesh cup
56 Mainshaft circlip
57 3rd speed gear thrustwasher
58 3rd speed gear
59 Gear bush
60 Selector washer
61 Gear bush
62 2nd speed gear
63 Thrustwasher
64 Synchromesh cup
65 Ball
66 Spring
67 1st/2nd speed synchromesh hub
68 Mainshaft reverse gear
69 Synchromesh cup
70 Split collar
71 1st speed gear
72 Thrustwasher
73 Mainshaft centre bearing
74 Snap ring
75 Selective washer
76 Circlip
77 Speedometer wheel
78 Oil flinger
79 Front thrustwasher
80 Bearing outer retaining ring
81 1st motion shaft (input shaft)
82 Needle roller bearing
83 Mainshaft
84 Washer
85 Ball-bearing
86 Drive flange
87 Washer
88 Self-locking nut
89 Laygear cluster
90 Bearing retaining inner ring
91 Needle rollers
92 Rear thrustwasher
93 Layshaft
94 Layshaft dowel
95 Laygear preload springs

from the front of the gearbox casing. This is because there is a layshaft restraining pin (94) at the rear to stop it rotating.

28 Locate the gearbox to allow the laygear cluster (89) to drop into the bottom of the casing.

29 Using a small drift placed on the bearing outer track, tap out the gearbox input shaft (81). If necessary, recover the caged needle roller bearing (82) from the end of the shaft. Also recover the 4th gear synchromesh cup.

30 The mainshaft may now be drifted rearwards slightly, sufficiently to move the bearing and locating circlip (73). Using a screwdriver between the circlip (74) and casing, ease the bearing out of its bore and from its locating shoulder on the mainshaft. Lift away the bearing from the end of the mainshaft.

31 The complete mainshaft may now be lifted away through the top of the gearbox main casing.

32 Unscrew the dowel bolt (38) that locks the reverse idler shaft (39) to the gearbox casing. Lift away the bolt and spring washer.

33 Using a small drift, tap the reverse idler shaft rearwards noting the hole in the shaft into which the dowel bolt locates.

34 Note which way round the reverse idler is fitted and lift it from the casing.

35 Lift out the laygear cluster noting which way round it is fitted.

36 Recover the two thrust washers, noting that the tags locate in grooves in the gearbox casing.

4 Gearbox – examination

1 The gearbox has been stripped, presumably, because of wear or malfunction, possibly excessive noise, ineffective synchromesh or failure to stay in a selected gear. The cause of most gearbox ailments is failure of the ball-bearings on the input or mainshaft and wear on the synchro rings, both the core surfaces and dogs. The nose of the mainshaft which runs in the needle roller bearing in the input shaft is also subject to wear. This can prove very expensive as the mainshaft would need replacement and this represents about 20% of the total cost of a new gearbox.

2 Examine the teeth of all gears for signs of uneven or excessive wear and, of course, chipping. If a gear on the mainshaft requires replacement, check that the corresponding laygear is not equally damaged. If it is, the whole laygear may need replacing also.

3 All gears should be a good running fit on the shaft with no signs of rocking. The hubs should not be a sloppy fit on the splines.

4 Selector forks should be examined for signs of wear or ridging on the faces which are in contact with the operating sleeve.

5 Check for wear on the selector rod and interlock spool.

6 The ball-bearings may not be obviously worn but if one has gone to the trouble of dismantling the gearbox it would be short sighted not to renew them. The same applies to the four synchronizer rings, although for these the mainshaft has to be completely dismantled for the new ones to be fitted.

7 The input shaft bearing retainer is fitted with an oil seal and this should be removed if there are any signs that oil has leaked past it into the clutch housing or, of course, if it is obviously damaged. The rear extension has an oil seal at the rear as well as a ball-bearing race. If either have worn or oil has leaked past the seal the parts should be renewed.

8 Before finally deciding to dismantle the mainshaft and replace parts, it is advisable to make enquiries regarding the availability of parts and their cost. It may still be worth considering an exchange gearbox even at this stage. The gearbox should be reassembled before exchange.

5 Input shaft – dismantling and reassembly

1 Place the input shaft in a vice, splined end upwards, and with a pair of circlip pliers, remove the circlip which retains the ball-bearing in place. Lift away the spacer.

2 With the bearing resting on the top of the open jaws of the vice and with the splined end upwards, tap the shaft through the bearing with a soft-faced hammer. Note that the offset circlip groove in the outer track of the bearing is towards the front of the input shaft.

3 Lift away the oil flinger.

4 Remove the oil caged needle roller bearing from the centre of the rear of the input shaft if it is still in place.

5 Remove the circlip from the old bearing outer track and transfer it to the new bearing.
6 Refit the oil flinger and, with the aid of a block of wood and vice, tap the bearing into place. Make sure it is the right way round.
7 Finally refit the spacer and bearing retaining circlip.

6 Mainshaft – dismantling and reassembly

1 The component parts of the mainshaft are shown in Fig. 6.4.
2 Lift the 3rd and 4th gear synchromesh hub and operating sleeve assembly from the end of the mainshaft.

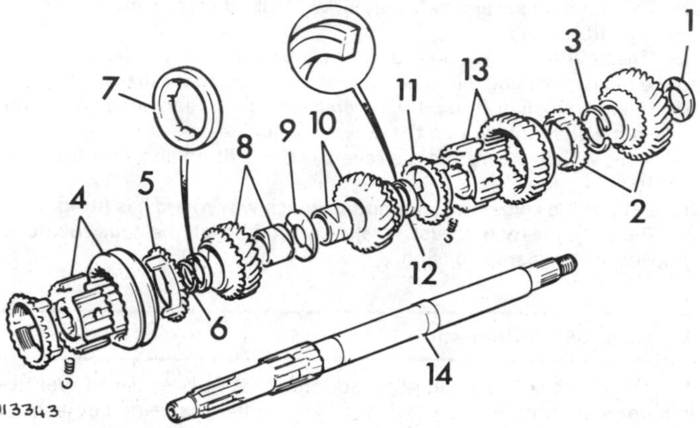

Fig. 6.4 Exploded view of mainshaft assembly (Sec 6)

1	*Thrustwasher*	7	*3rd speed gear thrustwasher*
2	*1st speed gear and synchro cup*	8	*3rd speed gear and bush*
		9	*Selective washer*
3	*Split collar*	10	*2nd speed gear and bush*
4	*3rd and 4th speed synchro unit*	11	*Synchromesh cup*
		12	*Thrustwasher*
5	*Synchro cup*	13	*1st/2nd speed synchro*
6	*Mainshaft circlip*	14	*Mainshaft*

3 Remove the 3rd gear synchromesh cup.
4 Using a small screwdriver, ease the 3rd gear retaining circlip from its groove in the mainshaft. Lift away the circlip.
5 Lift away the 3rd gear thrustwasher.
6 Slide the 3rd gear and bush from the mainshaft followed by the thrustwasher. Note this is a selective thrustwasher.
7 Slide the 2nd gear and bush from the mainshaft followed by the grooved washer. Note which way round it is fitted.
8 Detach the 2nd gear synchromesh cup from inside the 2nd and 1st gear synchromesh hub and lift away.
9 Slide the 2nd and 1st gear synchromesh hub and reverse gear sleeve assembly from the mainshaft. Recover the 1st gear synchromesh cup.
10 Using a small electricians screwdriver, lift out the two split collars from their groove in the mainshaft.
11 Slide the 1st gear mainshaft washer from the mainshaft and follow this with the 1st gear.
12 The mainshaft is now completely dismantled.
13 Before reassembling, refer to Fig. 6.5 and measure the endfloat of the 2nd and 3rd gears on their respective bushes. The endfloat should be within the limits quoted in the Specifications. Obtain a new bush, if necessary, to achieve the correct endfloat.
14 Temporarily refit the 2nd gear washer, oil grooved face away from the mainshaft shoulder, to the mainshaft. Assemble to the mainshaft the 3rd gear bush, selective washer, 2nd gear bush, 3rd gear thrustwasher with its oil grooved face to the bush, and fit the 3rd gear mainshaft circlip. Measure the endfloat of the bushes on the mainshaft, which should be within the limits quoted in the Specifications. Obtain a new selective washer to obtain the correct endfloat. Remove the parts from the mainshaft.
15 To reassemble, slide the 1st gear onto the mainshaft followed by the washer (photo).
16 Fit the two halves of the split collar into the groove in the mainshaft and push the 1st gear hard up against the collar (photo). If the split collars are marked on one side with orange dye (as is the case on some gearboxes), then the orange marked side must face away from the 1st speed gear.
17 Fit the synchromesh cup onto the core of the 1st gear (photo).
18 Slide the 1st and 3nd gear synchromesh hub and reverse gear sleeve on the mainshaft and engage it with the synchromesh cup (photo).
19 Fit the 2nd gear synchromesh cup to the synchromesh hub (photo).
20 Fit the 2nd gear washer onto the end of the mainshaft splines so that the oil grooved face is towards the front of the mainshaft (photo).

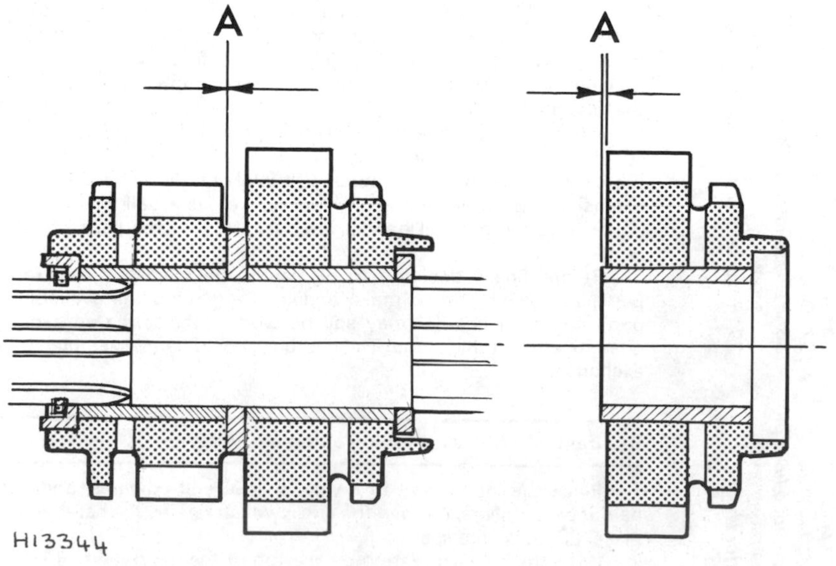

Fig. 6.5 Mainshaft gear endfloat (Sec 6)

A – 0.002 to 0.006 in (0.050 to 0.152 mm)

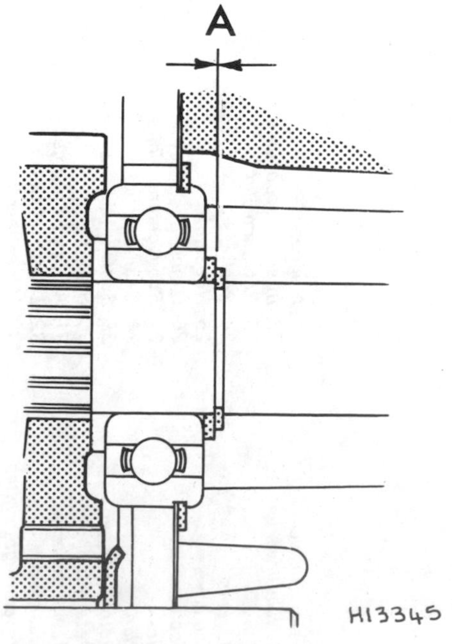

Fig. 6.6 Selective washer thickness 'A' (Sec 6)

6.15 Fitting 1st gear and large washer onto mainshaft

6.16 Inserting split collar into groove

6.17 Synchromesh cone for 1st gear

6.18 Fitting synchromesh hub and reverse gear sleeve

6.19 2nd gear synchromesh cup placement on synchromesh hub

6.20 The oil grooved face must face towards front of mainshaft

6.21 Sliding 2nd gear bush onto mainshaft

6.22 Fitting 2nd gear onto bush

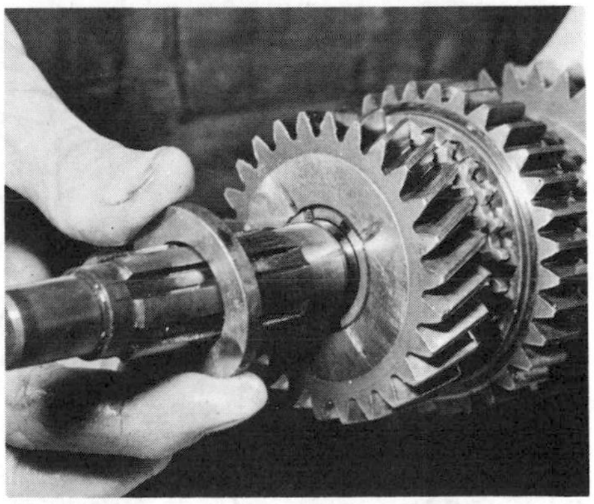

6.23 2nd and 3rd gear selective washer

6.24 Sliding 3rd gear bush onto mainshaft

6.25 Fitting 3rd gear onto bush

6.26 Sliding 3rd gear thrustwasher onto mainshaft

6.27 The circlip must be correctly located in its groove

6.28 Fitting 3rd gear synchromesh cup into synchromesh hub

6.29 Sliding 3rd and 4th gear synchromesh hub assembly onto mainshaft

21 Slide the 2nd gear bush onto the mainshaft (photo).
22 Fit the 2nd gear onto the bush on the mainshaft and engage the taper with the internal taper of the synchromesh cup (photo).
23 Fit the 2nd and 3rd gear selective washer (photo).
24 Slide the 3rd gear bush onto the mainshaft (photo).
25 Fit the 3rd gear onto the bush on the mainshaft, the cone facing the front of the mainshaft (photo).
26 Slide the 3rd gear thrustwasher onto the mainshaft splines (photo).
27 Ease the 3rd gear retaining circlip into its groove in the mainshaft. Make quite sure it is fully seated (photo).
28 Fit the 3rd gear synchromesh cup onto the cone of the 3rd gear (photo).
29 Finally slide on the 3rd and 4th gear synchromesh hub and operating sleeve assembly and engage it with the synchromesh cup (photo).

7 Gearbox – reassembly

1 Check that the magnet is in position in the gearbox casing (photo).
2 Position the laygear needle bearing roller bearing inner retainers into the laygear bore. Apply multi-purpose grease to the ends of the laygear and fit the needle rollers. Retain the needle rollers in position with the outer retainers (photo).
3 Make up a piece of tube the same diameter as the layshaft and the length of the laygear plus thrustwashers. Slide the tube into the laygear. This will retain the needle rollers in position. Apply grease to the thrustwashers and fit to the ends of the laygear. The tags must face outwards (photo).
4 Carefully lower the laygear into the bottom of the gearbox casing (photo).

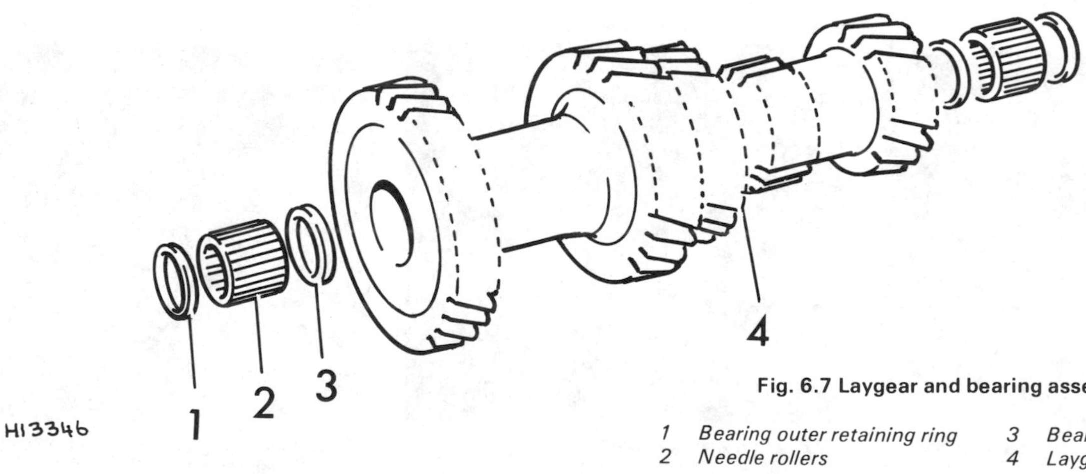

Fig. 6.7 Laygear and bearing assemblies (Sec 7)

1 Bearing outer retaining ring
2 Needle rollers
3 Bearing inner retaining ring
4 Laygear

7.1 Location of magnet

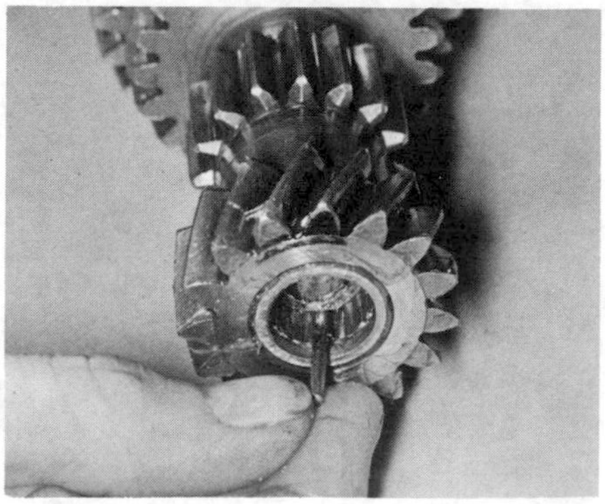
7.2 Inserting needle bearing rollers

7.3 Thrustwasher with tag facing outwards

7.4 Lowering laygear into gearbox casing

7.5 Sliding idler shaft into position

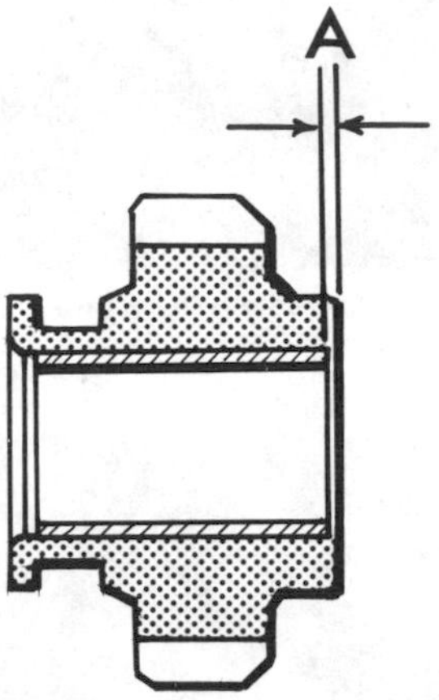
Fig. 6.8 Reverse idler gear bush location (Sec 7)

A – 0.000 to 0.010 in (0.000 to 0.254 mm)

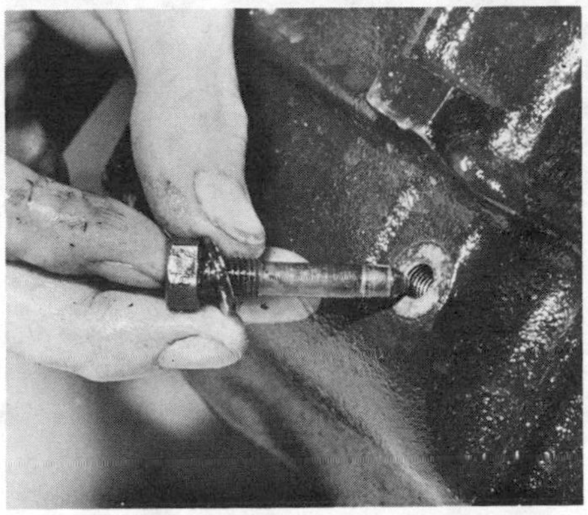
7.6 Special dowel bolt with shaped end

7.7 Fitting mainshaft into gearbox casing

7.8 Sliding bearing up mainshaft

7.9 Supporting mainshaft spigot

7.10 Drifting bearing into position

5 Fit the reverse gear operating lever to the operating lever pivot. Hold the reverse idler in its approximate fitted position and slide in the idler shaft, drilled end first (photo).

6 Carefully line up the drilled hole in the idler shaft and gearbox casing and refit the dowel bolt and spring washer (photo).

7 The assembled mainshaft may now be fitted into the gearbox casing (photo).

8 Ease the mainshaft bearing up the mainshaft, circlip offset on the outer track towards the rear (photo).

9 Place a metal lever in the position shown in the photo, so supporting the mainshaft spigot (photo).

10 Using a suitable diameter tube, carefully drift the mainshaft bearing into position in the rear casing (photo).

11 Fit the 4th gear synchromesh cup onto the end of the input shaft.

12 Lubricate the needle roller bearing and fit it into the end of the input shaft (photo). Don't forget first to fit the spacer.

13 Fit the input shaft to the front of the gearbox casing, taking care to engage the synchromesh cup with the synchromesh hub (photo).

14 Tap the input bearing until the circlip is hard up against the front gearbox casing. Check that the mainshaft bearing outer track circlip is hard up against the rear casing. Refit the washer and circlip.

15 Invert the gearbox. Fit the pin into the drilled hole in the layshaft and carefully insert the layshaft from the rear of the main casing. This will push out the previously inserted tube. The pin must be to the rear of the main casing (photo).

16 Line up the layshaft pin with the groove in the rear face and push the layshaft fully home (photo).

17 Fit the 3rd and 4th gear selector fork to the synchromesh sleeve (photo).

18 Fit the 1st and 2nd gear selector fork to the synchromesh sleeve (photo).

19 Slide the selector fork shaft from the front through the two selector forks and into the rear of the main casing (photo).

20 Fit a new gasket to the rear face of the main casing and retain it in position with a little grease (photo).

21 Place the speedometer drivegear onto the mainshaft and using a

7.12 Inserting the needle roller bearing into input shaft

7.13 Fitting input shaft

7.15 The pin must be to the rear of the gearbox casing

7.16 Lining up pin with groove

7.17 Fitting 3rd and 4th gear selector fork to synchromesh sleeve

7.18 Fitting 1st and 2nd gear selector fork to synchromesh sleeve

7.19 Inserting selector fork shaft through selector forks

7.20 Fitting new gasket to gearbox casing rear face

7.21 Drift speedometer drivegear up to previously made mark

7.22 Slide washer onto mainshaft

7.23 The bearing code letters must face outwards

7.24 Using socket and hammer to tap bearing into position

tube, drive the gear into its previously noted position (photo).
22 Slide the washer onto the mainshaft (photo).
23 Place the rear extension bearing into its bore, letters facing outwards (photo).
24 Tap the bearing into position using a suitable diameter socket (photo).
25 Fit a new rear extension oil seal and tap it into position with the previously used socket. The lip must face inwards (photo).
26 Slide the interlock spool onto the selector shaft, making sure it is the correct way round as shown. This is to give an idea of the final fitted position. Remove the interlock spool again (photo).
27 Place the interlock spool on the selector forks with the flanges correctly engaged (photo).
28 Offer up the gearbox rear extension to the rear of the main casing, at the same time feeding the selector shaft through the interlock spool. It will be necessary to rotate the selector shaft to obtain correct engagement (photo).
29 Secure the rear extension with the eight bolts and spring washers (photo).
30 Refit the spring-pin into the end of the selector shaft, ensuring the ends are equidistant from the shaft (photo).
31 This photo shows the interlock spool and selector shaft correctly aligned with the pegs engaged (photo).
32 Insert the speedometer driven gear and housing into the rear extension (photo).
33 Fit the driveflange onto the mainshaft splines (photo).
34 Hold the driveflange and tighten the retaining nut and washer fully (photo).
35 Refit the reverse lift plate and secure it with the bolt and spring washer (photo).
36 Refit the rear extension end cover and tap it into position with the end of the lip flush with the end of the casing (photo).
37 Refit the interlock spool plate in the same position as was noted before removal (photo).
38 Fit a new gasket to the top of the gearbox casing and refit the top cover (photo).
39 Secure the top cover with the nine bolts and spring washers. These should be progressively tightened in a diagonal manner (photo).

7.25 Fitting rear extension oil seal with lip facing inwards

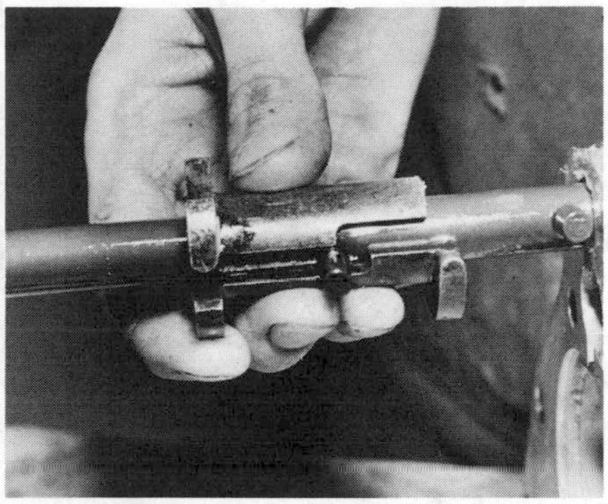

7.26 Trial fitting of interlock spool

7.27 Fitting interlock spool to selector forks

7.28 Rotate the selector shaft to obtain correct engagement

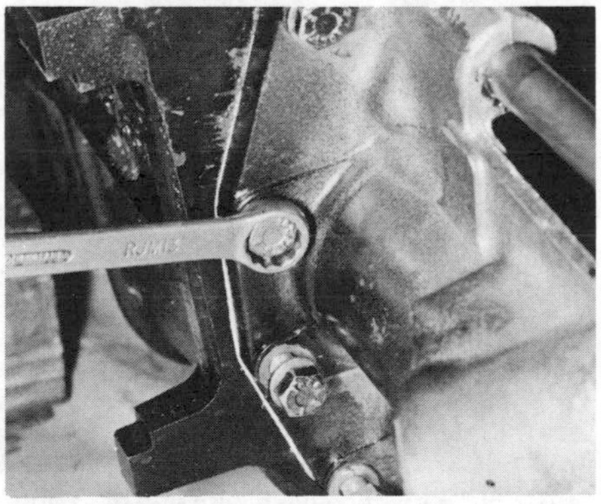

7.29 Tightening rear extension securing bolts in a progressive manner

7.30 Refitting spring pin to end of selector shaft

7.31 Interlock spool and selector shaft aligned with pegs engaged

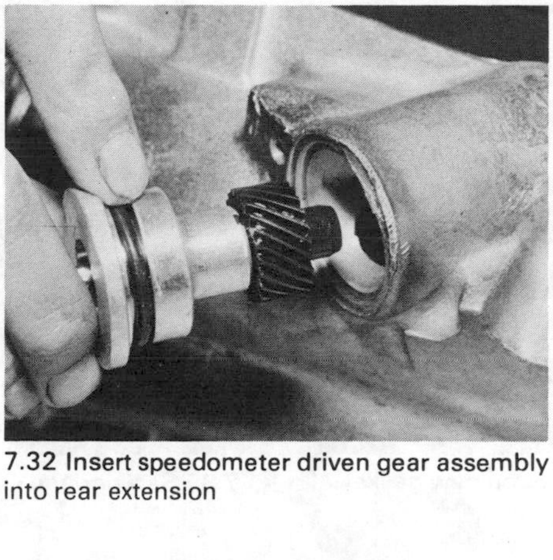

7.32 Insert speedometer driven gear assembly into rear extension

Fig. 6.9 Speedometer drive assembly (Secs 2, 3 and 7)

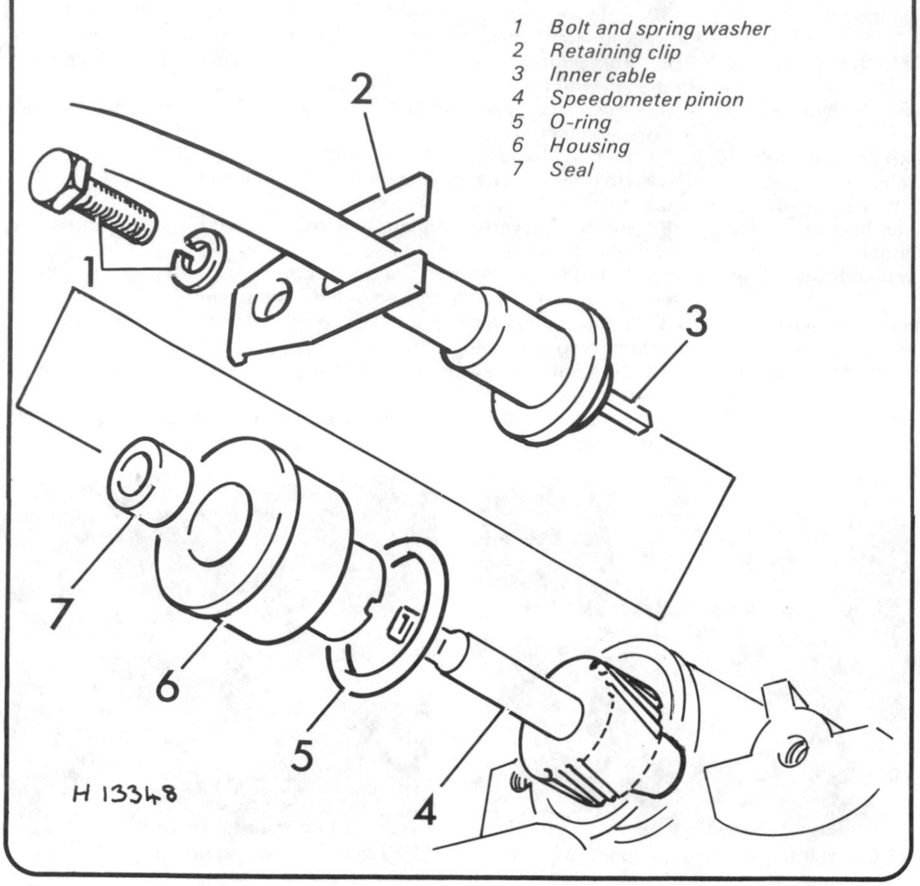

7.33 Offering up the flange to mainshaft

7.34 Tightening flange retaining nut

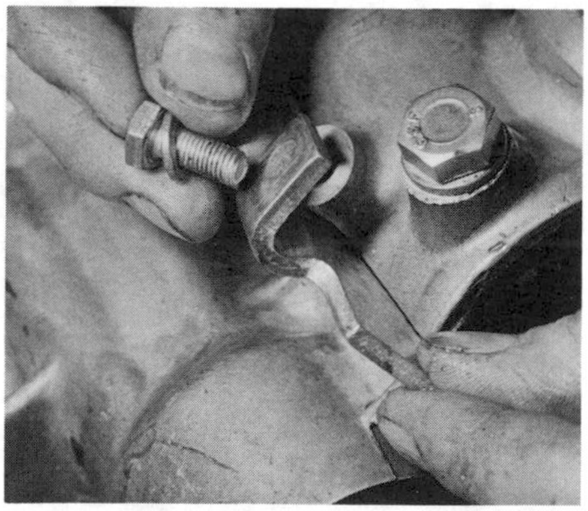
7.35 Refitting reverse lift plate

7.36 The lip must be flush with end of casing

7.37 Refitting interlock spool plate

7.38 Positioning top cover on new gasket

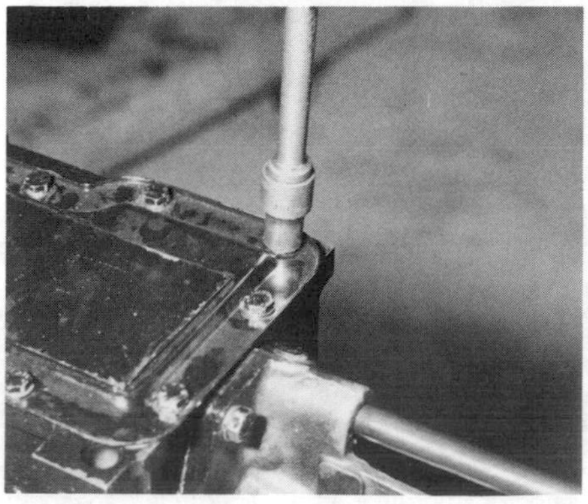
7.39 Securing top cover to gearbox casing

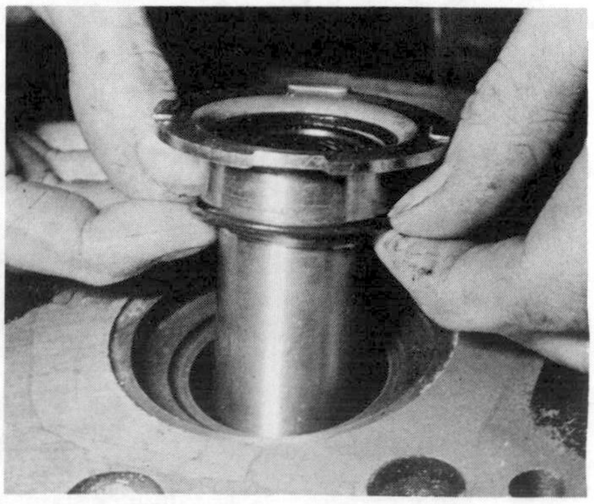
7.40 Fitting new O-ring to input shaft retainer

7.41 Fit a new gasket to gearbox casing front face

7.42 Positioning clutch bellhousing

7.43 This bolt requires a copper washer

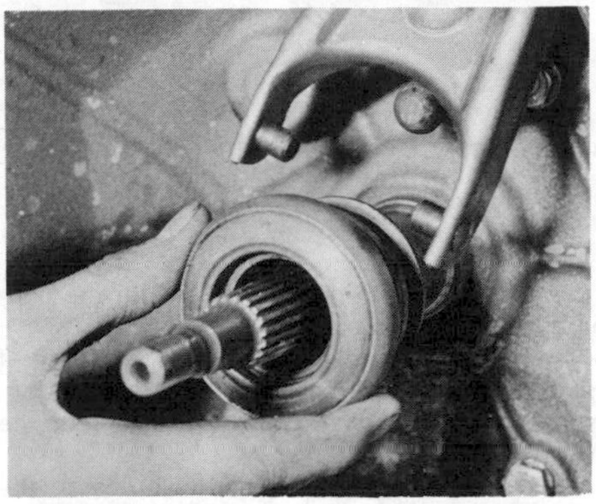

7.44 Sliding clutch release bearing assembly onto its guide

7.45 Clutch bellhousing upper flange bolts – refitting

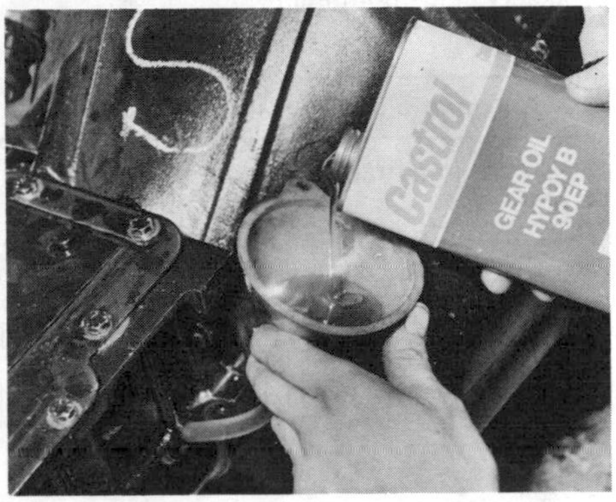

7.46 Use a funnel if possible – flexitops will work

40 Fit a new O-ring to the input shaft retainer and refit the retainer (photo).
41 Fit a new gearbox casing front face gasket and retain it in position with a little grease (photo). Fit the three layshaft preload springs in their locations on the front of the gearbox casing.
42 Move the gearbox to the end of the bench and offer up the clutch bellhousing (photo).
43 Refit the five bolts that secure the bellhousing to the main casing. Note that four bolts have spring washers and the fifth (lowermost) has a copper washer (photo).
44 Slide the clutch release bearing assembly onto its guide, at the same time engaging the release lever (photo).
45 If the gearbox was removed in unit with the engine, it may now be reattached. Secure it in position with the retaining nuts, bolts and spring washers (photo).
46 Refill the gearbox with $1\frac{1}{2}$ pints (0.85 litres) SAE 90 EP gear oil (photo).

8 Fault diagnosis

Symptom	Reason/s
Weak or ineffective synchromesh	Synchronising cones worn, split or damaged Synchromesh dogs worn or damaged
Jumps out of gear	Broken gearchange fork rod spring Gearbox coupling dogs badly worn Selector fork rod groove badly worn
Excessive noise	Incorrect grade of oil in gearbox or oil level too low Bush or needle roller bearings worn or damaged Gearteeth excessively worn or damaged Laygear thrustwashers worn, allowing excessive endplay
Excessive difficulty in engaging gear	Clutch fault

PART B: AUTOMATIC TRANSMISSION

9 General description

Borg-Warner automatic transmission is fitted.

The automatic transmission system comprises two main components: A three-element hydrokinetic torque converter coupling capable of torque multiplication at an infinitely variable ratio between 2:1 and 1:1 and a torque speed responsive and hydraulically operated epicyclic gearbox, comprising a planetary gear set providing three forward ratios and reverse gear.

Due to the complexity of the automatic transmission unit, if performance is not up to standard, or overhaul is necessary, it is imperative that this is undertaken by BL main agents who will have special equipment for accurate fault diagnosis and rectification.

The contents of the following Sections are therefore solely general and servicing information.

10 Automatic transmission – fluid level

1 It is important that the transmission fluid is of the correct specification, use automatic transmission fluid type G to specification MC2 33G. The capacity of the unit is approximately 11.5 pints (6.54 litres) when dry. For a drain and refill, which is not normally necessary except during repair, the capacity will be approximately 6 pints (3.4 litres) as the converter cannot be completely drained.

2 The location of the dipstick is shown in Fig. 6.11. Full information on checking the fluid level will be found in the Routine Maintenance Section at the beginning of this manual.

11 Automatic transmission – removal and refitting

1 Any suspected faults must be referred to the main agent before unit removal, as with this type of transmission its fault must be confirmed using special equipment before it is removed from the car.

2 As the automatic transmission is relatively heavy it is best if the car is raised from the ground on ramps, but it is possible to remove the unit if the car is placed on high axle-stands.

3 Disconnect the battery.

4 Disconnect the downshift cable from the throttle linkage at the side of the carburettor.

5 Remove the dipstick from its guide tube.

6 Disconnect the breather pipe from the guide tube.

7 Remove the dipstick guide tube after removing the securing bolt.

8 Detach the exhaust pipe from the manifold.

9 Place a clean container of at least 8 pints (4.5 litres) capacity under the sump drain plug, remove the drain plug and drain the fluid into the container. Refit the drain plug.

10 Disconnect the manual selector rod from both the gearbox lever and gearchange lever. Lift away the manual selector rod.

11 Disconnect the inhibitor/reverse lamp switch at the multi-connector plugs.

12 Using a scriber or file, mark the propeller shaft and gearbox flanges so that they may be refitted in their original positions.

13 Undo and remove the four locknuts and bolts that secure the two flanges together.

14 Lift the front end of the propeller shaft away from the rear of the gearbox and tie it to the torsion bar with wire or strong string.

15 Undo and remove the speedometer cable clamp bolt and spring washer on the gearbox extension. Lift away the clamp and withdraw the cable.

16 Using a hoist take the weight of the complete power unit or alternatively, position a hydraulic jack under the torque converter housing to take the weight of the unit.

17 Undo and remove the two bolts, spring and plain washers that secure the rear mounting crossmember to the underside of the body.

18 Carefully lower the gearbox so as to provide access to the top.

19 Using a second jack, support the weight of the gearbox.

20 Undo and remove the two bolts that secure the sump connecting plate to the converter housing.

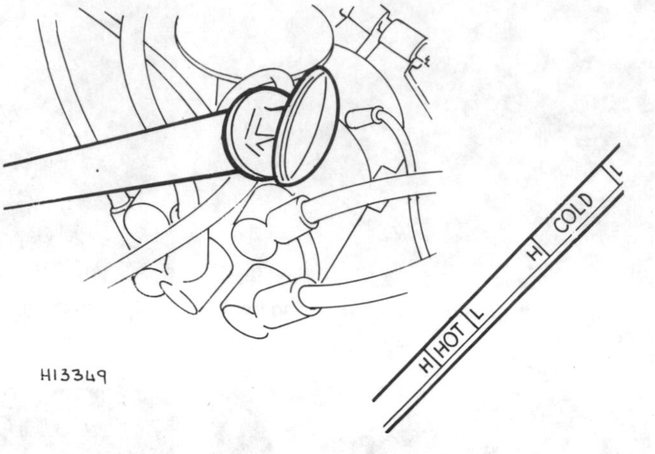

Fig. 6.10 Location of dipstick and filler tube (Sec 10)

21 Undo and remove the remaining nine bolts that secure the converter housing to the gearbox adaptor plate.

22 Very carefully withdraw the gearbox rearwards until it is clear of the torque converter and then lift it away from under the car. It is very important that the weight of the gearbox is not allowed to hang on the input shaft as the shaft may become damaged.

23 Refitting is the reverse of the removal procedure, but note the following points:

24 Carefully align the converter and front pump driving dogs and slots in the horizontal plane.

25 Carefully align and then locate the input shaft and drive dogs.

26 Ensure that the marks on the propeller shaft and gearbox flanges are correctly aligned.

27 Tighten the gearbox securing bolts to the specified torque.

28 Refill the gearbox with automatic transmission fluid type G to specification MC2 33G to the correct level, refer to Section 10.

12 Torque converter – removal and refitting

1 Remove the gearbox as described in Section 11.

2 Remove the cover from the access aperture on the gearbox adaptor plate (Fig. 6.12).

3 With a scriber, mark the relative positions of the driveplate and torque converter if these are to be refitted. This will ensure that they are refitted in their original positions relative to each other.

4 Working through the aperture in the adaptor plate, turn the converter and then progressively slacken the four securing bolts.

5 Support the torque converter and completely remove the securing bolts and washers. Lift away the torque converter. Be prepared to mop up fluid that will issue from the torque converter as it cannot be completely drained.

6 Remove the abutment ring from the centre spigot of the converter.

7 Refitting is the reverse of the removal procedure. If the original parts are being refitted, ensure that the marks made previously at removal are in alignment. Tighten the four securing bolts to the specified torque.

13 Starter inhibitor/reverse lamp switch – removal and refitting

1 Firmly chock all wheels and apply the handbrake. This is to prevent the car moving should the starter be operated whilst the inhibitor switch is disconnected or the manual selector lever be left in any of the drive positions, ie R, D, 2 or 1.

2 Disconnect the wiring from the switch at the multi-pin connectors.

3 Undo the screw that secures the switch.

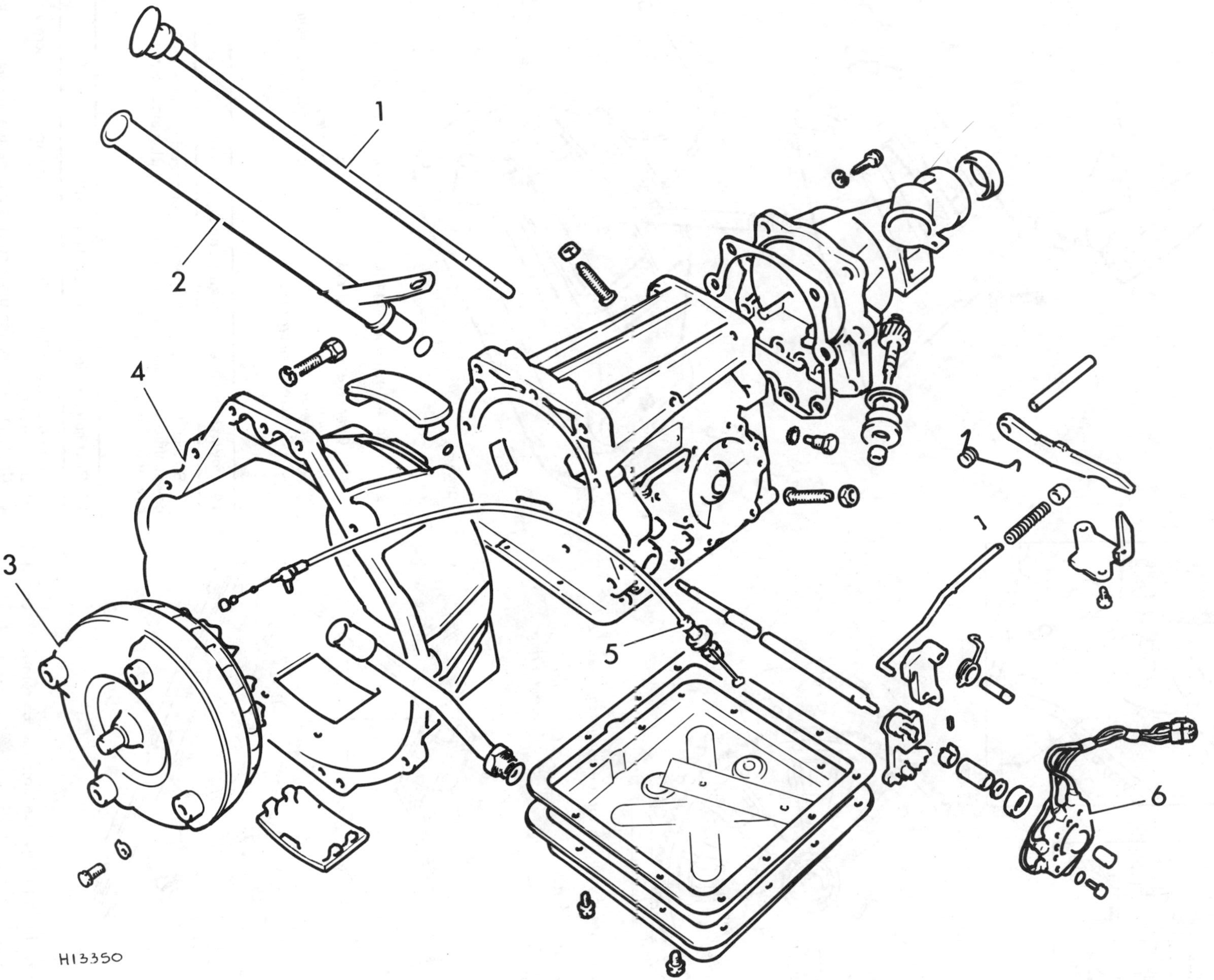

Fig. 6.11 Exploded view of automatic transmission casing and external components (Sec 9)

1 *Dipstick*
2 *Filler tube*
3 *Torque converter*
4 *Converter housing*
5 *Downshift cable*
6 *Inhibitor switch*

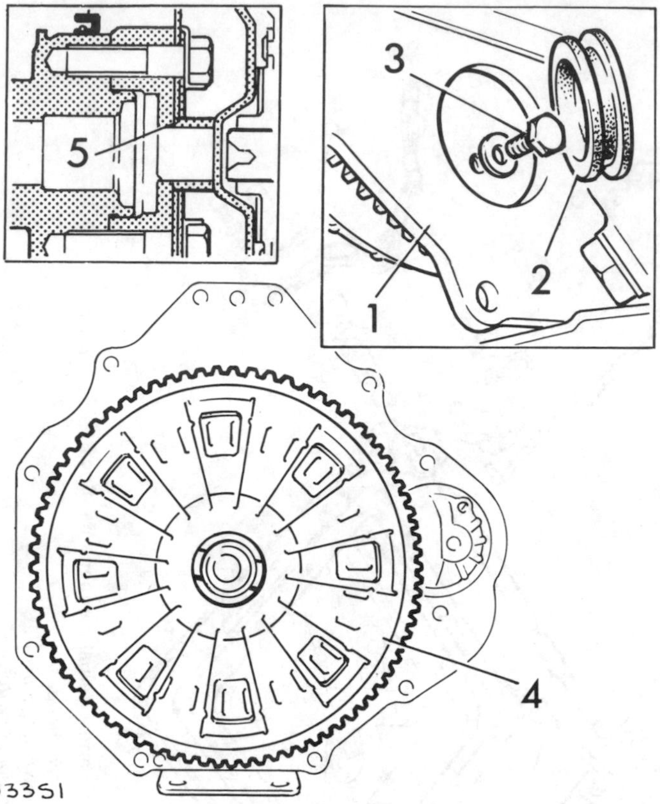

Fig. 6.12 Removing the torque converter (Sec 12)

1 Adaptor plate
2 Cover
3 Bolt
4 Converter
5 Abutment ring

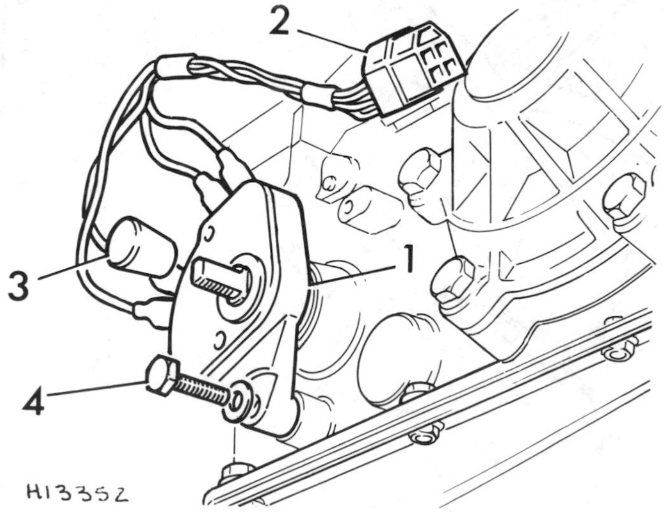

Fig. 6.13 Starter inhibitor/reverse lamp switch removal (Sec 13)

1 Switch
2 Multi-pin connectors
3 Cap
4 Bolt

4 Remove the starter/inhibitor reverse lamp switch from the gearbox.
5 Refitting is the reverse of the removal procedure.
6 Check that the starter motor only operates in the P and N positions of the selector and that the reverse light only comes on in the R position.

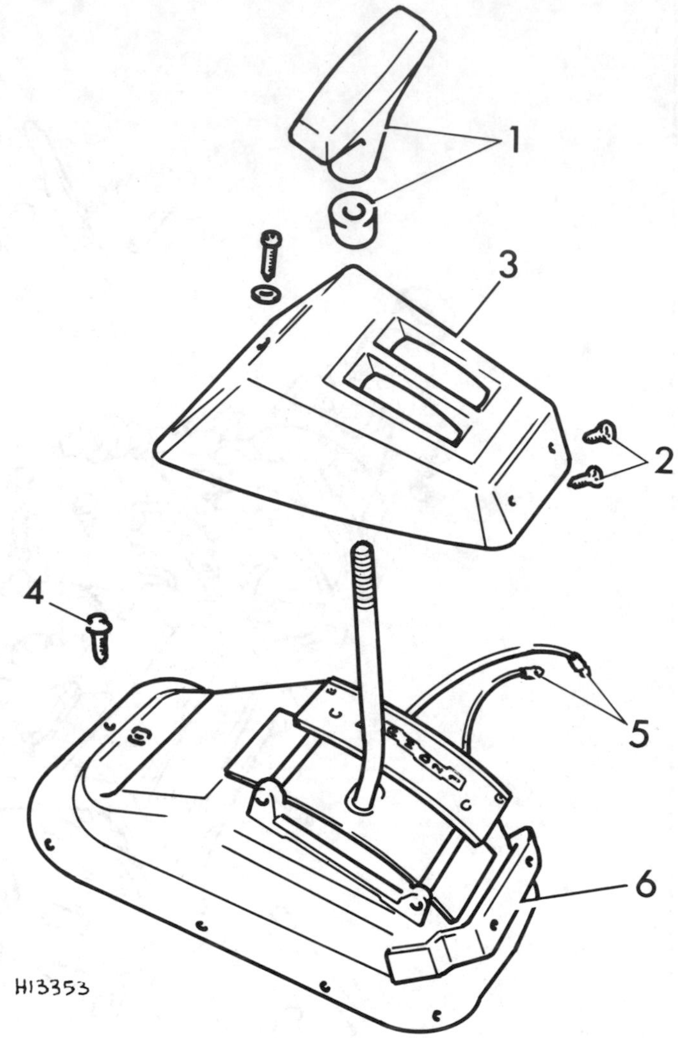

Fig. 6.14 Removing the gear selector lever (Sec 15)

1 Gear lever knob and lock collar
2 Screws
3 Selector gate cover
4 Screw
5 Lamp leads
6 Selector gate housing

14 Downshift cable – checking and adjustment

Special test equipment is required for checking and adjusting the downshift cable. This should be carried out by BL main agents who will have the equipment required.

15 Gear selector lever – removal and refitting

1 Undo the lock collar on the selector lever and then unscrew the selector lever knob.
2 Undo and remove the three screws that secure the selector gate cover.
3 Remove the selector gate cover and then the carpet.
4 Undo and remove the screws that secure the selector gate housing to the floor.
5 Disconnect the wiring from the selector gate housing at the push-in connectors.
6 Lift the selector gate housing and release the selector rod to selector lever retaining clip. Disconnect the selector lever from the selector rod. Remove the gear selector lever.

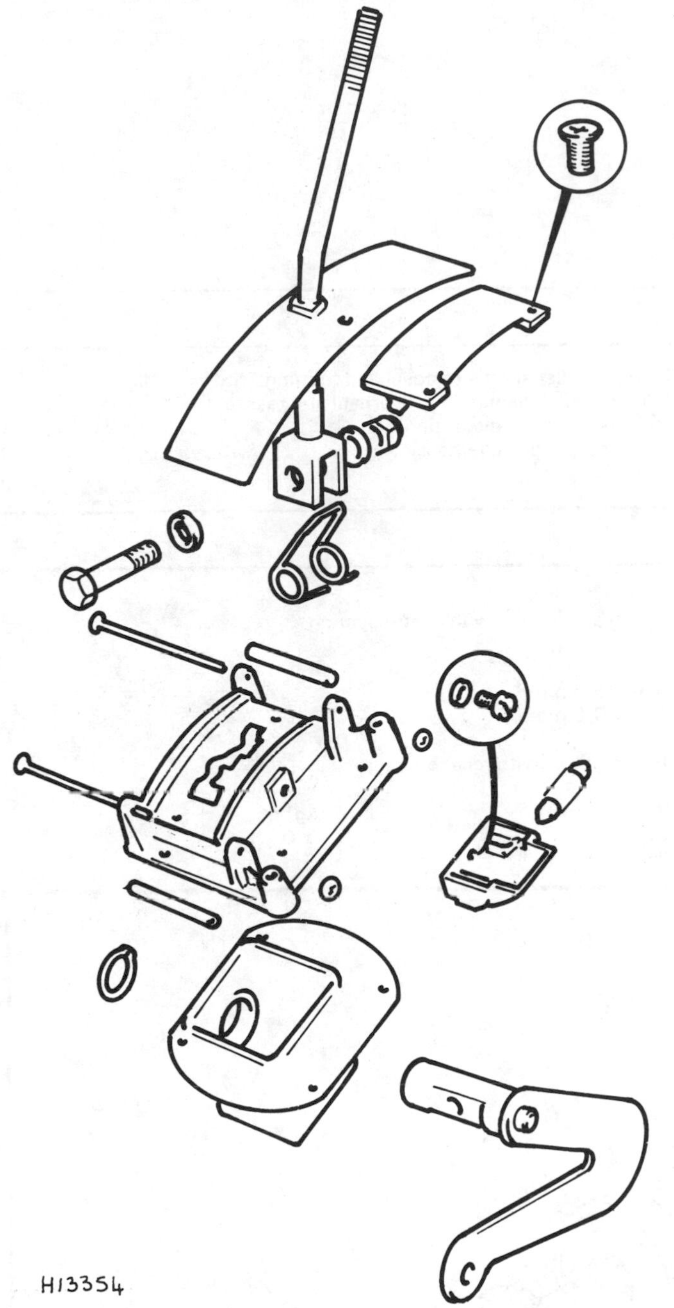

Fig. 6.15 Exploded view of gear selector lever assembly (Sec 15)

7 Refitting is the reverse of the removal procedure. Check the operation of the gear selector lever and if necessary, adjust the selector linkage as described in Section 16.

16 Selector linkage – adjustment

1 Apply the handbrake and move the gear selector lever to the N position.
2 Refer to Fig. 6.16. Slacken the selector rod locknut.
3 Release the clip that secures the selector rod to the gearbox selector lever and separate the rod from the lever.
4 Ensure that the gearbox selector lever is in the N position, which is the third detent from the fully forward position.

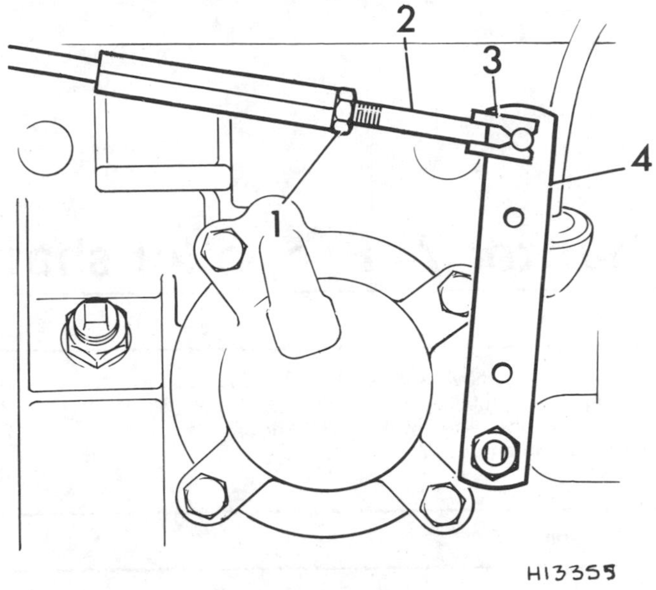

Fig. 6.16 Selector linkage adjustment (Sec 16)

1 Locknut
2 Threaded rod
3 Clip
4 Gearbox selector lever

5 The length of the selector rod is adjusted by screwing the threaded part of the selector rod in or out until the selector rod can be connected to the gearbox lever without any tension on the rod.
6 Secure the selector rod to the gearbox lever with the retaining clip and tighten the locknut on the selector rod.
7 Check the operation of the gear selector lever in all positions. Check that the starter motor only operates when the selector lever is in the P and N positions.

17 Fault diagnosis

Stall test procedure

The function of a stall test is to determine that the torque converter and gearbox are operating satisfactorily.
1 Check the condition of the engine. An engine which is not developing full power will affect the stall test readings.
2 Allow the engine and transmission to reach correct working temperatures.
3 Connect a tachometer to the vehicle.
4 Chock the wheels and apply the handbrake and footbrake.
5 Select R and depress the throttle to the kickdown position. Note the reading on the tachometer which should be 1800 to 2000 rpm. If the reading is below 1150 rpm, suspect the converter for stator slip. If the reading is down to 1400 rpm, the engine is not developing full power. If the reading is in excess of 2200 rpm, suspect the gearbox for brake band or clutch slip.
Note: *Do not carry out a stall test for a longer period than 10 seconds, otherwise the transmission will overheat*

Converter diagnosis

Inability to start on steep gradients, combined with poor acceleration from rest and low stall speed (1150 rpm), indicates that the converter stator uni-directional clutch is slipping. This condition permits the stator to rotate in an opposite direction to the impeller and turbine, and torque multiplication cannot occur.

Poor acceleration in third gear above 30 mph and reduced maximum speed, indicates that the stator uni-directional clutch has seized. The stator will not rotate with the turbine and impeller and the 'fluid flywheel' phase cannot occur. This condition will also be indicated by excessive overheating of the transmission although the stall speed will be correct.

Chapter 7 Propeller shaft

Contents

Specifications

Type	Two piece tubular with centre bearing	
Diameter		
Front	3 in (76·2 mm)	
Rear	2 in (50·8 mm)	
Universal joints	Hardy Spicer with roller bearings	
Torque wrench settings	**lbf ft**	**kgf m**
Centre bearing mounting bolts	22	3·0
Front and rear flange bolts	28	3·8

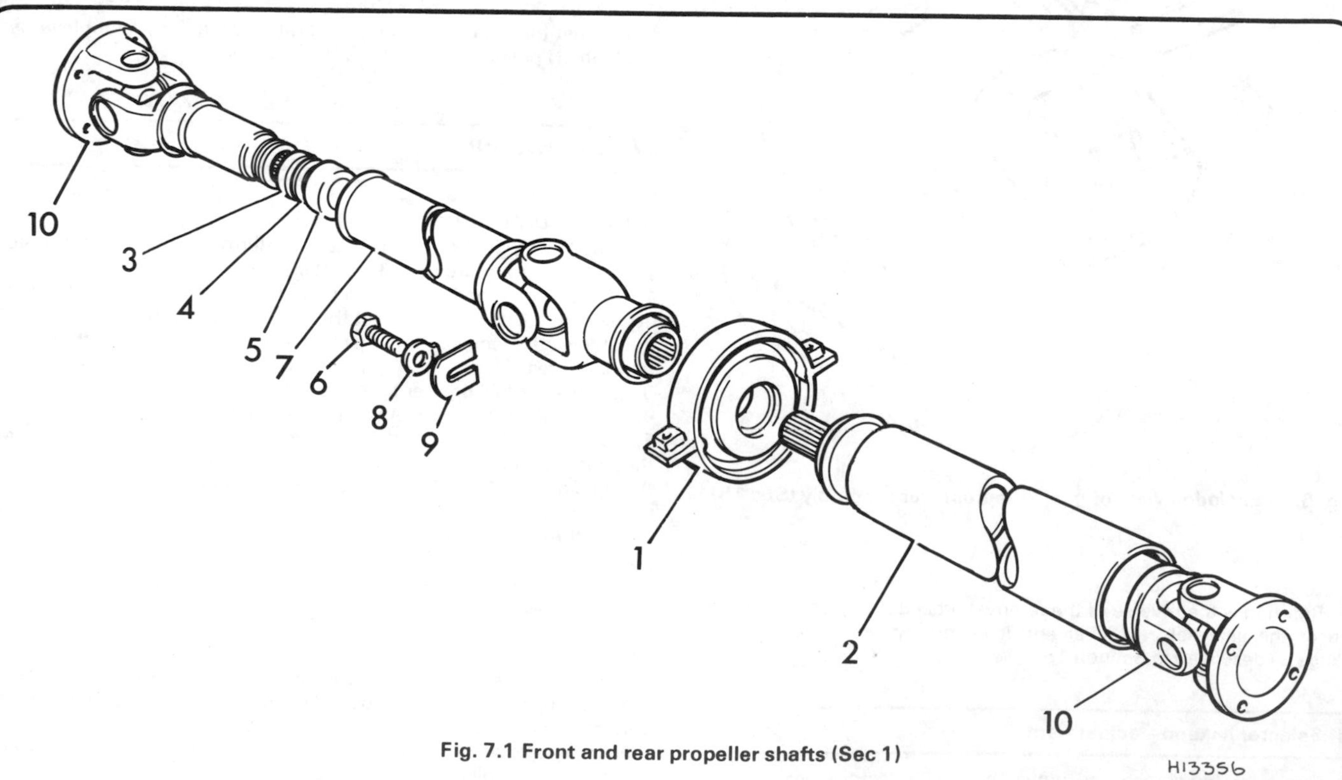

Fig. 7.1 Front and rear propeller shafts (Sec 1)

1 Centre bearing mounting
2 Front shaft
3 Seal
4 Seal retainer
5 Screw cap
6 Retaining bolt
7 Rear shaft
8 Tab washer
9 C-washer
10 Universal joints

1 General description

Drive is transmitted from the gearbox to the rear axle by means of a finely balanced Hardy Spicer tubular propeller shaft, split into two halves and supported at the centre by a rubber mounted bearing.

Fitted to the front, centre and rear of the propeller shaft assembly are universal joints which allow for vertical movement of the rear axle and slight movement of the complete power unit on its rubber mountings. Each universal joint comprises a four legged centre spider, four needle roller bearings and two yokes.

Fore and aft movement of the rear axle is absorbed by a sliding spline at the rear of the propeller shaft assembly. This is splined and mates with a sleeve and yoke assembly. When assembled, a dust cap, steel washer and cork washer seal the end of the sleeve and sliding joint.

The yoke flange of the front universal joint is fitted to the gearbox mainshaft flange with four bolts, spring washers and nuts. The yoke flange on the rear universal joint is secured to the pinion flange on the rear axle in the same way.

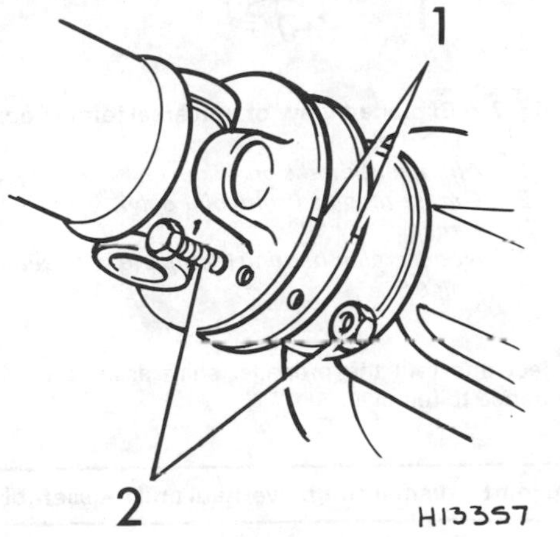

Fig. 7.2 Propeller shaft flange attachments (Sec 2)

1 Front/rear propeller shaft and gearbox/rear axle flanges
2 Securing bolt and locknut

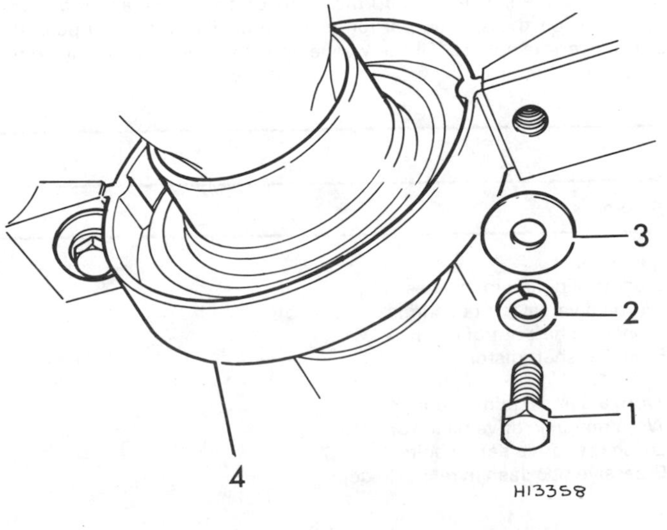

Fig. 7.3 Centre bearing attachment (Sec 2)

1 Bolt | *3 Plain washer*
2 Spring washer | *4 Centre bearing assembly*

2 Front propeller shaft – removal and refitting

1 Jack up the rear of the car and support on firmly based axle-stands. Alternatively, position the rear of the car on a ramp. Chock the front wheels.
2 The propeller shaft is carefully balanced to fine limits and it is important that it is refitted in exactly the same position as fitted prior to its removal. Scratch marks on the gearbox, differential pinion and propeller driveshaft flanges for correct re-alignment when refitting.
3 Support the weight of the front propeller shaft. Undo and remove the four gearbox end flange nuts and bolts (Fig. 7.2).
4 Support the weight of the rear propeller shaft. Undo and remove the four axle end flange nuts and bolts.
5 Undo and remove the two bolts, spring and plain washers that retain the centre bearing mounting to the body brackets (Fig. 7.3).
6 Lift away the propeller shaft assembly from the underside of the car.
7 To separate the two halves of the propeller shaft assembly, first bend back the locking washer tab and undo and remove the retaining bolt. Lift away the C-washer and tab washer (Fig. 7.4).
8 Draw the front propeller shaft away from the rear propeller shaft universal joint splines.
9 Reconnection and refitting the two propeller shaft halves is the reverse sequence to removal but the following additional points should be noted:

(a) Ensure that the mating marks scratched on the propeller shaft, gearbox and differential pinion flanges are lined up
(b) Tighten the centre bearing mounting bolts to a torque wrench setting of 22 lbf ft (3.0 kgf m)
(c) Tighten the front and rear flange retaining nuts to a torque wrench setting of 28 lbf ft (3.8 kgf m)

3 Rear propeller shaft – removal and refitting

The sequence for removing the rear propeller shaft is the same as for removing the front propeller shaft, in that the complete assembly must be removed first and then the two halves parted. See Section 2.

4 Universal joints – inspection

1 Wear in the needle roller bearings is characterised by vibration in

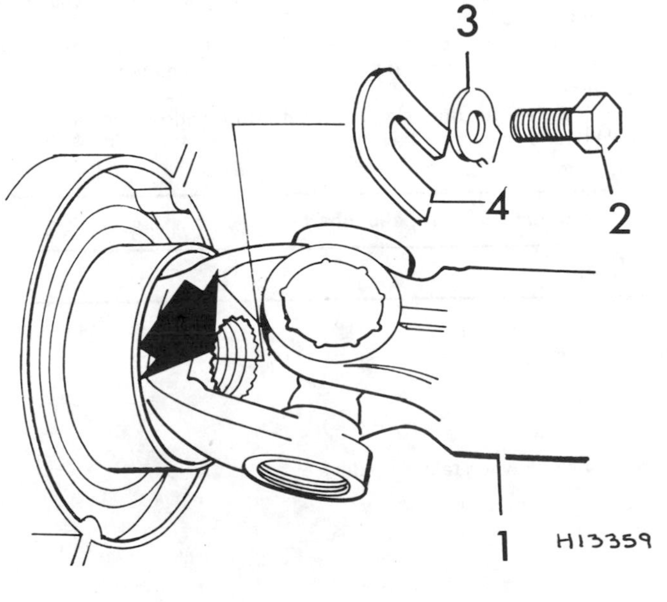

Fig. 7.4 Rear propeller shaft to centre bearing attachment (Sec 2)

1 Rear propeller shaft | *3 Lockwasher*
2 Bolt | *4 C-washer*

the transmission, 'clonks' on taking up the drive and in extreme cases of lack of lubrication, metallic squeaking and ultimately grating and shrieking sounds as the bearings break up.

2 It is easy to check if the needle roller bearings are worn with the propeller shaft in position, by trying to turn the shaft with one hand whilst the other hand holds the rear axle flange when the rear universal joint is being checked, and the front half coupling when the front universal joint is being checked. Any movement between the propeller shaft and the front, centre or rear half couplings is indicative of considerable wear. Check also by trying to lift the shaft and noticing any movement in the splines.

3 Test the propeller shaft for wear and if worn, it will be necessary to purchase a new rear half coupling, or if the yokes are badly worn, an exchange half propeller shaft. It is not possible to fit oversize bearings and journals to the trunnion bearing holes.

5 Universal joints – overhaul

1 The universal joints are retained by circlips. Clean away all traces of dirt then mark the yokes so that they can be fitted in the same relative positions (Fig. 7.5).

2 Remove the grease nipple (front joint only), then remove the circlips using long-nosed pliers or a suitable screwdriver. If difficult to remove, apply some penetrating oil and tap the bearing housing away from the circlip to alleviate the pressure.

3 It should be possible to drive out the bearings, but if they will not move by this method, mount the assembly in a vice and press them out using a socket on one bearing housing and another socket on the yoke to allow the opposite housing to slide out.

4 Repeat this operation as necessary for the other bearings, then remove the spider from the yoke. On front joints where a grease nipple is fitted, some leverage may be required.

5 Clean the yokes carefully then fit new seals to the new spider. Pack each bearing housing with a lithium based grease, place the spider in the yoke and press in the bearing. Ensure that the needle rollers are not dislodged.

6 Ensure that the oil seals register over the bearing housing lips then fit the circlips.

7 On the front joint, fit the grease nipple and apply 3 strokes from a grease gun containing a lithium based grease.

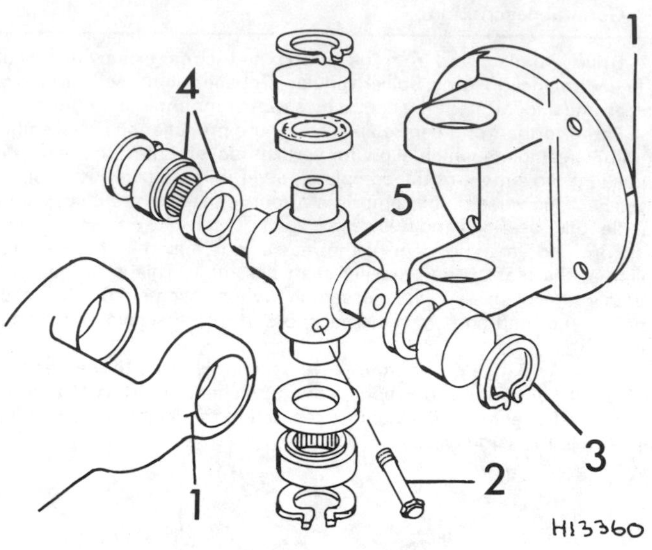

Fig. 7.5 Exploded view of universal joint (Sec 5)

1 *Alignment marks on yokes*
2 *Grease nipples (front joint only)*
3 *Circlip*
4 *Needle roller bearing (complete with seal)*
5 *Spider*

6 Centre bearing – removal and refitting

1 Refer to Section 2 and remove the propeller shaft assembly. Separate the two halves.

2 Using a universal puller and suitable thrust block (a suitable size bolt will do), draw the centre bearing from the end of the front propeller shaft.

3 To fit a new bearing, simply drift it into position using a piece of suitable diameter metal tube.

4 Reconnect and refit the propeller shaft assembly, this being the reverse sequence to removal.

7 Sliding joint – dismantling, overhaul and reassembly

1 Refer to Section 2 and remove the propeller shaft assembly.

2 Unscrew the dust cap from the sleeve and then slide the sleeve from the shaft. Take off the steel washer and the cork washer.

3 With the sleeve separated from the shaft assembly, the splines can be inspected. If worn, it will be necessary to purchase a new sleeve assembly.

4 To reassemble, fit the dust cap, steel washer and a new cork gasket over the splined part of the propeller shaft.

5 Grease the splines, line up the arrow on the sleeve assembly with the arrow on the splined portion of the propeller shaft and push the sleeve over the splines. Fit the washers to the sleeve and screw up the dust cap.

8 Fault diagnosis – propeller shaft

Symptom	Reason/s
Vibration	Wear in sliding joint splines Worn universal joint bearings Propeller shaft out of balance Propeller shaft distorted
Knock, or clunk when taking up drive	Worn universal joint bearings Worn rear axle drive pinion splines Loose rear drive flange bolts Excessive backlash in rear axle gears

Chapter 8 Rear axle

Contents

Specifications

Type — Hypoid, semi-floating

Ratio — 3·636 : 1

Tolerances

Distance of bearing from threaded end of axle shaft	2·84 in (72·14mm)
Differential bearing shims	0·003 in (0·076mm) 0·005 in (0·127mm) 0·010 in (0·254mm) 0·020 in (0·508mm)
Differential pinion gears thrustwasher range	8 by 0·002 in (0·05mm) increments
Thrustwasher range	0·027 in (0·685mm) to 0·043 in (1·092mm)
Crown wheel runout (max)	0·003 in (0·076mm)
Crown wheel backlash, optimum setting	0·005 in (0·127mm)
Pinion bearing preload	15 to 18 lbf ft (0·17 to 0·21 kgf m)
Pinion head washer sizes:	
Standard	0·077 in (1·956 mm)
Alternative	0·075 in (1·905mm) to 0·096 in (2·438mm) in a range of 21 increments
Pinion bearing shims	0·003 in (0·76mm) 0·005 in (0·127mm) 0·010 in (0·254mm) 0·020 in (0·508mm) 0·030 in (0·762mm)

Oil capacity — 1·25 pints (0·71 litres)

Torque wrench settings

	lbf ft	kgf m
Backplate securing nuts	18	2·5
Axle shaft nut	110	15·2
Differential case to axle retaining nuts	20	2·7
Pinion bearing pre-load	1·25 to 1·50	0·17 to 0·21
Driveflange nut	90	12·4
Axle to spring U-bolt nuts	15 to 18	2·0 to 2·4
Propeller shaft flange retaining nuts	28	3·8
Bearing cap to case bolts	39	5·4

1 General description

The rear axle is of the semi-floating type and is held in place by semi-elliptic springs. These springs provide the necessary lateral and longitudinal location of the axle. The rear axle incorporates a hypoid crownwheel and pinion and a two-pinion differential. All repairs can be carried out to the component parts of the rear axle without removing the axle casing from the car.

The crownwheel and pinion, together with the differential gears, are mounted on the differential unit which is bolted to the front face of the banjo type axle casing.

Adjustments are provided for the crownwheel and pinion backlash, pinion depth of mesh, pinion shaft bearing preload, and backlash between the differential gears. All these adjustments may be made by varying the thickness of the various shims and thrustwashers.

The axle or halfshafts are easily withdrawn and are splined at their inner ends to fit into the splines in the differential gears.

The wheel bearings are mounted on the outer ends of the axleshafts and are retained by the oil seal housings to the rear axle casing.

2 Rear axle – removal and refitting

1 Chock the front wheels, jack up the rear of the car and place it on firmly based axle-stands located under the body and forward of the rear axle.

2 With a scriber or file, mark the pinion and propeller shaft driveflange so that they may be refitted in their original positions.

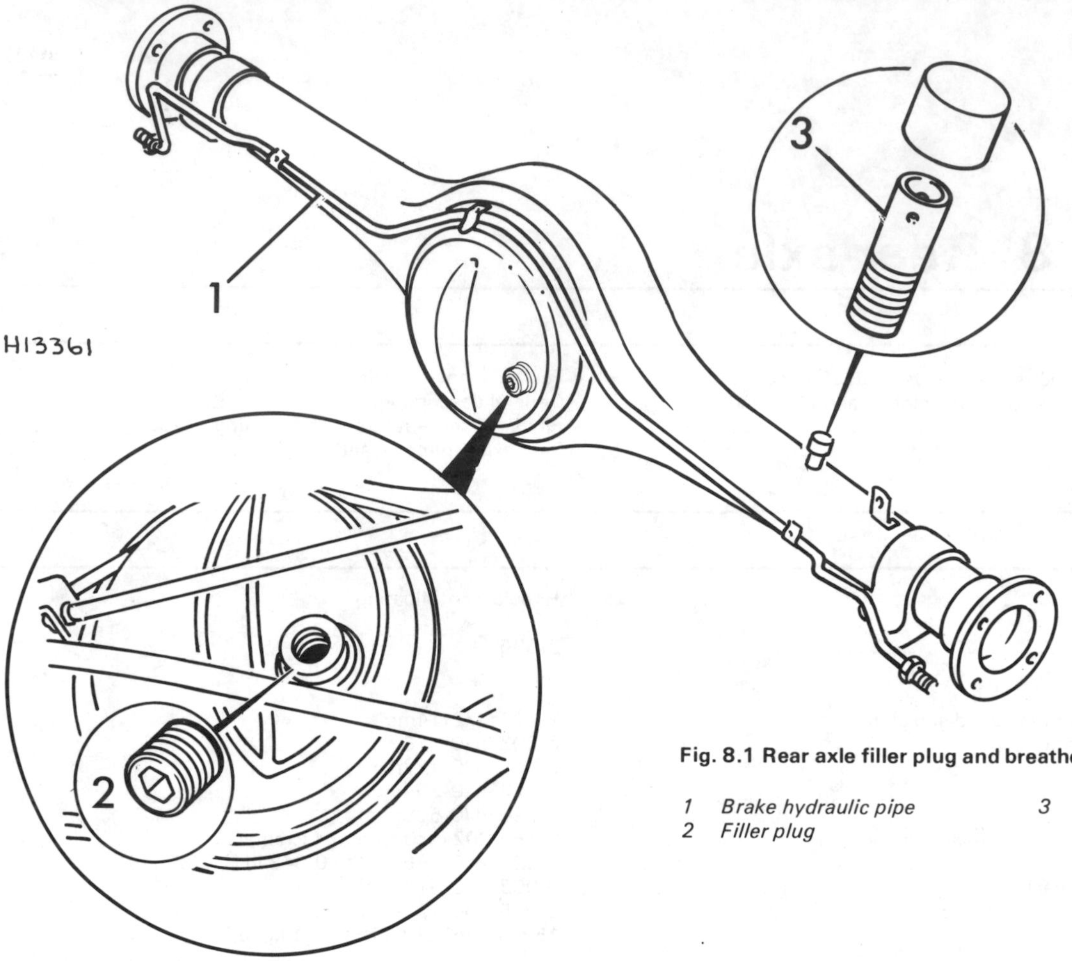

Fig. 8.1 Rear axle filler plug and breather (Sec 1)

1 Brake hydraulic pipe
2 Filler plug
3 Breather

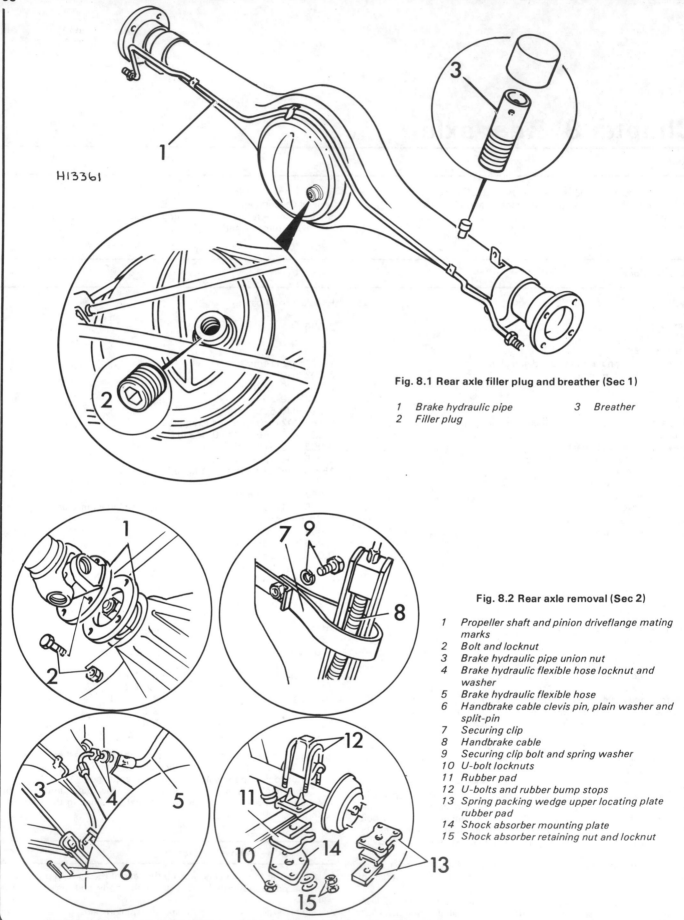

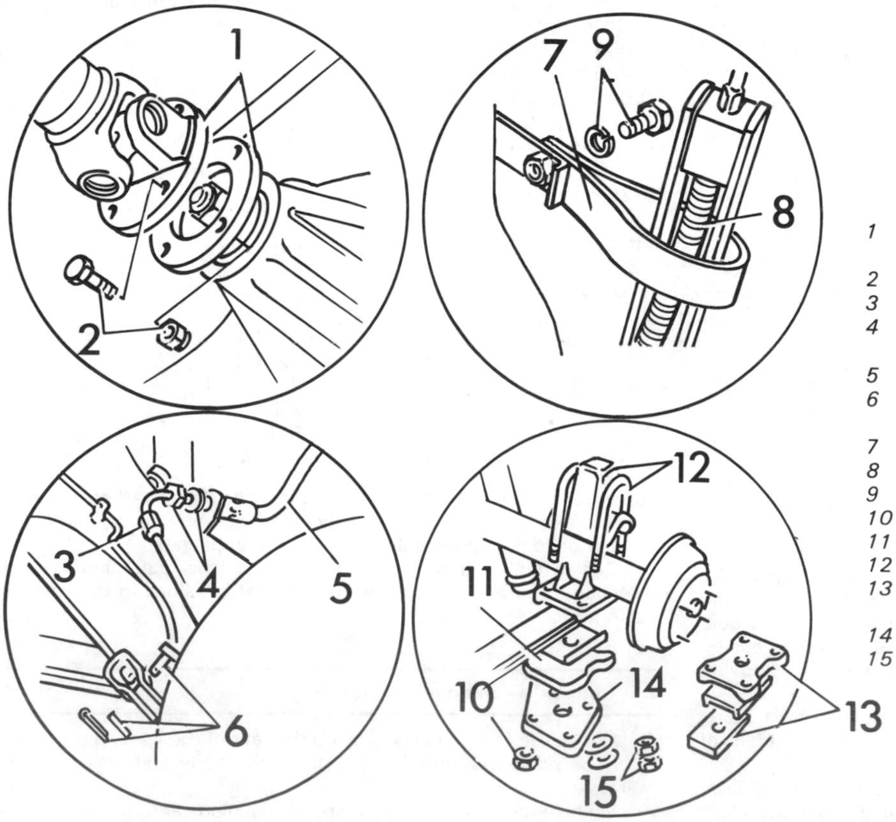

Fig. 8.2 Rear axle removal (Sec 2)

1 Propeller shaft and pinion driveflange mating marks
2 Bolt and locknut
3 Brake hydraulic pipe union nut
4 Brake hydraulic flexible hose locknut and washer
5 Brake hydraulic flexible hose
6 Handbrake cable clevis pin, plain washer and split-pin
7 Securing clip
8 Handbrake cable
9 Securing clip bolt and spring washer
10 U-bolt locknuts
11 Rubber pad
12 U-bolts and rubber bump stops
13 Spring packing wedge upper locating plate rubber pad
14 Shock absorber mounting plate
15 Shock absorber retaining nut and locknut

3 Undo and remove the four nuts and bolts that secure the rear propeller shaft flange to the pinion flange. Lower the propeller shaft.
4 Remove the wheel trims, undo and remove the wheel nuts and lift away the road wheels.
5 Wipe the top of the brake master cylinder reservoir, unscrew the cap and place a piece of thin polythene sheet over the top of the reservoir. Refit the cap. This will prevent hydraulic fluid syphoning out during subsequent operations.
6 Wipe the area around the union nut on the brake feed pipe at the axle bracket. Unscrew the union nut.
7 Undo and remove the locknut and washer from the flexible hose.
8 Detach the flexible hose from its support bracket.
9 Extract the split-pins that lock the brake lever clevis pins. Lift away the plain washers and withdraw the clevis pins.
10 Undo and remove the bolt and spring washer that secure the handbrake cable clip to the axle casing.
11 Using axle-stands or other suitable means, support the weight of the rear axle.
12 Undo and remove the eight U-bolt locknuts.
13 Detach the shock absorber mounting brackets and move them back to one side. If necessary, tie them back with string or wire.
14 Undo and remove tha anti-roll bar clamp securing nuts and bolts and detach the clamps.
15 Detach the lower mounting plates and rubber pads.
16 Lift away the U-bolts and rubber bump stops.
17 The rear axle may now be lifted over the rear springs and drawn away from one side of the car. Make a special note of the location of the spring packing wedge, upper locating plates and rubber pad. Refitting the rear axle is the reverse sequence to removal. The following additional points should be noted:

- a) *Make sure that the spring packing wedges are refitted in their original positions.*
- b) *Inspect the rubber mounting pads and if they show signs of deterioration fit new ones.*
- c) *Tighten the U bolt nuts to a torque wrench setting of 15 to 18 lbf ft (2.0 to 2.4 kfg m).*
- d) *It will be necessary to bleed the brake hydraulic system. See Chapter 9.*

3 Axleshaft, bearing and oil seal – removal and refitting

1 Chock the front wheels, jack up the rear of the car and place on firmly based axle-stands. Remove the rear wheel.
2 Undo and remove the axleshaft nut and plain washer.
3 Undo and remove the two screws that secure the brake drum to the hub flange. Lift away the brake drum.
4 Should it be tight to remove, back off the brake adjusters and using a soft-faced hammer, tap outwards on the circumference of the brake drum.
5 Using a heavy duty puller and suitable thrust block over the end of the axleshaft draw off the rear hub from the axleshaft.
6 Extract the split-pin from the handbrake lever clevis pin at the rear of the brake backplate. Lift away the plain washer and withdraw the clevis pin so separating the handbrake cable yoke from the handbrake lever.
7 Wipe the top of the brake master cylinder reservoir, unscrew the cap and place a piece of thin polythene sheet over the top of the reservoir. Refit the cap. This will prevent hydraulic fluid syphoning out during subsequent operations.
8 Wipe the area of the brake pipe union/s at the rear of the wheel cylinder and unscrew the union/s from the wheel cylinder.
9 Undo and remove the four nuts, spring washers and bolts that secure the brake backplate to the axle casing.
10 Make a note of the fitted position of the drip lip relative to the brake slave cylinder and remove the oil catcher.
11 The brake backplate assembly may now be lifted away.
12 Remove the rear hub oil seal and retainer assembly: the seal should be renewed if it is worn or has any signs of leaking. To remove the old seal, carefully ease it out using a screwdriver. Once removed an old seal must never be refitted.
13 Place a clean container under the end of the axle banjo to catch any oil that will issue from the end.
14 Using a screwdriver or a pair of pliers, remove the axleshaft key and put it in a safe place where it will not be lost.
15 Using either an impact slide hammer or if not available, mole grips, on the end of the axleshaft (with the nut refitted), draw the

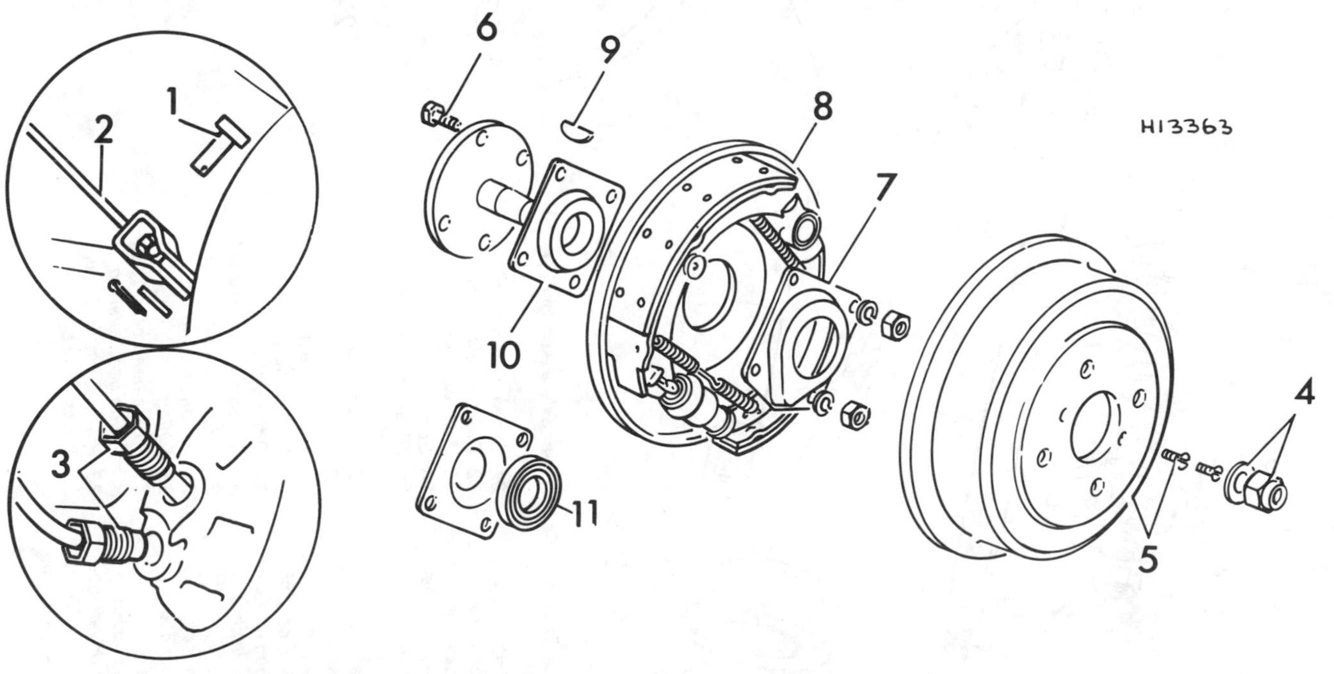

Fig. 8.3 Hub and axleshaft removal (Sec 3)

1 *Clevis pin, plain washer and split-pin*
2 *Handbrake cable*
3 *Brake hydraulic pipe*
4 *Axleshaft nut and plain washer*
5 *Brake drum and retaining screw*
6 *Brake backplate securing screw*
7 *Oil catcher*
8 *Brake backplate assembly*
9 *Woodruff key*
10 *Oil seal housing*
11 *Oil seal*

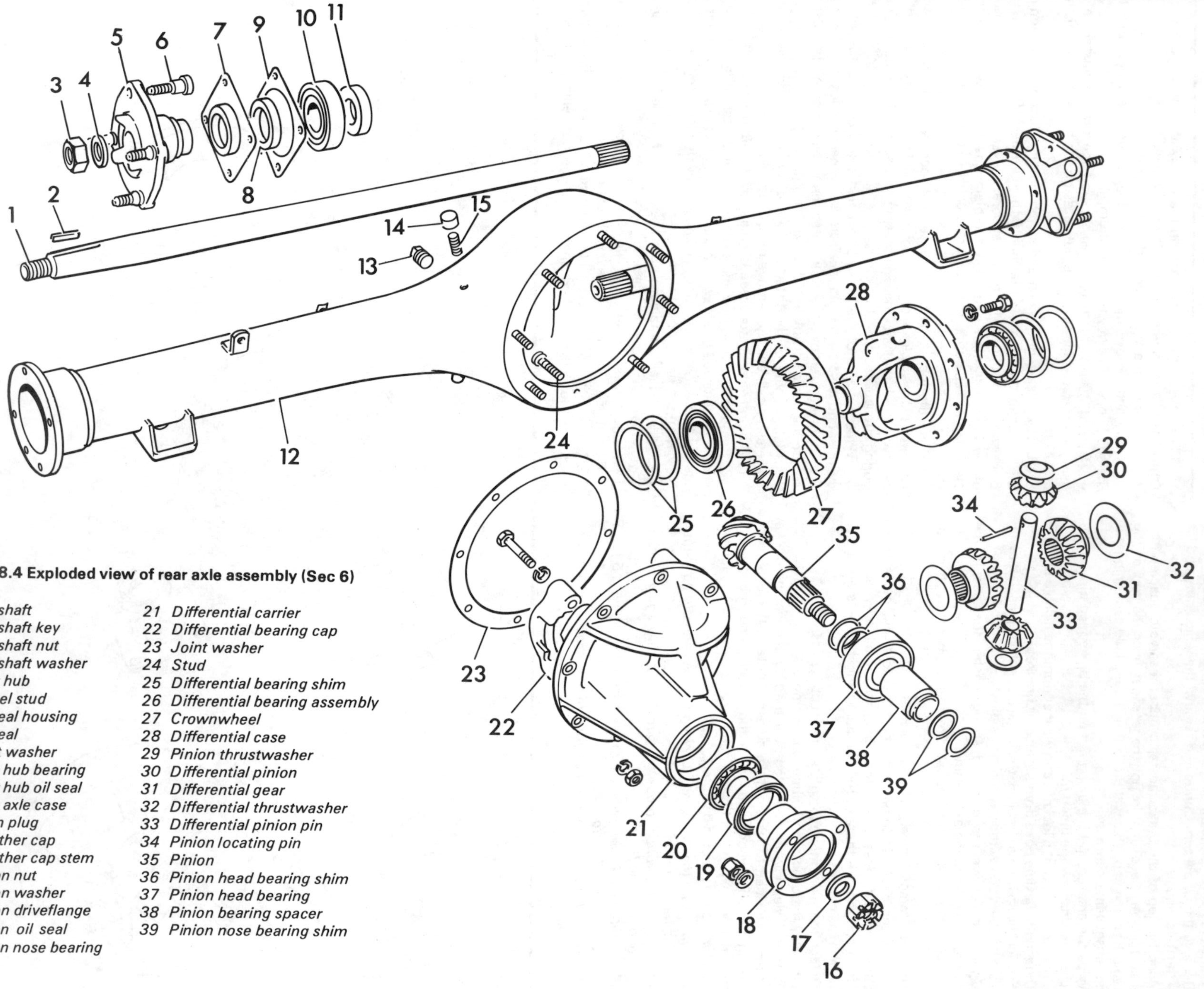

Fig. 8.4 Exploded view of rear axle assembly (Sec 6)

1 *Axleshaft*
2 *Axleshaft key*
3 *Axleshaft nut*
4 *Axleshaft washer*
5 *Rear hub*
6 *Wheel stud*
7 *Oil seal housing*
8 *Oil seal*
9 *Joint washer*
10 *Rear hub bearing*
11 *Rear hub oil seal*
12 *Rear axle case*
13 *Drain plug*
14 *Breather cap*
15 *Breather cap stem*
16 *Pinion nut*
17 *Pinion washer*
18 *Pinion driveflange*
19 *Pinion oil seal*
20 *Pinion nose bearing*
21 *Differential carrier*
22 *Differential bearing cap*
23 *Joint washer*
24 *Stud*
25 *Differential bearing shim*
26 *Differential bearing assembly*
27 *Crownwheel*
28 *Differential case*
29 *Pinion thrustwasher*
30 *Differential pinion*
31 *Differential gear*
32 *Differential thrustwasher*
33 *Differential pinion pin*
34 *Pinion locating pin*
35 *Pinion*
36 *Pinion head bearing shim*
37 *Pinion head bearing*
38 *Pinion bearing spacer*
39 *Pinion nose bearing shim*

axleshaft from the casing.

16 The inner oil seal may be removed by using a piece of metal bar shaped in the form of a hook. Pull on the seal and draw it from inside the axle casing.

17 If required, the bearing may be removed from the axleshaft by placing the bearing on the top of the jaws of a vice and driving the axleshaft through using a soft-faced hammer on the end of the axleshaft nut.

18 Refitting the oil seal, bearing and axleshaft is the reverse sequence to removal but the following additional points should be noted:

(a) Pack the bearing with a lithium based grease. Lubricate the new oil seal with a little engine oil
(b) When fitting the oil seal the lip must face inwards
(c) Using a suitable diameter tube, drift the bearing onto the axleshaft until the distance of the bearing to the threaded end of the axleshaft is 2.84 in (72.14 mm)
(d) Always use a new rear hub joint washer
(e) The backplate securing nut should be tightened to a torque wrench setting of 18 lbf ft (2.5 kgf m)
(f) To ensure positive locking of the axleshaft nut, always apply a little thread locking compound to the axleshaft thread
(g) Tighten the axleshaft nut to a torque wrench setting of 110 lbf ft (15.2 kgf m)
(h) Top up the rear axle oil level
(i) Bleed the brake hydraulic system as described in Chapter 9

4 Pinion oil seal – removal and refitting

1 Chock the front wheels, jack up the rear of the car and support it on firmly based axlestands.

2 With a scriber or file, mark the propeller shaft and pinion flanges so that they may be refitted correctly in their original positions.

3 Undo and remove the four nuts, bolts and spring washers that secure the pinion flange to the propeller shaft flange. Lower the propeller shaft to the floor.

4 Apply the handbrake firmly. Using a pair of pliers, extract the flange nut locking split-pin.

5 With a socket wrench, undo the flange nut. Lift away the nut and plain washer, (see Fig. 8.4).

6 Place a container under the pinion end of the rear axle to catch any oil that seeps out.

7 Using a universal puller and suitable thrust block, draw the pinion flange from the pinion.

8 The old oil seal may now be prised out using a screwdriver or thin piece of metal bar with a small hook on one end.

9 Refitting the new oil seal is the reverse sequence to removal but the following additional points should be noted:

(a) Soak the new oil seal in engine oil for 1 hour prior to fitting
(b) Fit the new seal with the lip facing inwards using a tubular drift
(c) Tighten the driveflange nut to a torque wrench setting of 90 lb ft (12.4 kgf m). Lock with a new split-pin.
(d) Top up the rear axle oil level as necessary

5 Differential unit – removal and refitting

1 If it is wished to overhaul the differential carrier assembly or to exchange it for a factory reconditioned unit, first remove the axleshafts as described in Section 3.

2 Mark the propeller shaft and pinion flanges to ensure their refitment in the same relative position.

3 Undo and remove the four nuts and bolts from the flanges. Separate the two parts and lower the propeller shaft to the ground.

4 Place a container under the differential unit assembly to catch oil that will drain out during subsequent operations.

5 Undo and remove the eight nuts and spring washers that secure the differential unit assembly to the axle casing.

6 Draw the assembly forwards from over the studs on the axle casing. Lift away from under the car. Remove the paper joint washer.

7 Refitting the differential assembly is the reverse sequence to removal. The following additional points should be noted:

(a) Always use a new joint washer and make sure the mating faces are clean, then apply a non-setting jointing compound
(b) Tighten the differential retaining nuts to a torque wrench setting of 20 lbf ft (2.7 kgf m)
(c) Refill the axle with 1.25 pints (0.71 litres) of high quality EP 90 gear oil.

6 Differential unit – dismantling, inspection, reassembly and adjustment

Make sure before attempting to dismantle the differential that it is both necessary and economical. A special tool is needed and whilst not difficult, some critical measurements have to be taken, It may well be cheaper to exchange the final drive assembly as a complete unit at the outset.

1 Obtain a special tool called an axle stretcher before commencing to dismantle the unit. It has a BL part number of 18G131C with adaptor plates 18G131E (Fig. 8.5).

2 Hold the differential unit vertically in a vice and then using a scriber or dot punch, mark the bearing caps and adjacent side of the differential carrier so that the bearing caps are refitted to their original positions.

3 Undo and remove the four bolts and spring washers that secure the end caps (Fig. 8.6).

4 Assemble the axle housing stretcher adaptor plates on the differential unit casing. Next fit the stretcher to the adaptor plates.

5 The differential unit case should now be stretched by tightening the nut three or four flats until the differential carrier can be levered out and the bearing shims and caps removed. **Note:** *To avoid damage to the case, do not attempt to spread it any more than is necessary.* To assist, each flat on the nut is numbered to give a check on the amount turned. The maximum stretch is 0.008 In (0.20 mm). When removing the differential carrier do not lever against the spreader. Remove the bearing inner races with a suitable puller.

6 Mark the relative positions of the crownwheel and differential carrier to ensure correct refitting.

7 Undo and remove the eight bolts and spring washers that secure the crownwheel to the differential carrier.

8 Separate the crownwheel from the differential carrier. If a little tight, tap with a soft-faced hammer.

9 Using a parallel pin punch, carefully drive out the differential pinion pin locking peg.

10 With a suitable diameter soft metal drift, remove the differential pinion pin.

11 Rotate the differential gear wheels until the differential pinions are opposite the openings in the differential gearcase, remove the differential pinions and their selective thrustwashers. Keep the pinions and respective thrustwashers together.

12 Remove the differential gear wheels and their thrustwashers.

13 Transfer the differential unit casing from its position in the vice and hold the driveflange firmly in the jaws.

14 Extract the driveflange nut split-pin and using a socket, undo and

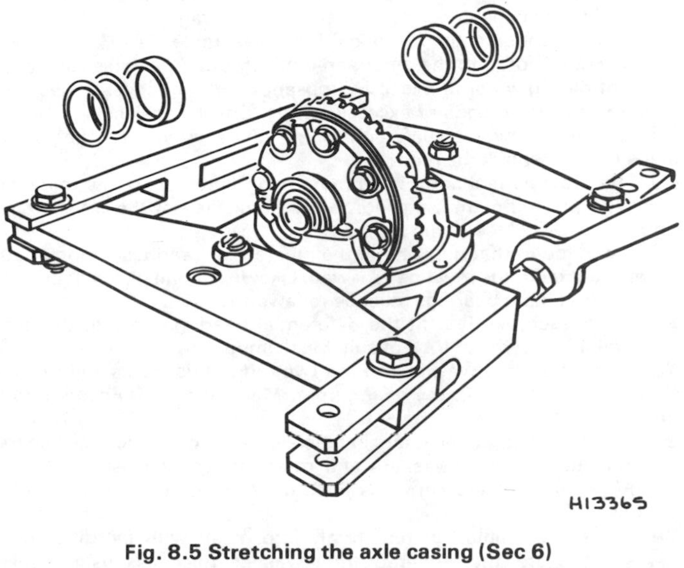

Fig. 8.5 Stretching the axle casing (Sec 6)

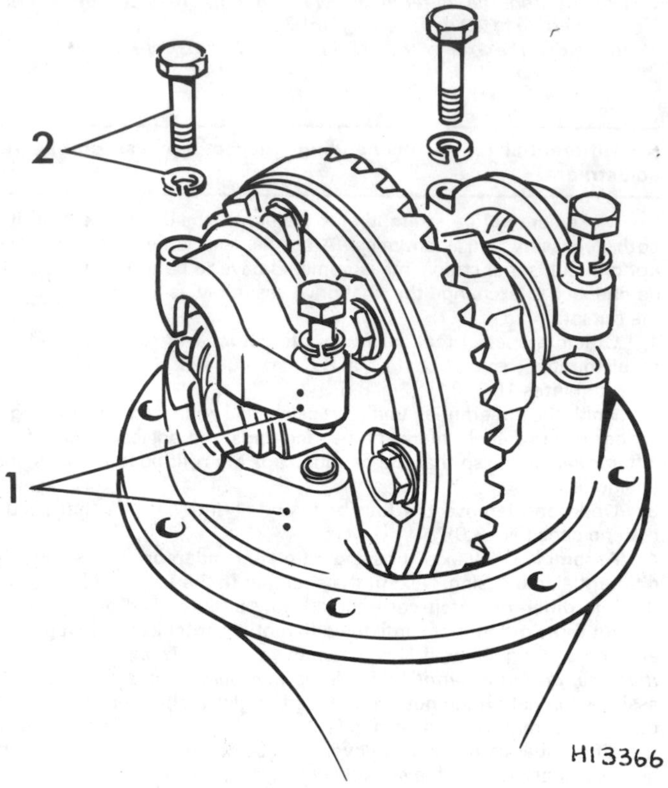

Fig. 8.6 Differential bearing cap removal (Sec 6)

1 *Bearing cap identification marks*
2 *Bearing cap securing bolt and washer*

remove the driveflange nut and washer.

15 Using a universal puller and suitable thrust block, remove the driveflange from the pinion.

16 The pinion may next be removed. To do this, carefully drive it out using a hard wood block and hammer.

17 Remove the pinion bearing shims and spacer.

18 If the pinion bearings are to be renewed, the inner bearings should be drawn off the pinion using a universal puller with long legs.

19 Lift the pinion head washer away from behind the pinion head.

20 Using a tapered soft-metal drift, carefully drift out the pinion outer bearing cup, bearing and oil seal. Also remove the pinion inner bearing cup.

21 Dismantling is now complete. Thoroughly wash all parts in petrol or paraffin and wipe dry using a clean non-fluffy rag.

22 Lightly lubricate the bearings and reassemble. Test for signs of roughness by rotating the inner and outer tracks. Check the rollers for signs of pitting, wear or excessive looseness in their cage. Inspect the thrustwashers for signs of excessive wear. Check for signs of wear on the differential pinion shaft and pinion gears. Any parts that show signs of wear should be renewed.

23 The crownwheel and pinion must only be replaced as a matched pair. The pair number is etched on the outer face of the crownwheel and the forward face of the pinion.

24 If it is found that only one of the differential bearings is worn, both differential bearings must be renewed. Likewise if one pinion bearing is worn, both pinion bearings must be renewed.

25 To reassemble, first fit the differential bearing cones to the gear carrier using a piece of tube of suitable diameter.

26 Place the thrustwashers on the two differential gears and then fit them to their bores in the gear carrier. Make sure these gears rotate easily.

27 Place the two pinion gears in mesh with the two differential gears, leaving out the thrustwashers and rotate the gear cluster until the pinion pin hole in the carrier is lined up with the pinions. Insert the pinion pin.

28 Press each pinion in turn firmly into mesh with the differential gears. Measure the required thrustwasher thickness using feeler gauges so that no backlash exists.

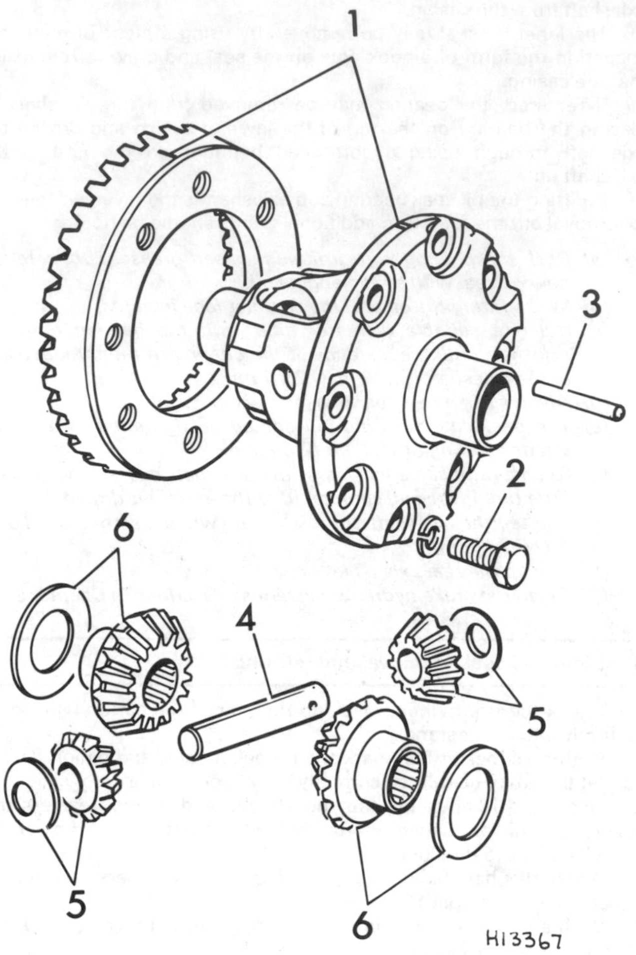

Fig. 8.7 Exploded view of differential unit (Sec 6)

1 *Crownwheel and differential carrier mating marks*
2 *Crownwheel securing bolt and spring washer*
3 *Locking peg*
4 *Differential pinion pin*
5 *Differential pinion and thrustwasher*
6 *Differential gearwheel and thrustwasher*

29 Remove the pinion gears, keeping them in their respective positions, and select a thrustwasher that has a thickness the same as that determined by the feeler gauges. Eight thrustwashers are available in 0.002 in (0.05 mm) increments from 0.027 in to 0.041 in (0.685 to 1.03mm).

30 Lubricate the thrustwashers, pinions and pinion pin and reassemble into the differential carrier. Check that there is no backlash. When this condition exists the gears will be stiff to rotate by hand.

31 Lock the pinion pin using the locking peg. Secure the peg by peening the metal of the differential carrier.

32 Carefully clean the crownwheel and gear carrier mating faces and fit the crownwheel. Any burrs can be removed with a fine oilstone. If the original parts are being used, line up the previously made marks.

33 Secure the crownwheel with the eight bolts and spring washers which should be tightened in a diagonal and progressive manner.

34 Fit the carrier bearing cups to the bearings and place the assembly in the case. Leave out the shims at this stage.

35 Refit the bearing caps in their original positions and tighten the retaining bolts. Using either a dial indicator gauge or feeler gauges, check the run-out of the crownwheel and carrier. This must not exceed 0.003 in (0.076mm).

36 If a reading in excess is obtained, check for dirt on the crownwheel or carrier mating faces or under the bearing cups.

37 Remove the bearing caps again and lift out the assembly.

38 If the pinion bearing cups were removed for the fitting of new bearings, these should next be refitted. For this, use a tube of suitable

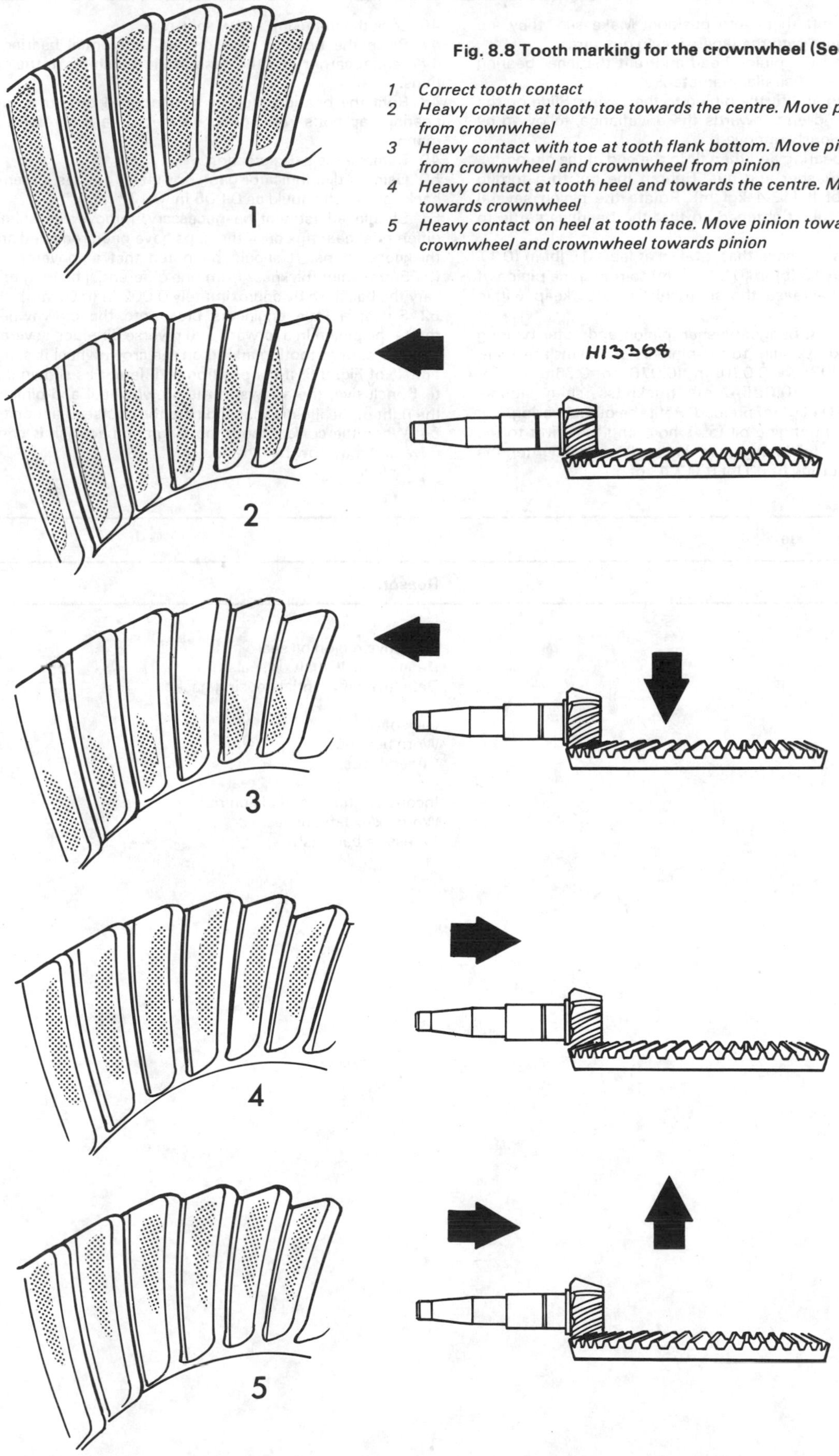

Fig. 8.8 Tooth marking for the crownwheel (Sec 6)

1. *Correct tooth contact*
2. *Heavy contact at tooth toe towards the centre. Move pinion away from crownwheel*
3. *Heavy contact with toe at tooth flank bottom. Move pinion away from crownwheel and crownwheel from pinion*
4. *Heavy contact at tooth heel and towards the centre. Move pinion towards crownwheel*
5. *Heavy contact on heel at tooth face. Move pinion towards crownwheel and crownwheel towards pinion*

diameter and carefully drift them into position. Make sure they are fitted the correct way round with the tapers facing outwards.
39 Fit the spacer behind the pinion head and refit the inner bearing using a piece of metal tube of similar diameter.
40 Lubricate the bearing and fit the pinion to the casing. Slide on the bearing spacer, chamfered end towards the driveflange, followed by the shims that were previously removed.
41 Lubricate the outer bearing and then fit to the end of the pinion.
42 Fit the driveflange washer and nut. Tighten the nut to a torque wrench setting of 90 lbf ft (12.4 kgf m). Rotate the pinion several times before the nut is fully tightened so that the bearings settle to their running positions.
43 Using a torque wrench, check that it takes at least 15 lbf in (0.17 kgf m) but not more than 18 lbf in (0.21 kgf m) to rotate the pinion. If necessary, increase or decrease the shim thickness to keep within these limits as follows:
44 Remove the pinion nut, flange, washer, pinion and outer bearing and fit the required thickness shim to the pinion. Four shims are available in sizes from 0.003 to 0.030 in (0.076 to 0.76mm). For assistance, a 0.001 in (0.0254mm) thickness shim equals approximately 4 lbf in (0.04 kgf m) preload. Refit the outer bearing.
45 Soak a new oil seal in engine oil for 1 hour and then fit it to the differential case. Refit the driveflange, washer and nut and tighten the nut to a torque wrench setting of 90 lbf ft (12.4 kgf m).
46 Lock the nut using a new split-pin.
47 Place the bearing cups on the differential bearings and fit the differential carrier into the case. Refit the shims in their original positions.
48 Refit the bearing caps in their original positions and tighten the bearing cap bolts with spring washers in a progressive and diagonal manner.
49 Remove the axle spreader.
50 Using a dial indicator gauge or feeler gauges, determine the total backlash which should be 0.005 in (0.127 mm).
51 Should adjustment be necessary, remove the shims behind the differential bearings once the caps have been removed and fit different thickness shims. It should be noted that a movement of 0.002 in (0.05mm) shim thickness from one differential bearing to the other will vary the backlash by approximately 0.002 in (0.05mm).
52 Smear a little engineers blue onto the crownwheel teeth and rotate the pinion in a forward and reverse direction several times.
53 The correct tooth marking on the crownwheel is shown in illustration 1, of Fig. 8.8. If the position is different as shown in illustration 2 to 5 inclusive, the arrows in the crownwheel and pinion diagrams to the right of the illustrations indicate the course of action to be taken.
54 When the correct tooth marking and backlash is correct, the final drive unit may now be refitted.

7 Fault diagnosis – rear axle

Symptom	Reason/s
Oil leakage	Defective pinion oil seal Defective axleshaft oil seals Defective differential housing gasket
Noise	Lack of oil Worn bearings General wear
Clonk on taking up drive	Incorrectly tightened pinion nut Worn axleshaft splines Excessive backlash

Chapter 9 Braking system

Contents

Specifications

Front disc brakes

Disc diameter	9·785 in (248·5 mm)
Disc run-out	0·006 in (0·152 mm)
Total pad area	17·4 in^2 (286·77 cm^2)
Total swept area	182·8 in^2 (2671 cm^2)
Pad lining material (service replacement)	Ferodo 2445
Minimum pad thickness	$\frac{1}{8}$ in (3 mm)

Rear drum brakes

Drum diameter	8 in (203·2 mm)
Lining dimensions	8 x 1·5 x 0·1875 in (203·2 x 38·1 x 4·76 mm)
Total swept area	76 in^2 (490·2 cm^2)
Lining material	Ferodo 2626

Servo unit

Girling Supervac

Torque wrench settings

	lbf ft	kgf m
Bleed screw	4 to 6	0·5 to 0·8
Master cylinder retaining nuts	17 to 19	2·3 to 2·6
Caliper retaining bolts	50	6·9
Wheel cylinder retaining bolts	4 to 5	0·55 to 0·7
Backplate securing nuts and bolts	15 to 18	2·0 to 2·5
Brake disc securing bolts	38 to 45	5·25 to 6·22

1 General description

The dual circuit hydraulic system is of the split front/rear type. A pressure failure warning light is located on the facia.

Disc brakes are fitted to the front and drum brakes to the the rear. They are operated by the hydraulic pressure created in the master cylinder when the brake pedal is depressed. This pressure is transferred to the wheel and caliper cylinders by a system of metal pipes and flexible hoses.

The rear drum brakes are of the internally expanding type whereby the shoes and linings are moved outwards into contact with the rotating brake drum. One wheel cylinder is fitted.

The handbrake operates on the rear brakes only, using a system of links and cables.

The front disc brakes are of the conventional fixed caliper design. Each half of the caliper contains a piston which operates in a bore, both being interconnected so that under hydraulic pressure their pistons move towards each other. By this action they clamp the rotating disc between two friction pads to slow rotational movement of the disc. Special seals are fitted between the piston and bore and these seals are able to stretch slightly when the piston moves to apply the brake. When the hydraulic pressure is released, the seals return to their natural shape and draw the pistons back slightly so giving a running clearance between the pads and disc. As the pads wear, the piston is able to slide through the seal allowing wear to be taken up.

The front disc brakes are self-adjusting; the rear drum brakes are of the manually adjusted type.

A brake servo unit is fitted between the brake pedal and master cylinder to add pressure on the master cylinder pushrod when the brake pedal is being depressed. This therefore reduces driver foot effort.

2 Rear drum brakes – adjustment

1 Chock the front wheels, release the handbrake completely, jack up the rear of the car and support it on firmly based stands.
2 A single adjuster for each side is located on the rear of the back-plate near the top, see Fig. 9.1. The surrounding area should be cleaned of all dirt and a small amount of engine oil smeared onto the adjuster threads.
3 Turn the adjuster clockwise (as viewed from the centre of the rear axle), preferably with a square adjuster spanner, until the brake drum is locked. Then back off the adjuster until the drum rotates without any signs of binding. About two clicks is normal.
4 Repeat the procedure given in paragraph 3 for the remaining wheel.
5 Finally lower the car to the ground.

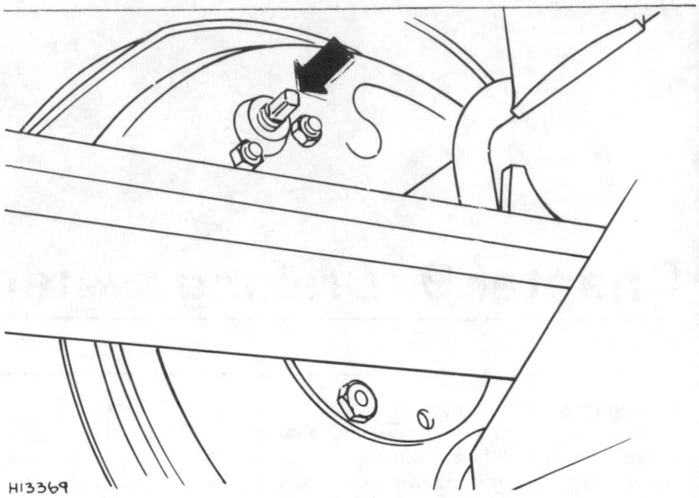

Fig. 9.1 Location of rear brake adjuster (Sec 2)

3 Bleeding the hydraulic system

Whenever the brake hydraulic system has been overhauled, a part renewed, or the level in the reservoir becomes too low, air will have entered the system necessitating its bleeding. During the operation the level of hydraulic fluid in the reservoir should not be allowed to fall below half full, otherwise air will be drawn in again.
1 Obtain two glass jars, two pieces of plastic tubing approximately 15 in long and of suitable diameter to fit tightly over the bleed screws and a supply of brake fluid.
2 Ensure that all hydraulic system connections are tight and all bleed screws closed.
3 Top up the master cylinder reservoir and fill the bottom inch of the jars with hydraulic fluid. Take extreme care that no brake fluid is allowed to come into contact with the paintwork as it acts as a solvent and will damage the finish.
4 Attach the bleed tubes to both the left-hand front and left-hand rear brake bleed screws and bleed both at the same time. Note that the left-hand rear bleed nipple controls the bleeding operation for both rear wheels.
5 Insert the ends of the bleed tubes in the jars, ensuring that the ends of the tubes are below the surface of the fluid.
6 Use a suitable open-ended spanner and unscrew the bleed screws about half a turn.
7 An assistant should now pump the brake pedal using firm full strokes. Carefully watch the flow of fluid into the jars and when air bubbles cease to emerge with the fluid, during the next downstroke tighten the bleed screws.
8 Finally bleed the right-hand front brake and top up the hydraulic fluid level in the reservoir as necessary with fresh hydraulic fluid; never re-use brake fluid that has been bled from the system.
9 After completing the bleeding operations, check the operation of the brakes and the brake pressure failure warning system.

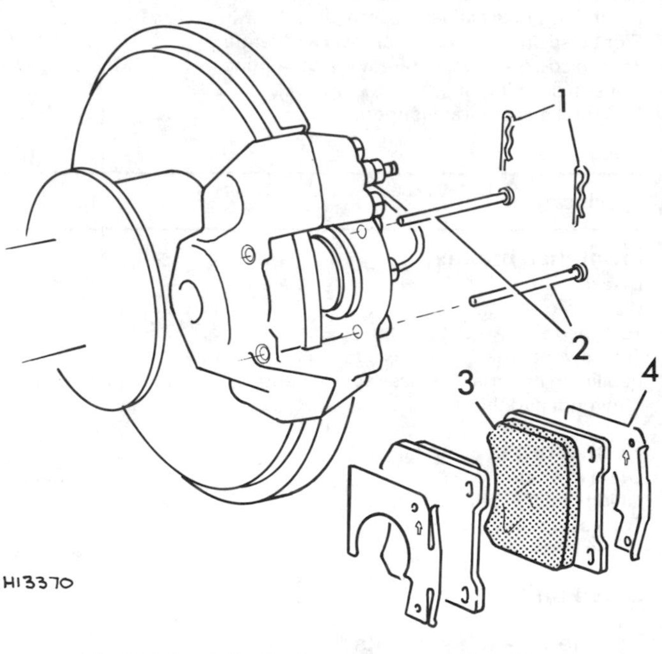

Fig. 9.2 Front disc brake pad removal (Sec 4)

1 Spring clip
2 Retaining pin
3 Pad
4 Anti-squeak shim

4 Front disc brake pads – removal and refitting

1 Chock the rear wheels and apply the handbrake. Jack up the front of the car and support it on firmly based axle-stands or other suitable supports. Remove the roadwheels.
2 Extract the two pad retaining pin spring clips and withdraw the two retaining pins (photos).
3 Lift away the brake pads and anti-squeak shims noting which way round the shims are fitted (photo).
4 Inspect the thickness of the lining material and, if it is less than $\frac{1}{8}$ in (3mm), the pads must be renewed. If one of the pads is slightly more worn than the other, it is permissible to change these round.
5 If new pads are being fitted, always use those manufactured to the recommended specifications given at the beginning of this Chapter.
6 To refit the pads, it is first necessary to extract a little brake fluid from the system. To do this, fit a plastic bleed tube to the bleed screw and immerse the free end in 1 in of hydraulic fluid in a jar. Slacken off the bleed screw one complete turn and press back the pistons into their bores. Tighten the bleed screw and remove the bleed tube.
7 Wipe the exposed end of the pistons and the recesses of the caliper free of dust or road dirt.
8 Refitting the pads is now the reverse sequence to removal but the following points should be noted:

(a) The anti-squeak shims are fitted with the arrows pointing upwards
(b) If it is suspected that air has entered through the system during the operation described in paragraph 6, the system must be bled as described in Section 3

9 Wipe the top of the hydraulic fluid reservoir and remove the cap. Top up and depress the brake pedal several times to settle the pads. Recheck the hydraulic fluid level.

5 Front disc brake caliper – removal and refitting

1 Apply the handbrake, chock the rear wheels, jack up the front of the car and support it on firmly based stands. Remove the roadwheel.
2 Wipe the top of the master cylinder reservoir, unscrew the cap and place a piece of thin polythene sheet over the top. Refit the cap. This is to stop hydraulic fluid syphoning out during subsequent operations.
3 Wipe the area around the caliper flexible hose to metal pipe union

4.2a Remove the spring clips (arrowed) ...

4.2b ... and withdraw the pad retaining pins

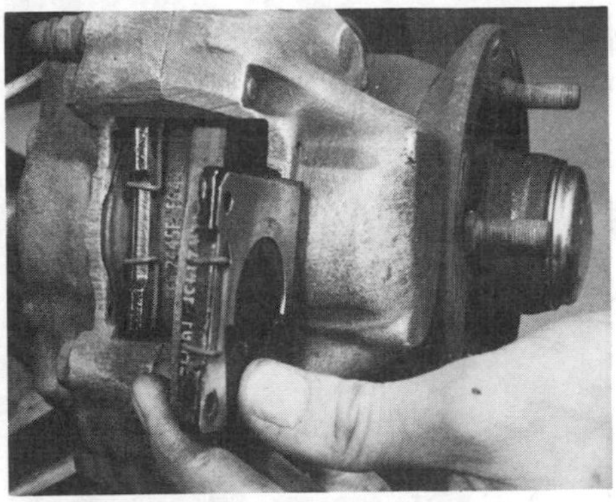

4.3 Removing the brake pads

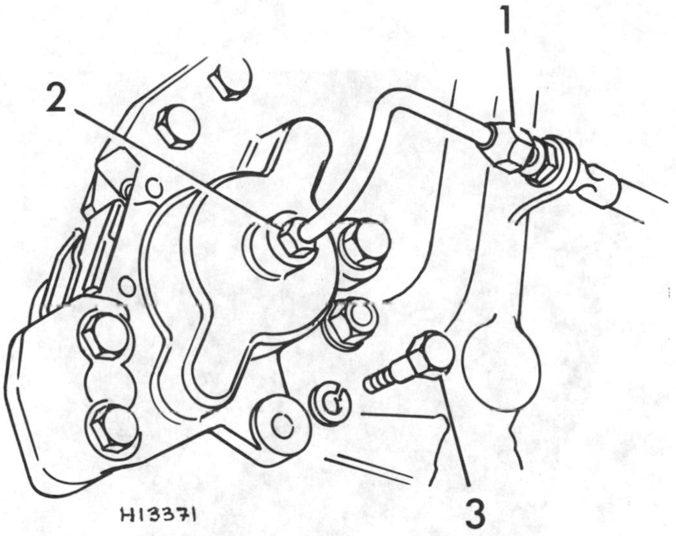

Fig. 9.3 Front disc brake caliper removal (Sec 5)

1 Metal pipe union nut
2 Metal pipe union
3 Securing bolt and spring washer

and the metal pipe to caliper connection. Unscrew the union nuts and lift away the metal pipe (Fig. 9.3).

4 Undo and remove the two bolts and spring washers that secure the caliper to the steering swivel. Lift the caliper from the disc.

5 Refitting the caliper is the reverse sequence to removal but the following additional points should be noted:

(a) The two caliper securing bolts should be tightened to a torque wrench setting of 50 lbf ft (6.9 kgf m)
(b) Bleed the brake hydraulic system as described in Section 3
(c) Depress the brake pedal several times to reset the pads in their correct operating position

6 Front brake disc – removal and refitting

1 Chock the rear wheels, apply the handbrake, jack up the front of the car and support it on firmly based axle-stands. Remove the road wheel.

2 Refer to Section 5 of this Chapter and remove the caliper assembly.

3 Using a wide blade screwdriver, ease off the hub grease cap.

4 Straighten the hub nut retainer locking split-pin legs and extract the split-pin. Remove the nut retainer and then undo and remove the nut-splined washer (Fig. 9.4).

5 Withdraw the complete front hub assembly from the spindle.

6 To separate the disc from the hub, first mark the relative position of the hub and disc. Undo and remove the four bolts that secure the hub to the disc and separate the two parts.

7 Should the disc surfaces be grooved and a new disc not obtainable, it is permissible to have the two faces ground by an engineering works. Score marks are not serious provided that they are concentric but not excessively deep. It is however, far better to fit a new disc rather than to re-grind the original one.

8 To refit the disc to the hub, make sure that the mating faces are very clean and then line up the previously made alignment marks if the original parts are to be used. Secure with the four bolts which should be tightened in a progressive and diagonal manner to a final torque wrench setting of 38 to 45 lbf ft (5.25 to 6.22 kgf m).

9 Refitting is now the reverse sequence to removal but the following additional points should be noted:

(a) Before refitting the caliper, check the disc runout at a 4.75 in (120.7mm) radius of the disc. The runout must not exceed 0.006 in (0.152mm). If necessary remove the disc and check for dirt on the mating faces. Should these be clean reposition the disc on the hub
(b) The hub bearing endfloat must be adjusted as described in Chapter 11

7 Rear drum brake shoes – inspection, removal and refitting

After high mileages, it will be necessary to fit replacement shoes with new linings. Refitting new brake linings to shoes is not considered economic or possible without the use of special equipment. However, if the services of a local garage or workshop having brake re-lining equipment are available, there is no reason why the original shoes should not be successfully relined. Ensure that the correct specification linings are fitted to the shoes.

1 Chock the front wheels, jack up the rear of the car and place it on firmly based axle-stands. Remove the roadwheel.

2 Undo and remove the two brake drum retaining screws and carefully pull off the brake drum. If it is tight it may be tapped outwards using a soft-faced hammer (photos).

3 The brake linings should be renewed if they are so worn that the rivet heads are flush with the surface of the lining. If bonded linings are fitted, they must be renewed when the lining material has worn down to $\frac{1}{16}$ in (1.6 mm) at its thinnest point.

4 Using a pair of pliers, release the steady springs and pins from the shoes by rotating through 90°. Lift away the steady spring, pin and cup washers from each brake shoe web (Fig. 9.5).

5 With a screwdriver, ease the trailing shoe from its adjuster link and the wheel cylinder adjuster slot.

6 Lift away the brake shoes complete with return springs. It will be necessary to ease the leading shoe from the handbrake operating lever.

7 Detach the pull-off springs and remove the retaining spring and support plate from the leading shoe.

8 If the shoes are to be left off for a while, do not depress the brake pedal otherwise the pistons will be ejected from the cylinders causing unnecessary work. Retain the pistons with strong elastic bands.

9 Thoroughly clean all traces of dust from the shoes, backplate and

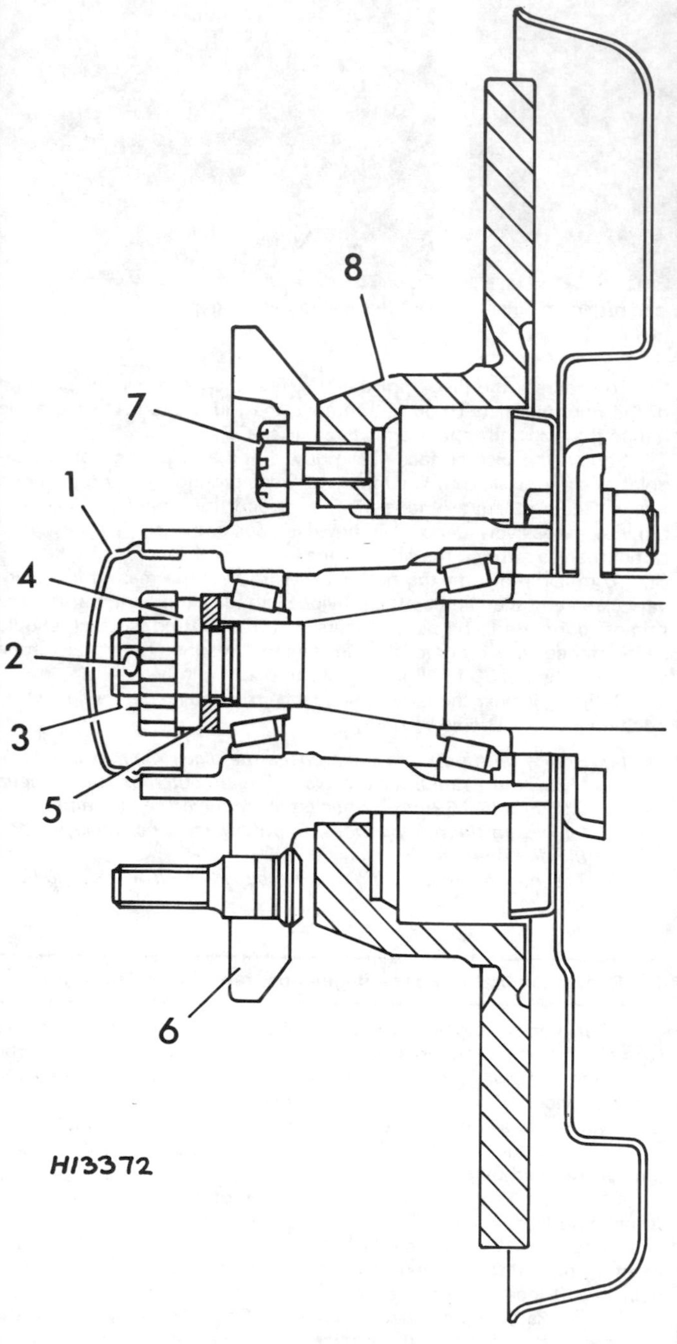

Fig. 9.4 Cross-section view of front brake disc (Sec 6)

1 *Grease cap*
2 *Split-pin*
3 *Nut retainer*
4 *Nut*
5 *Splined washer*
6 *Hub*
7 *Disc securing bolt*
8 *Disc*

7.2a Unscrew the rear brake drum retaining screws ...

7.2b ... and pull off the brake drum

7.2c Rear brake shoes assembly with drum removed

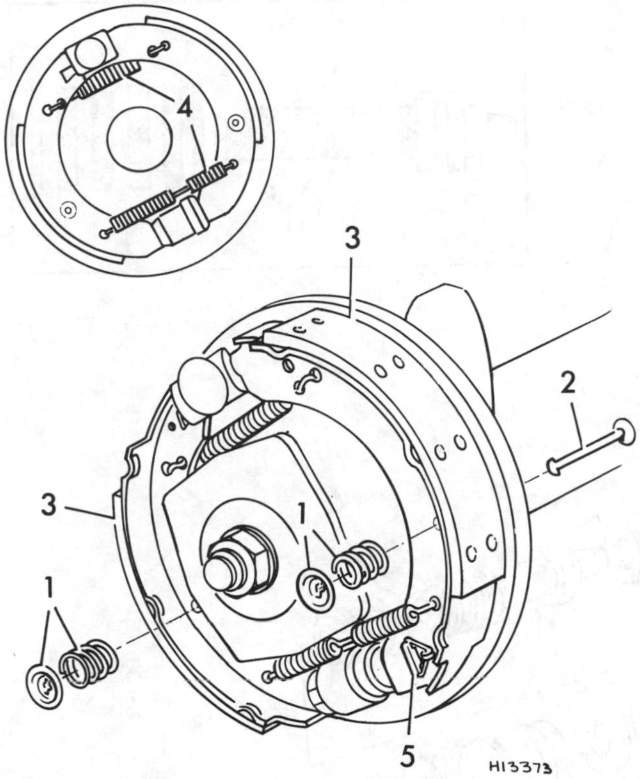

Fig. 9.5 Rear drum brake assembly (Sec 7)

1 Cup washer and spring
2 Steady pin
3 Brake shoe
4 Brake shoe pull-off springs
5 Shoe retaining spring and support plate

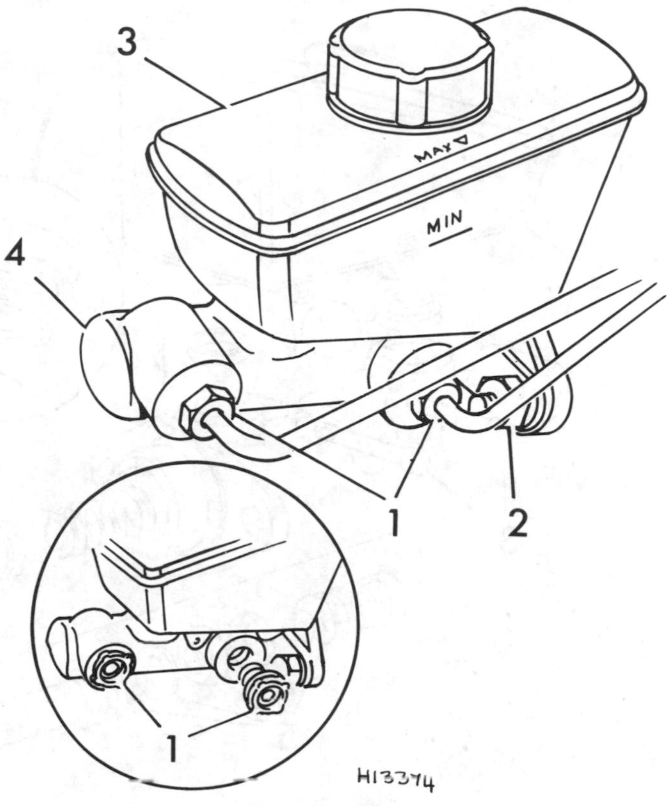

Fig. 9.6 Master cylinder removal (Sec 9)

1 Hydraulic pipe unions
2 Securing nut
3 Fluid reservoir
4 Master cylinder

drum using a stiff brush. **Caution:** *Do not use compressed air as it blows up dust which must not be inhaled as it is of an asbestos nature.* Brake dust can cause judder or squeal and therefore it is important to clean away all traces.

10 Check that each piston is free in its cylinder, the rubber dust covers are undamaged and in position and that there are no hydraulic fluid leaks.

11 Check that the wheel cylinder is free to move in its slot in the backplate.

12 Prior to reassembly, smear a trace of brake grease to the ends of the brake shoes, steady platforms anchor posts and the thread of the adjuster. Do not allow any grease to come into contact with the linings or rubber parts. Refit the shoes in the reverse sequence to removal. The two pull-off springs should preferably be renewed every time new shoes are fitted and must be in their original web holes, see Fig. 9.5.

13 Refit the brake drum and secure with the two retaining screws.

14 Refit the roadwheel.

15 Adjust the rear brakes as described in Section 2. Road test the car to ensure that the brakes operate correctly.

8 Rear brake backplate – removal and refitting

For full information refer to Chapter 8, Section 3, which describes the removal of the axleshaft and hub assembly and includes the removal of the backplate.

9 Master cylinder – removal and refitting

1 Apply the handbrake and chock the front wheels. Drain the fluid from the master cylinder reservoir and master cylinder by attaching a plastic bleed tube to one of the front brake bleed screws. Undo the screw one turn and then pump the fluid out into a clean glass container by means of the brake pedal. Hold the brake pedal against the floor at the end of each stroke and tighten the bleed nipple. When the pedal has returned to its normal position, loosen the bleed nipple and repeat the process until the master cylinder reservoir is empty.

2 Wipe the area around the hydraulic pipe unions on the master cylinder. Undo the union nuts and lift out the hydraulic pipes (Fig. 9.6).

3 Undo and remove the two nuts and spring washers that secure the master cylinder to the servo unit.

4 Lift away the master cylinder, taking care not to allow any hydraulic fluid to drip onto the paintwork.

5 Remove the master cylinder O-ring seal.

6 Refitting is the reverse of the removal procedure. Always fit a new master cylinder O-ring and seal. Tighten the securing nuts to a torque of 17.0 to 19.0 lbf ft (2.3 to 2.6 kgf m).

7 Bleed the hydraulic system as described in Section 3.

8 Road test the car to check the operation of the brakes.

10 Master cylinder – dismantling and reassembly

If a replacement master cylinder is to be fitted, it will be necessary to lubricate the seals before fitting to the car as they have a protective coating when originally assembled. Remove the blanking plugs from the hydraulic pipe union seatings. Ease back and remove the plunger dust cover. Inject clean hydraulic fluid into the master cylinder and operate the piston several times so the fluid will spread over all the internal working surfaces.

If the master cylinder is to be dismantled after removal, proceed as follows:

1 Unscrew the filler cap, and lever out the plastic baffle and rubber washer (Fig. 9.7).

2 Extract the hairpin retaining clips and withdraw the two fluid reservoir securing pins, then lift the reservoir off the master cylinder.

3 Note the fitted position of the two reservoir seals and then extract them from the master cylinder body.

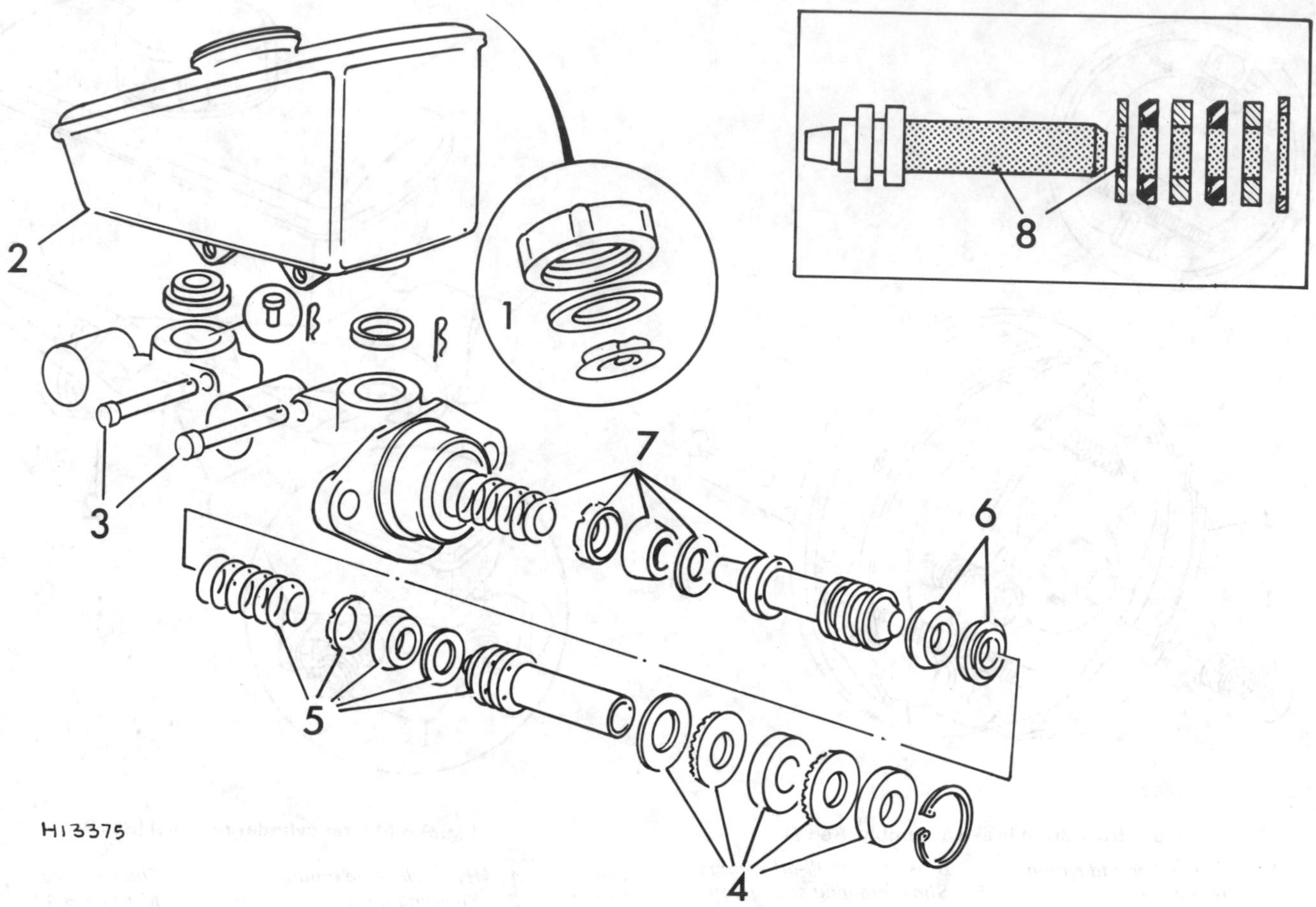

Fig. 9.7 Exploded view of brake master cylinder (Sec 10)

1 Filler cap assembly
2 Reservoir
3 Reservoir retaining pins
4 Primary plunger assembly
5 Primary plunger assembly
6 Intermediate seals (later models have only one)
7 Secondary plunger assembly
8 Lubrication points (shaded)

4 With the master cylinder mounted in a soft-jawed vice, use a suitable diameter rod to push the plunger fully down the cylinder bore and then extract the secondary plunger stop pin.

5 Using a pair of circlip pliers, extract the circlip from the end of the master cylinder bore and then remove the primary plunger assembly. Place each item on a clean surface in the exact order of removal, noting which way round the seals are fitted.

6 Shake out or alternatively blow out with a tyre pump or air line, the secondary plunger assembly; apply air pressure to the secondary outlet port.

7 Withdraw the two vacuum seals and spacers from the primary plunger tube end and detach the spring, retainer, seal and washer from the inner end of the primary plunger.

8 Similarly remove the spring and seals from the secondary plunger, again noting their fitted position.

9 Examine the bore of the cylinder carefully for any signs of scores or ridges. If this is found to be smooth all over, new seals can be fitted. If, however, there is any doubt of the condition of the bore, then a new cylinder must be fitted.

10 If examination of the seals shows them to be apparently oversize or swollen, or very loose on the plungers, then suspect oil contamination in the system. Oil will swell these rubber seals, and if one is found to be swollen, it is reasonable to assume that all seals in the braking system will need attention.

11 Thoroughly clean all parts in methylated spirit. Ensure that the bypass ports are clear.

12 All seals should be assembled wet by dipping in clean brake fluid. Using the fingers only, fit new seals to the primary and secondary plungers ensuring that they are the correct way round.

13 Reassembly is a reversal of the dismantling procedure but the following additional points should be noted:

(a) The secondary plunger return spring is larger than the primary plunger return spring
(b) The master cylinder bore should be smeared with clean brake fluid before inserting the plunger assemblies
(c) The primary plunger, vacuum seals and spacers should be lubricated at the friction areas shown in Fig. 9.7 using grease supplied with the overhaul kit

11 Rear drum brake wheel cylinder – removal and refitting

1 If hydraulic fluid is leaking from the brake wheel cylinder, it will be necessary to dismantle it and renew the seal. Should brake fluid be found running down the side of the wheel or if it is noted that a pool of liquid forms alongside one wheel and the level in the master cylinder has dropped, it is indicative of seal failure.

2 Remove the brake drum and brake shoes as described in Section 7.

3 Wipe the top of the brake master cylinder reservoir and unscrew the cap. Place a piece of thin polythene sheet over the top of the reservoir and refit the cap.

4 Using an open-ended spanner, carefully unscrew the hydraulic pipe connection union to the rear of the wheel cylinder. Note that the feed pipe from the left-hand and right-hand wheel cylinder locates in the lower opening (Fig. 9.8).

5 Again using an open-ended spanner, undo and remove the bridge

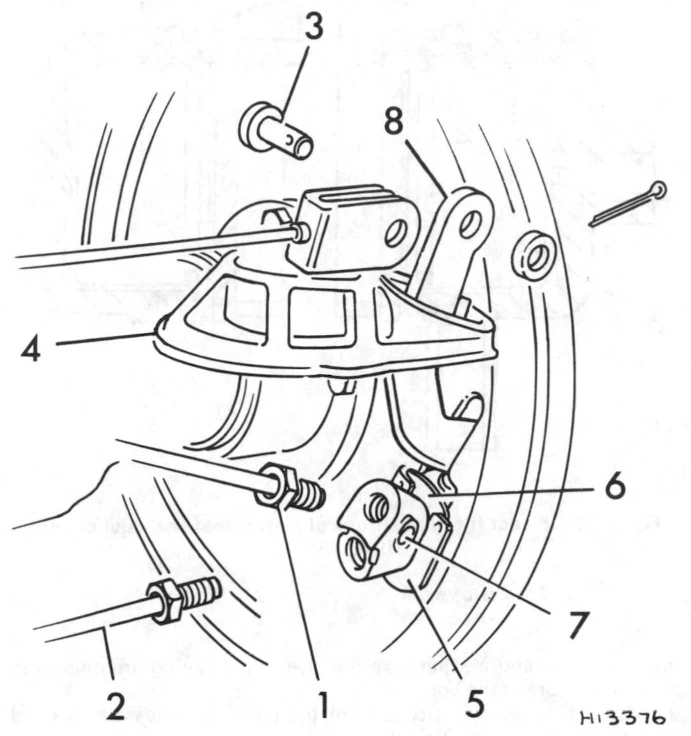

Fig. 9.8 Rear drum brake wheel cylinder removal (Sec 11)

1 Main hydraulic feed pipe
2 Bridge feed pipe
3 Clevis pin with plain washer and split-pin
4 Rubber boot
5 Retaining plate
6 Spring plate
7 Wheel cylinder
8 Wheel cylinder operating lever

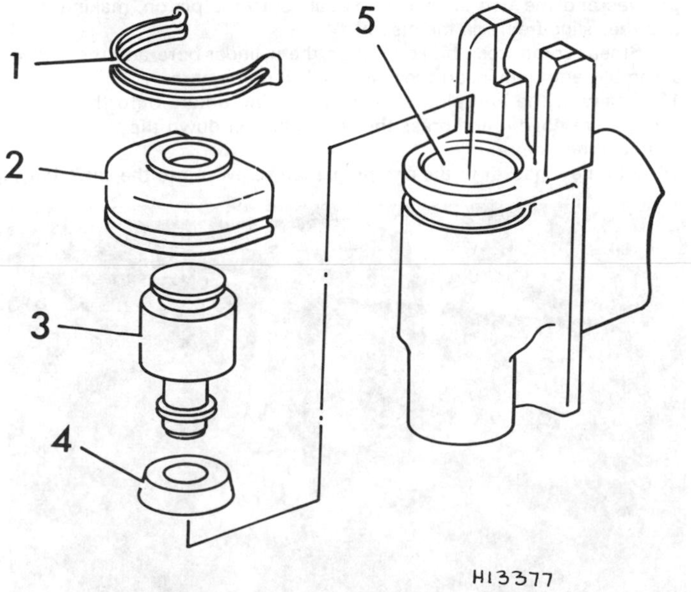

Fig. 9.9 Rear drum brake wheel cylinder component parts (Sec 12)

1 Clip
2 Dust cover
3 Piston
4 Seal
5 Wheel cylinder body

feed pipe from the right-hand wheel cylinder.

6 Extract the split-pin and lift away the washer and clevis pin that connects the handbrake cable yoke to the wheel cylinder operating lever.

7 Ease off the rubber boot from the rear of the wheel cylinder.

8 Using a screwdriver, carefully draw off the retaining plate and spring plate from the rear of the wheel cylinder.

9 The wheel cylinder may now be lifted away from the brake backplate. Detach the handbrake lever from the wheel cylinder.

10 To refit the wheel cylinder, first smear the backplate where the wheel cylinder slides with a little brake grease. Refit the handbrake lever on the wheel cylinder, ensuring that it is the correct way round. The spindles of the lever must engage in the recess on the cylinder arms.

11 Slide the spring plate between the wheel cylinder and backplate. The retaining plate may now be inserted between the spring plate and wheel cylinder, taking care the pips of the spring plate engage in the holes of the retaining plate.

12 Refit the rubber boot and reconnect the handbrake cable yoke to the handbrake lever. Insert the clevis pin, head upwards, and plain washer. Lock with a new split-pin.

13 Refitting the brake shoes and drum is the reverse sequence to removal. Adjust the brakes as described in Section 2 and finally bleed the hydraulic system following the instructions in Section 3.

12 Rear drum brake wheel cylinder – overhaul

1 Ease off the rubber dust cover that protects the open end of the cylinder bore (Fig. 9.9).

2 Withdraw the piston from the wheel cylinder body.

3 Using fingers only, carefully remove the piston seal from the piston, noting which way round it is fitted. Do not use a screwdriver as this could scratch the piston.

4 Inspect the inside of the cylinder for score marks caused by impurities in the hydraulic fluid. If any are found, the cylinder will require renewal. **Note:** *If the wheel cylinder is to be renewed, always ensure that the replacement is exactly similar to the one removed.*

5 If the cylinder is sound, thoroughly clean it out with fresh hydraulic fluid.

6 The old rubber seal will probably be swollen and visibly worn so it must be discarded. Smear the new rubber seal with hydraulic fluid and fit it to the piston so that the small diameter is towards the piston.

7 Carefully insert the piston and seal into the bore, making sure the fine edge lip does not roll or become trapped.

8 Refit the dust cover, engaging the lip with the groove in the outer surface of the wheel cylinder body. Refit the retaining ring.

13 Front disc brake caliper – overhaul

Note: *At no point in the overhaul procedure should the two halves of the caliper be separated.*

1 Extract the two pad retaining pin spring clips and withdraw the two retaining pins (Fig. 9.10).

2 Lift away the brake pads and anti-squeak shims, noting which way round the shims are fitted.

3 Remove the dust covers, then temporarily reconnect the caliper to the hydraulic system and support its weight. Do not allow the caliper to hang on the flexible hose but support its weight. Using a small G-clamp, hold the piston in the mounting half of the caliper. Carefully depress the footbrake pedal with the bleed nipple open so as to bleed the system, and then close the nipple. Depress the footbrake again and this will push the piston in the rim half of the caliper outwards. Release the dust cover retaining ring and the cover. Depress the footbrake again until the piston has been ejected sufficiently to continue removal by hand. It is advisable to have a container or tray available to catch any hydraulic fluid once the piston is removed.

4 Using a tapered wooden rod or an old plastic knitting needle, carefully extract the fluid seal from its bore in the caliper half.

5 Remove the G-clamp from the mounting half piston. Temporarily refit the rim half piston and repeat the operations in paragraph 3 and 4 of this Section.

6 Thoroughly clean the internal parts of the caliper using methylated spirit. Any other fluid cleaner will damage the internal seals between the two halves of the caliper.

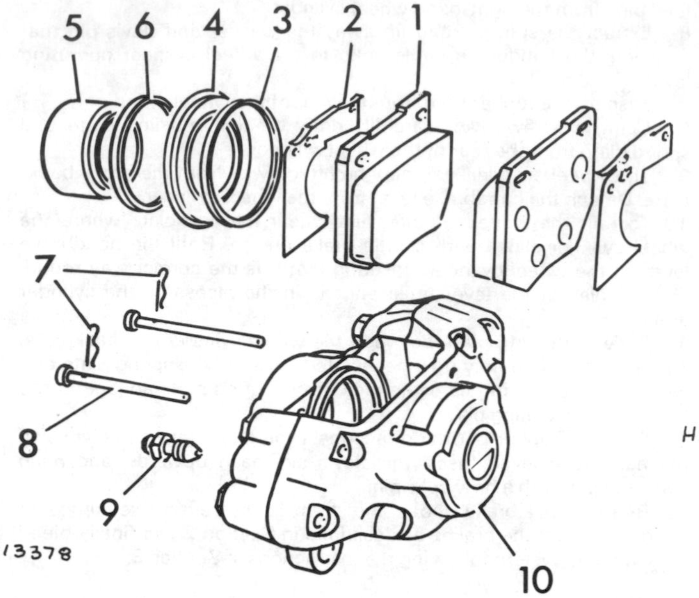

Fig. 9.10 Front disc brake caliper component parts (Sec 13)

1 *Pad*
2 *Anti-squeak shim*
3 *Dust cover retaining ring*
4 *Dust cover*
5 *Piston*
6 *Sealing ring*
7 *Spring clip*
8 *Retaining pin*
9 *Bleed screw*
10 *Caliper body*

7 Inspect the caliper bores and pistons for signs of scoring which if evident means that a new assembly should be fitted.
8 To reassemble the caliper, first wet a new fluid seal with brake fluid and carefully insert it into its groove in the rim half of the caliper seating, ensuring that it is correctly fitted. Refit the dust cover into its special groove in the cylinder (Fig. 9.11).
9 Release the bleed screw in the caliper one complete turn. Coat the side of the piston with hydraulic fluid and with it positioned squarely in the top of the cylinder bore, ease the piston in until 0.31 in (8mm) is left protruding. Engage the outer lip of the dust cover in the piston groove and push the piston into the cylinder as far as it will go. Fit the dust cover retaining ring.
10 Repeat the operations in paragraphs 8 and 9 for the mounting half of the caliper.
11 Fit the pads and anti-squeak shims into the caliper and retain in position with the two pins and spring clips.
12 The caliper is now ready for refitting.

14 Pressure differential warning actuator (PDWA) valve – removal, overhaul and refitting

1 Refer to Figs. 9.12 and 9.13.
2 Disconnect the battery negative terminal and the valve switch supply lead (photo).
3 Clean the top of the brake hydraulic fluid reservoir and unscrew the cap. Place a piece of polythene over the top and refit the cap. This is to prevent fluid syphoning out when the pipes are disconnected from the PDWA valve.
4 Clean the area around the PDWA valve assembly and, using an open-ended spanner, unscrew the union nut that secures the rear brake fluid pipe to the valve, followed by the master cylinder to valve rear brake pipe, front brake fluid pipes and finally, the master cylinder to valve front brake pipe.
5 Undo the PDWA valve securing screw and lift away the screw, spring washer and the valve.
6 Wipe down the outside of the valve and then unscrew and remove the switch from the top of the body.
7 Unscrew and remove the end plug and gasket; a new gasket must be fitted on reassembly.
8 Shake or tap out the piston components noting the order in which

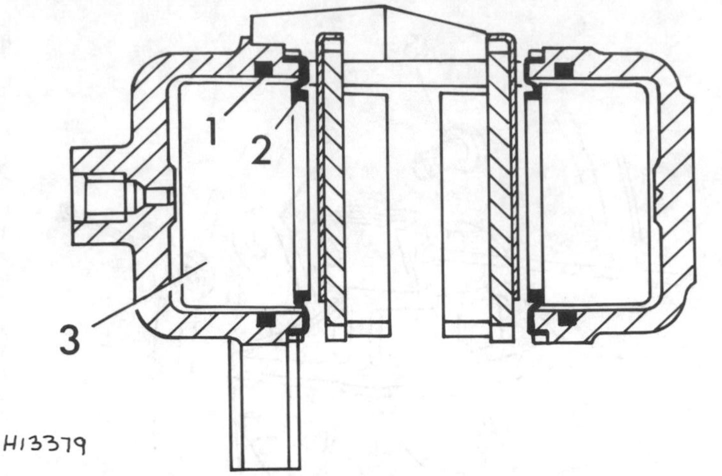

Fig. 9.11 Correct fitted positions of piston seal and dust cover (Sec 13)

1 *Piston seal*
2 *Dust cover*
3 *Piston*

they are fitted and making sure the sleeve and O-ring are recovered from the bottom of the bore.
9 Carefully prise the C-clips from the piston grooves and discard them together with the O-rings.
10 Examine the bore of the valve carefully for any signs of scores, ridges or corrosion. If this is found to be smooth all over, new seals can be fitted. If there is any doubt of the condition of the bore, a new valve must be obtained.
11 If examination of the two seals shows them to be apparently oversize or swollen, or very loose on the pistons, then suspect oil contamination in the system. Oil will swell these rubber seals and if one is found to be swollen, it is reasonable to assume that all seals in the braking system will need attention.
12 Thoroughly clean all parts in methylated spirit.
13 To reassemble the valve, first fit the piston C-clips into their grooves and the two sleeves and seals onto the piston, making sure that they slide freely on the piston.
14 Smear some fresh brake fluid on the cylinder bore and the piston assembly, and then insert the piston fully into the bore.
15 Screw in the end plug so that the O-ring enters onto the piston and then remove it and press the O-ring further down the bore until it contacts the sleeve.
16 Fit a new gasket to the end plug, then screw it into the valve body

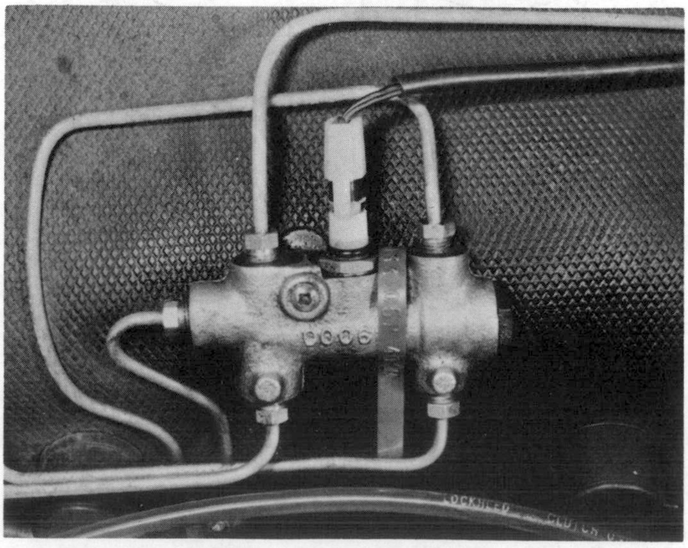

14.2 The PDWA valve is mounted on the bulkhead

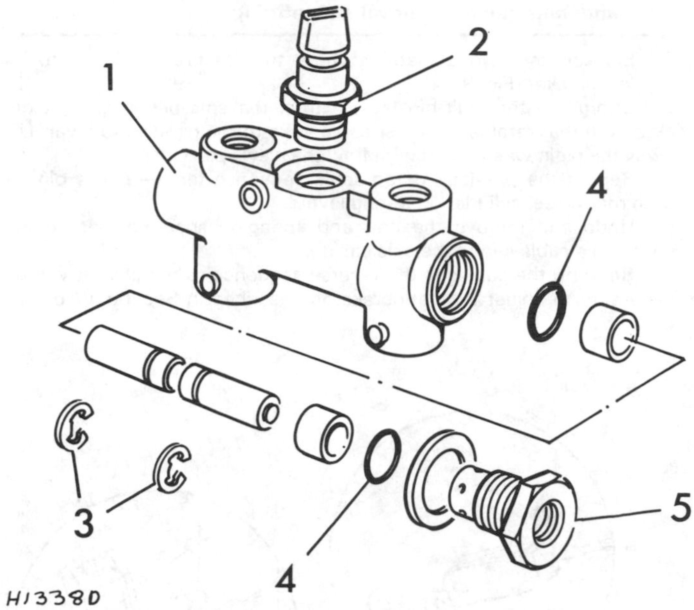

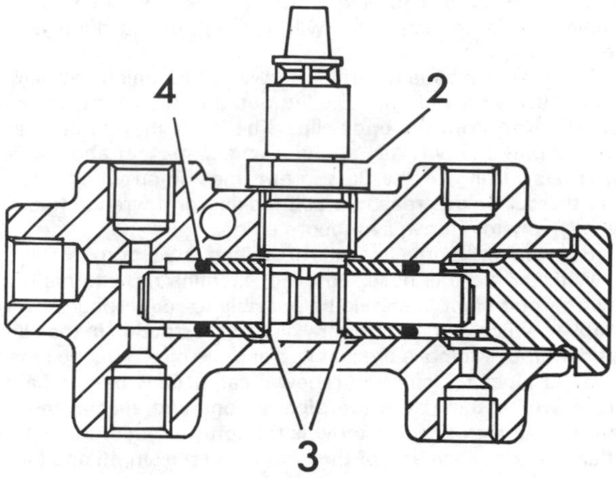

Fig. 9.12 Pressure differential warning actuator valve component parts (Sec 14)

1 Valve body
2 Warning lamp switch
3 Circlips
4 O-ring seals
5 End plug

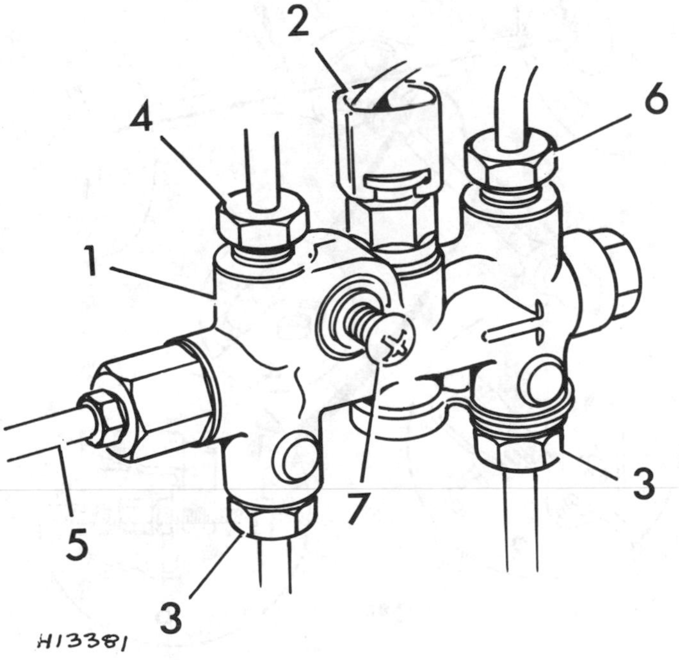

Fig. 9.13 Pressure differential warning actuator valve removal (Sec 14)

1 PDWA valve
2 Switch
3 Master cylinder to valve brake pipes
4 Valve to left-hand front wheel caliper pipe
5 Valve to right-hand front wheel caliper pipe
6 Valve to rear brakes pipe
7 Valve securing screw

and tighten it down to a torque wrench setting of 38 lbf ft (5.2 kgf m).
17 Using a screwdriver through the switch aperture, move the piston to its central position, then screw in the switch and tighten it to a torque wrench setting of 3.5 lbf ft (0.48 kgf m). Make sure that the piston is central, otherwise the switch may foul the two sleeves.
18 Refitting is a reversal of the removal procedure but it will be necessary to bleed the complete hydraulic system as described in Section 3.

15 Flexible hoses – inspection, removal and refitting

1 Inspect the condition of the flexible hydraulic hoses leading from the brake and clutch metal pipes. If any are swollen, damaged, cut or chafed, they must be renewed.
2 Unscrew the metal pipe union nut from its connection to the flexible hose and then, holding the hexagon on the hose with a spanner, unscrew the attachment nut and washer.
3 The end of the flexible hose can now be withdrawn from the mounting bracket and will be quite free.
4 Disconnect the flexible hose from the slave or wheel cylinder by unscrewing it, using a spanner.
5 Refitting is the reverse sequence to removal. It will now be necessary to bleed the brake hydraulic system as described in Section 3.

16 Hydraulic pipes and hoses – general

1 Periodically, all brake pipes, pipe connections and unions should be carefully examined.
2 First examine for signs of leakage, where the pipe unions occur. Then examine the flexible hoses for signs of chafing and fraying and , of course, leakage. This is only a preliminary part of the flexible hose inspection as exterior condition does not necessarily indicate the interior condition, which will be considered later.
3 The steel pipes must be examined carefully and methodically. They must be cleaned off and examined for any signs of dents, or other damage, rust and corrosion. Rust and corrosion should be scraped off, and if the depth of pitting in the pipes is significant, they will need renewing. This is particularly likely in those areas underneath the body

and along the rear axle where the pipes are exposed to the full force of road and weather conditions.
4 If any section of pipe is to be taken off, first wipe and then remove the fluid reservoir cap and place a piece of polythene over the reservoir. Refit the cap. This will stop syphoning during subsequent operations.
5 Rigid pipe is usually straightforward. The unions at each end are undone, the pipe and union pulled out, and the centre sections of the pipe removed from the body clips. The joints may sometimes be very tight. As one can only use an open-ended spanner and the unions are not large, burring of the flats is not uncommon when attempting to undo them. For this reason a self-locking grip wrench (mole) is often the only way to remove a stubborn union.
6 Removal of flexible hoses is described in Section 15.
7 With the flexible hose removed, examine the internal bore. If it is blown through first, it should be possible to see through it. Any specks of rubber which come out or any signs of restriction in the bore, means that the rubber lining is breaking up and the pipe must be renewed.
8 Rigid pipes which need renewal can usually be purchased at any garage where they have the pipe, unions and special tools to make them up. All they need to know is the total length of the pipe, the type of flare used at each end of the union, and the length and thread of the union.
9 Refitting the pipe is a straightforward reversal of the removal procedure. If the rigid pipes have been made up, it is best to get all the sets (bends) in them before trying to fit them. Also if there are any acute bends, ask your supplier to put these in for you on a special tube bender. Otherwise you may kink the pipe and thereby decrease the bore area and fluid flow.
10 With the pipes refitted, remove the polythene from the reservoir cap and bleed the system as described in Section 3.

17 Handbrake cable – adjustment

1 Refer to Section 2 and adjust the rear brakes.
2 Chock the front wheels and completely release the handbrake. Pull up the handbrake four clicks on the ratchet.
3 Jack up the rear of the car and support it on firmly based stands.
4 Check the cable adjustments by attempting to rotate the rear wheels. If this is possible, the cable may be adjusted as described in the subsequent paragraphs.
5 Refer to Fig. 9.14 and slacken the adjuster locknut (photo).
6 Turn the adjustment nut clockwise whilst the outer cable is held with an open-ended spanner until the correct adjustment is obtained. Retighten the locknut.
7 Release the handbrake and check that the rear wheels can be rotated freely.
8 Lower the rear of the car to the ground.

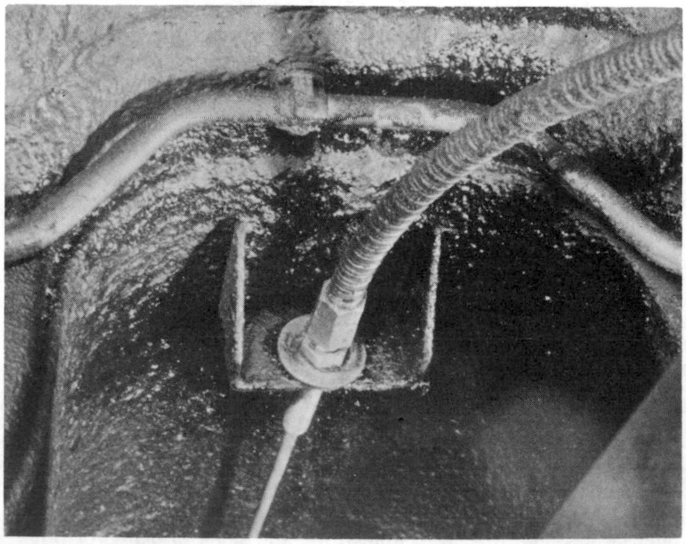

17.5 The handbrake cable support bracket attached to the floor pan

18 Handbrake cable – removal and refitting

1 Slacken the two adjustment nuts that secure the cable to its support bracket (Fig. 9.14).
2 Straighten the split-pin legs, extract the split-pin that locks the clevis pin that retains the inner cable yoke to the handbrake lever. Lift away the plain washer and withdraw the clevis pin.
3 Repeat the previous paragraph's sequence for the clevis pin on both rear wheel cylinder operating levers.
4 Undo and remove the bolt and spring washer that secure the handbrake cable clip to the axle casing.
5 Refitting the cables is the reverse sequence to removal. It will be necessary to adjust the handbrake as described in Section 17 of this Chapter.

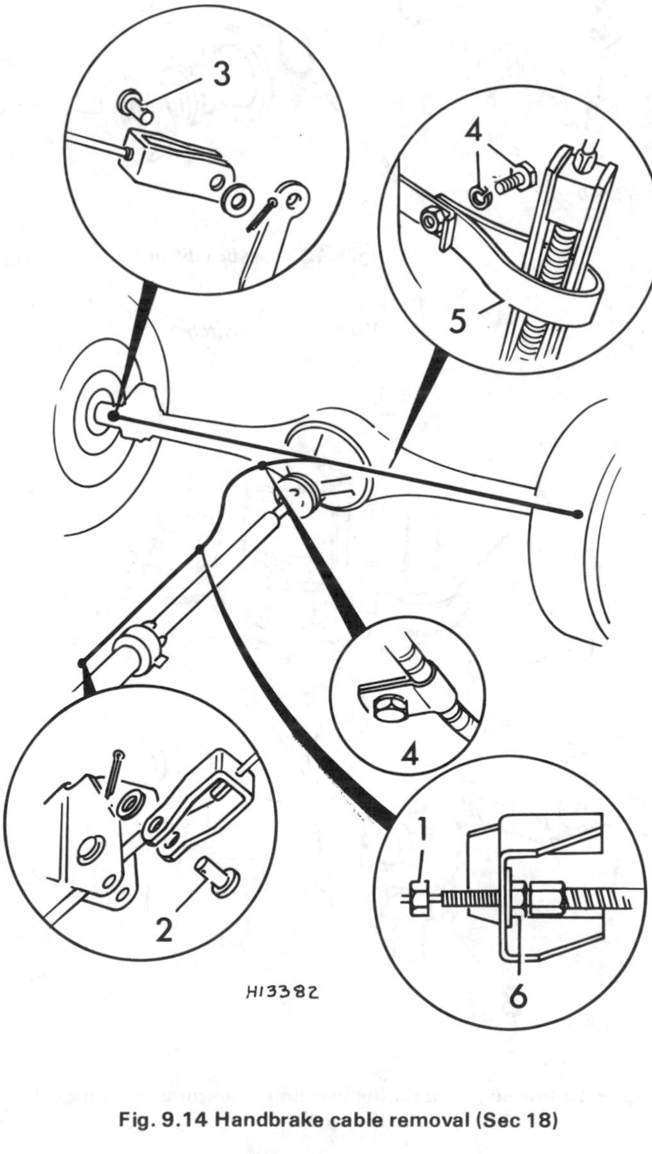

Fig. 9.14 Handbrake cable removal (Sec 18)

1 *Locknut*
2 *Cable to handbrake lever clevis pin*
3 *Cable to brake operating lever clevis pin*
4 *Clip securing bolt and washer*
5 *Handbrake cable clip*
6 *Adjustment nut*

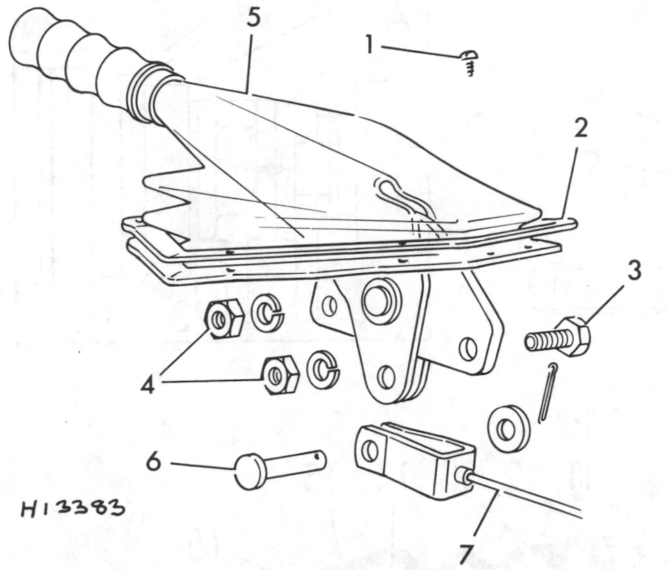

Fig. 9.15 Handbrake lever assembly removal (Sec 19)

1 *Self-tapping screw*
2 *Gaiter and metal plate*
3 *Mounting bolt*
4 *Nut and spring washer*
5 *Handbrake lever*
6 *Clevis pin with plain washer and split-pin*
7 *Handbrake cable*

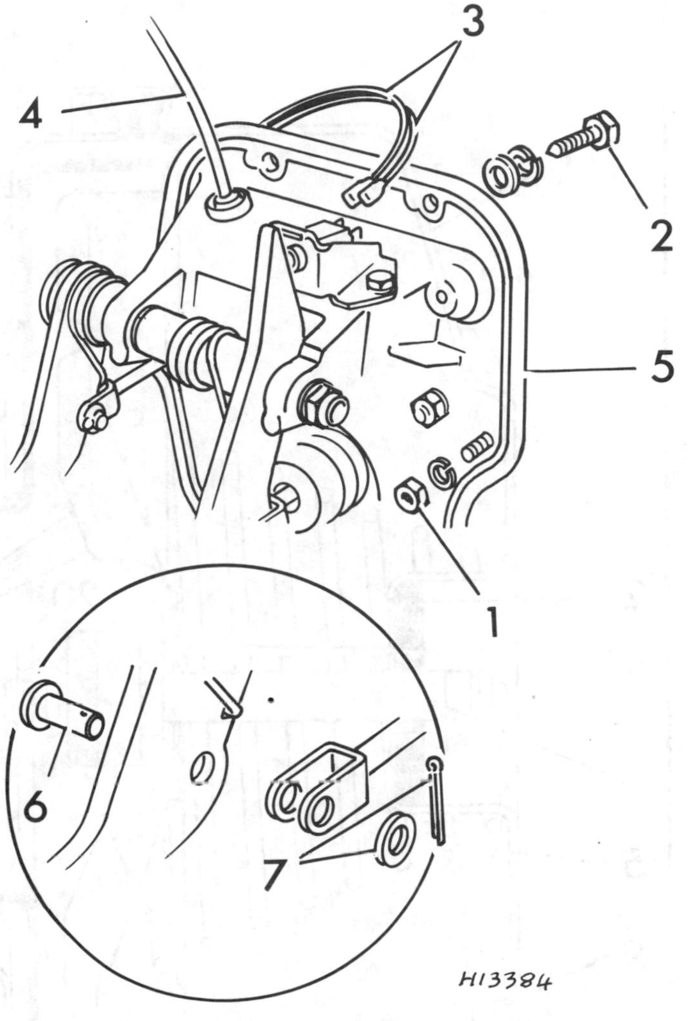

Fig. 9.16 Brake pedal assembly removal (Sec 20)

1 *Nut*
2 *Bolt*
3 *Stop light cables*
4 *Speedometer cable*
5 *Pedal mounting bracket*
6 *Clutch operating rod to clutch pedal clevis pin*
7 *Split-pin and washer*

19 Handbrake lever assembly – removal and refitting

1 Remove the centre console (if fitted) as described in Chapter 12, Section 33.
2 Draw back the floor covering from around the handbrake lever. Undo and remove the four self-tapping screws that retain the handbrake lever gaiter to the floor panel. It may be necessary to make cuts at each corner of the carpet to gain access to the retaining screws.
3 Straighten the split-pin legs and extract the split-pin that retains the handbrake cable to lever clevis pin. Lift away the plain washer and withdraw the clevis pin.
4 Slide the gaiter up the handbrake lever. Undo and remove the two nuts, spring washers and bolts that secure the handbrake lever assembly to its mounting bracket. If a handbrake switch is fitted, disconnect the supply lead from the switch.
5 Lift away the handbrake lever assembly.
6 Refitting the handbrake lever assembly is the reverse of the removal procedure. Lubricate all pivots with engine oil.

20 Brake pedal assembly – removal and refitting

1 Refer to Chapter 12 and remove the face vent hose.
2 Refer to Chapter 3 and remove the throttle pedal.
3 Remove the split-pin that retains the clutch master cylinder operating rod clevis pin and withdraw the clevis pin.
4 Undo and remove the clutch master cylinder securing nuts and move the cylinder to one side; then tap out the master cylinder lower securing stud.
5 Clean the top of the brake master cylinder and remove the fluid reservoir cap. Place a piece of polythene over the top of the reservoir and refit the cap. This is to prevent the hydraulic fluid syphoning out during subsequent operations.
6 Disconnect the brake master cylinder fluid pipes from the PDWA valve on the bulkhead.
7 Slacken the retaining clip and disconnect the vacuum hose from the servo unit connector.
8 Make a note of the cable connections on the ignition coil. Detach the cables and remove the ignition coil.
9 Undo and remove the nuts, bolts, spring and plain washers that secure the pedal mounting assembly to the bulkhead (Fig. 9. 16).
10 Partially withdraw the pedal assembly and disconnect the electrical connections at the stop light switch, then remove the clevis pin from the brake pedal.
11 Carefully pull the speedometer cable through the grommet in the pedal mounting assembly after disconnecting it.
12 The pedal and mounting assembly may now be lifted away from inside the car.
13 Undo and remove the locknut and plain washer that retain the brake pedal pivot pin. Lift away the throttle pedal spring bracket, noting which way round it is fitted.
14 Withdraw the clutch pedal complete with pivot pin and remove the brake pedal and return springs.
15 Refitting is the reverse of the removal procedure. Ensure that the servo operating rod is connected to the lower hole in the brake pedal. Bleed the brake hydraulic system as described in Section 3.

21 Brake servo unit – description

The vacuum servo unit is fitted into the brake hydraulic circuit in series with the master cylinder to provide power assistance to the driver when the brake pedal is depressed.

The unit operates by vacuum obtained from the induction manifold and comprises basically a booster diaphragm and a non-return valve.

The servo unit and hydraulic master cylinder are connected

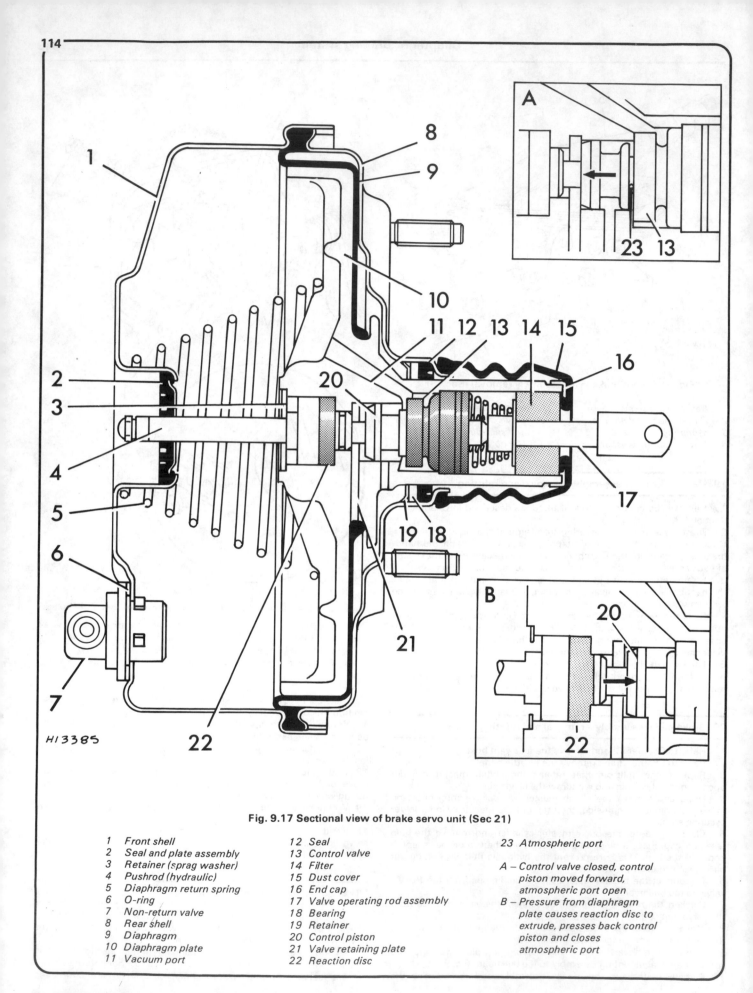

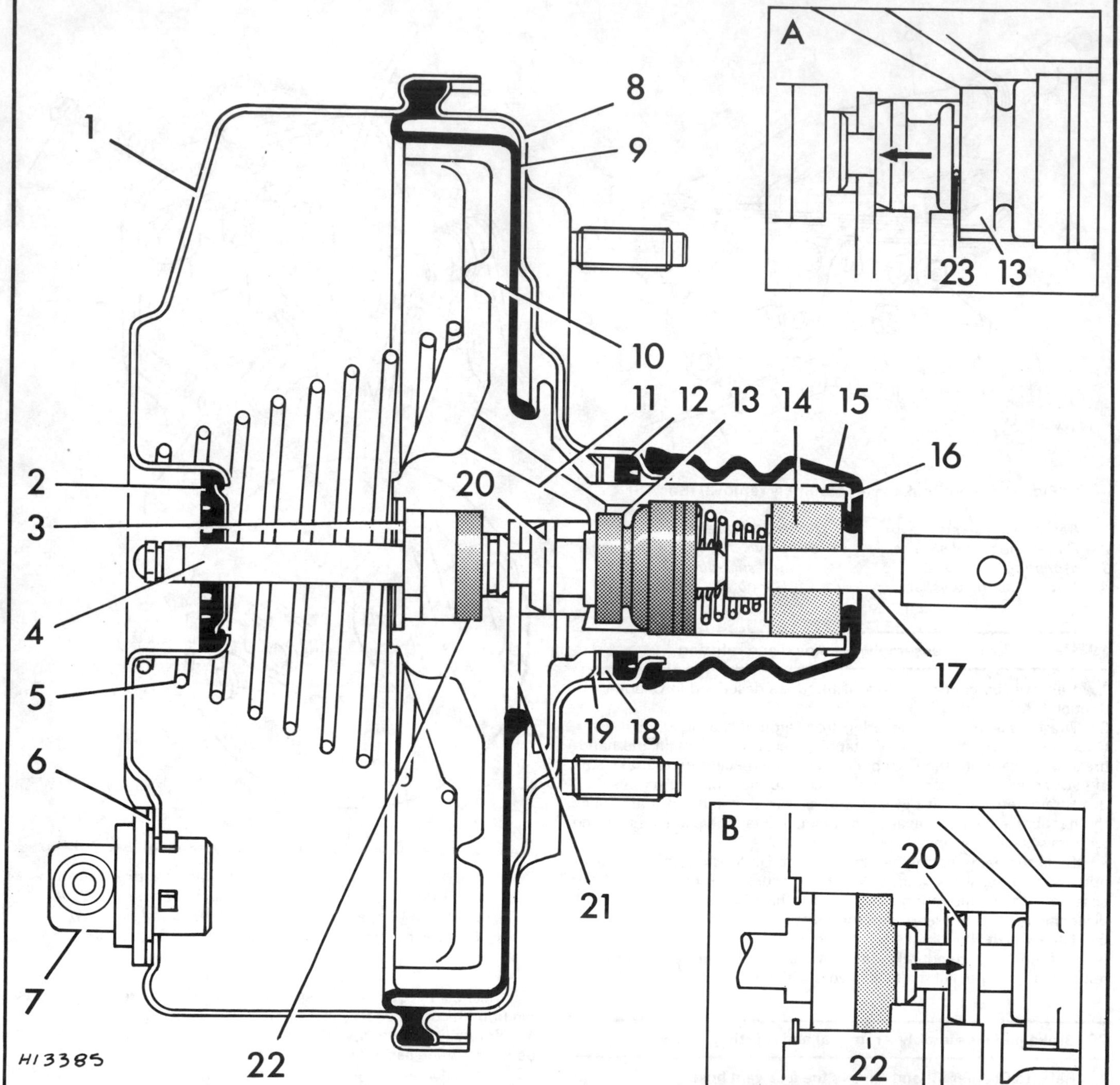

Fig. 9.17 Sectional view of brake servo unit (Sec 21)

1 *Front shell*
2 *Seal and plate assembly*
3 *Retainer (sprag washer)*
4 *Pushrod (hydraulic)*
5 *Diaphragm return spring*
6 *O-ring*
7 *Non-return valve*
8 *Rear shell*
9 *Diaphragm*
10 *Diaphragm plate*
11 *Vacuum port*
12 *Seal*
13 *Control valve*
14 *Filter*
15 *Dust cover*
16 *End cap*
17 *Valve operating rod assembly*
18 *Bearing*
19 *Retainer*
20 *Control piston*
21 *Valve retaining plate*
22 *Reaction disc*
23 *Atmospheric port*

A – Control valve closed, control piston moved forward, atmospheric port open
B – Pressure from diaphragm plate causes reaction disc to extrude, presses back control piston and closes atmospheric port

together so that the servo unit piston rod acts as the master cylinder pushrod. The driver's braking effort is transmitted through another pushrod to the servo unit piston and its built in control system. The servo unit piston does not fit tightly into the cylinder but has a strong diaphragm to keep its edges in constant contact with the cylinder walls, so assuring an air-tight seal between the two parts. The forward chamber is held under vacuum conditions created in the inlet manifold of the engine and, during periods when the brake pedal is not in use, the controls open a passage to the rear chamber, so placing it under vacuum. When the brake pedal is depressed, the vacuum passage to the rear chamber is cut off and the chamber opened to atmospheric pressure. The consequent rush of air pushes the servo piston forward in the vacuum chamber and operates the main pushrod to the master cylinder. The controls are designed so that assistance is given under all conditions and, when the brakes are not required, vacuum in the rear chamber is established when the brake pedal is released. Air from the atmosphere entering the rear chamber is passed through a small air filter.

22 Brake servo unit – removal and refitting

1 Refer to Section 9 and remove the brake master cylinder.
2 Slacken the hose clip and detach the vacuum hose from the servo connector.
3 Undo and remove the two nuts and spring washers that secure the throttle pedal bracket. Lift away the throttle pedal and bracket.
4 Straighten the legs of the split-pin that retains the servo to brake pedal pushrod clevis pin. Extract the split-pin, lift away the plain washer and withdraw the clevis pin.
5 Undo and remove the four nuts and spring washers that secure the servo unit to the mounting bracket. Lift away the servo unit and gasket.
6 Refitting the servo unit is the reverse of the removal procedure. Always use a new gasket. It is important that the servo operating rod is attached to the lower of the two holes in the brake pedal lever.
7 Bleed the brake hydraulic system as described in Section 3.

23 Brake servo unit air filter – renewal

Under normal operating conditions, the vacuum servo unit is very reliable and does not require overhaul except possibly at very high mileages. In this case it is far better to obtain a service exchange unit, rather than repair the original.

However, the air filter may be renewed and fitting details are given. This will not however, repair any fault.
1 Pull back the dust cover (Fig. 9.19) and slide up the pushrod.
2 Using a screwdriver, ease out the end cap and then with a pair of scissors cut off the old air filter.
3 Make a diagonal cut through the new air filter element and fit it over the pushrod. Hold it in position and refit the end cap.
4 Reposition the dust cover on the servo unit body.

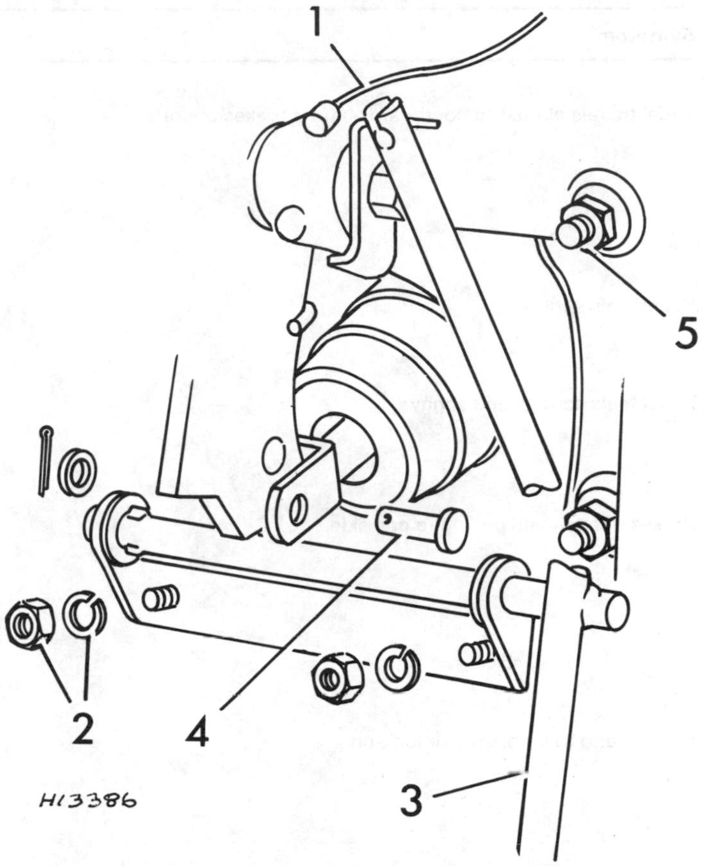

Fig. 9.18 Brake servo unit attachment – car interior (Sec 22)

1 *Throttle cable*
2 *Throttle pedal bracket securing nut and spring washer*
3 *Throttle pedal*
4 *Clevis pin with plain washer and split-pin*
5 *Servo unit securing nut and spring washer*

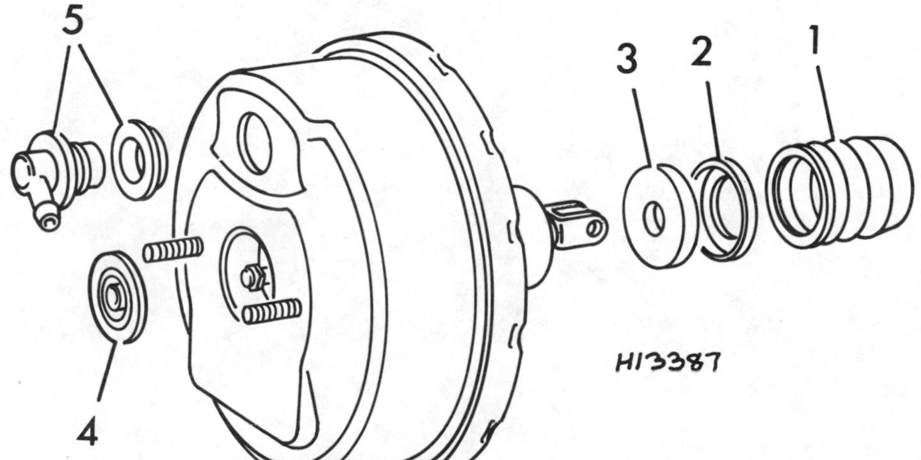

Fig. 9.19 Servo air filter renewal (Sec 23)

1 *Dust cover*
2 *End cap*
3 *Filter*
4 *Front seal*
5 *Non-return valve and grommet*

24 Fault diagnosis – braking system

Symptom	Reason/s
Pedal travels almost to floorboards before brakes operate	Brake fluid level too low Wheel cylinder or caliper leaking Master cylinder leaking (bubbles in master cylinder fluid) Brake flexible hose leaking Brake line fractured Brake system unions loose Linings over 75% worn
Pedal feels springy	New linings not yet bedded-in Brake drums or discs badly worn and weak or cracked Master cylinder securing nuts loose
Pedal feels spongy and soggy	Wheel cylinder or caliper leaking Master cylinder leaking (bubbles in master cylinder reservoir) Brake pipe line or flexible hose leaking Unions in brake system loose
Brakes uneven and pulling to one side	Linings and brake drums or discs contaminated with oil, grease, or hydraulic fluid Tyre pressures unequal Brake backplate, caliper or disc loose Brake shoes or pads fitted incorrectly Different type of linings fitted at each wheel Anchorages for front or rear suspension loose Brake drums or discs badly worn, cracked or distorted
Brakes tend to bind, drag or lock-on	Brake shoes adjusted too tightly Handbrake cable over-tightened Reservoir vent hole in cap blocked with dirt Master cylinder bypass port restricted, brakes seize in 'on' position Wheel cylinder seizes in 'on' position Drum brake shoe pull-off springs broken, stretched or loose Drum brake shoe pull-off spring fitted wrong way round, omitted, or wrong type used

Chapter 10 Electrical system

Contents

Specifications

System
12V negative earth

Battery

Type	Lucas A9, A11, A13 Exide 6VTP7 – BR, 6VTP9 – BR, 6VTPZ11 – BR
Capacity at 20 hr rate/maximum fast charge time:	
A9	40 Ah : 1½ hours
A11	50 Ah : 1½ hours
A13	60 Ah : 1 hour
6VTP7 – BR	30 Ah : 1½ hours
6VTP9 – BR	40 Ah : 1½ hours
6VTPZ11 – BR	50 Ah : 1½ hours

Alternator

Type	Lucas 16ACR, 17ACR, or 18ACR
Output at 14 volts and 6000 rpm:	
16 ACR	34 amps
17ACR	36 amps
18ACR	45 amps
Maximum permissible rotor speed	15 000 rpm
Stator phases	3
Rotor poles	12
Minimum brush length	0·2 in (5 mm) protruding beyond brush box moulding
Brush spring tension	9 to 13 oz (255 to 369 g) with brush face flush with brush box
Control unit	Integral with alternator

Rotor winding resistance at 20°C (68°F):

Alternator type	Resistance	Winding identification colour
16ACR	4·3 ohms ± 5%	Pink
16ACR	3·3 ohms ± 5%	Purple
17ACR	4·2 ohms ± 5%	Pink
17ACR	3·2 ohms ± 5%	Green
18ACR	3·2 ohms ± 5%	Green

Starter motor

Type	**Lucas M35J and M35J/PE**
Brush spring tension	28 oz (0·8 kg)
Minimum brush length	0·375 in (9·5 mm)
Lock torque	7 lbf ft (0·97 kgf m) with 350 to 375 amps
Torque at 1000 rpm	4·4 lbf ft (0·61 kgf m) with 260 to 275 amps
Light running current	65 amps at 8000 to 10 000 rpm
Type	**Lucas 2M100 pre-engaged**
Brush spring tension	36 oz (1·02 kg)
Minimum brush length	0·375 in (9·5 mm)
Lock torque	14·4 lbf ft (1·99 kgf m) with 463 amps
Torque at 1000 rpm	7·3 lbf ft (1·0 kgf m) with 300 amps
Light running current	40 amps at 6000 rpm

Wiper motor

Type	Lucas 14W, 2-speed
Armature endfloat	0·002 to 0·008 in (0·051 to 0·21 mm)
Light running current:	
Normal speed	1·5 amp
High speed	2·0 amp
Light running speed:	
Normal speed	46 to 52 rpm
High speed	60 to 70 rpm

Fuses

Fuse No	Rating	Circuit protected
1	17A	Direction indicators Stop lamps Reverse lamps Heated rear window and warning light Tailgate wiper and washer
2	8A	Side and tail lights Number plate lamp Panel lamps Glovebox light Automatic transmission selector light Rear fog guard lamps
3	17A	Windscreen wiper Windscreen washer Heater motor
4	8A	Hazard warning flashers
5	17A	Horn Headlight flasher Interior lamp Lighter Boot light Brake failure warning light Clock

Bulbs

	Wattage
Headlight:	
Sealed beam unit	60/55
Renewable bulb	60/55
Sidelamp	5
Front and rear flashers	21
Stop/tail lamp	6/21
Number plate	5
Interior	6
Panel and warning	2·2
Reverse lamps	21
Boot light	6
Automatic selector lever light	3
Heated backlight switch, brake failure light and hazard warning light	0·75
Glovebox light	6

Torque wrench settings

	lbf ft	kgf m
Starter motor bolts	27	3·8
Alternator adjusting link	9	1·3
Alternator pulley nut	20 to 30	2·7 to 4·2

1 General description

The electrical system is of the 12 volt type and the major components comprise a 12 volt battery of which the negative terminal is earthed, a Lucas alternator which is fitted to the front right-hand side of the engine and is driven from the pulley on the front of the crankshaft, and a starter motor which is mounted on the rear right-hand side of the engine.

The battery supplies a steady amount of current for the ignition, lighting and other electrical circuits, and provides a reserve of electricity when the current consumed by the electrical equipment exceeds that being produced by the alternator.

The battery is charged by a Lucas ACR alternator and information on this component will be found in Section 6.

When fitting electrical accessories to cars with a negative earth system, it is important if they contain silicone diodes or transistors, that they are connected correctly, otherwise serious damage may result to the component concerned. Items such as radios, tape recorders, electronic tachometer, automatic dipping, parking lamp and anti-dazzle mirrors should all be checked for correct polarity.

It is important that the battery negative lead is always disconnected if the battery is to be boost charged or if any body or mechanical repairs are to be carried out using electric arc welding equipment. Serious damage can be caused to the more delicate instruments, specially those containing semi-conductors. It is equally important to ensure that neither battery lead is disconnected whilst the engine is running and that the battery terminals are not inadvertently connected with the polarity reversed.

2 Battery – removal and refitting

1 The battery is in a special carrier fitted on the right-hand wing valance of the engine compartment. It should be removed once every three months for cleaning and testing. Disconnect the negative and then the positive leads from the battery terminals by slackening the clamp retaining nuts and bolts, or by unscrewing the retaining screws if terminal caps are fitted instead of clamps.

2 Unscrew the clamp bar retaining nuts and lower the clamp bar to the side of the battery. Carefully lift the battery from its carrier. Hold the battery vertical to ensure that none of the electrolyte is spilled.

3 Refitting is a direct reversal of this procedure. Ensure that the positive lead is fitted before the negative lead and smear the terminals with petroleum jelly (vaseline) to prevent corrosion. Do not use an ordinary grease as applied to other parts of the car.

3 Battery – maintenance and inspection

1 Normal weekly battery maintenance consists of checking the electrolyte level of cells to ensure that the separators are covered by ¼ in of electrolyte. If the level has fallen, top up the battery using distilled water only. Do not overfill. If the battery is overfilled or any electrolyte spilled, immediately wipe away excess as electrolyte attacks and corrodes any metal it comes into contact with very rapidly.

2 If the battery is of the Lucas Pacemaker design, a special topping up procedure is necessary as follows:

(a) The electrolyte levels are visible through the translucent battery case or may be checked by fully raising the vent cover and tilting to one side. The electrolyte level in each cell must be kept such that the separator plates are just covered. To avoid flooding, the battery must not be topped up within half an hour of it having been charged from any source other than from the generating system fitted to the car

(b) To top up the levels in each cell, raise the vent cover and pour distilled water into the trough until all the rectangular filling slots are full of distilled water and the bottom of the trough is just covered. Wipe the cover seating grooves dry and press the cover firmly into position. The correct quantity of distilled water will automatically be distributed to each cell

(c) The vent must be kept closed at all times except when being topped up

3 If the battery has the Auto-fill device fitted, a special topping up sequence is required. The white balls in the Auto-fill battery are part of the automatic topping up device which ensures correct electrolyte level. The vent chamber should remain in position at all times except when topping up or taking specific gravity readings. If the electrolyte level in any of the cells is below the bottom of the filling tube, top up as follows:

(a) Lift off the vent chamber cover

(b) With the battery level, pour distilled water into the trough until all the filling tubes and trough are full

(c) Immediately refit the cover to allow the water in the trough and tubes to flow into the cells. Each cell will automatically receive the correct amount of water

4 As well as keeping the terminals clean and covered with petroleum jelly, the top of the battery and especially the top of the cells, should be kept clean and dry. This helps to prevent corrosion and ensures that the battery does not become partially discharged by leakage through dampness and dirt.

5 Inspect the battery securing nuts, battery clamp plate, tray and battery leads for corrosion (white fluffy deposits on the metal which are brittle to touch). If any corrosion is found, clean off the deposit with ammonia and paint over the clean metal with an anti-rust, anti-acid paint.

6 At the same time, inspect the battery case for cracks. If a crack is found, clean and plug it with one of the proprietary compounds marketed for this purpose. If leakage through the crack has been excessive, then it will be necessary to refill the appropriate cell with fresh electrolyte as described later. Cracks are frequently caused at the top of the battery case by pouring in distilled water in the middle of winter after instead of before running the car. This gives water no chance to mix with the electrolyte and so the former freezes and splits the battery case.

7 If topping up becomes excessive and the case has been inspected for cracks that could cause leakage, but none are found, the battery is being overcharged and the voltage regulator will have to be checked and reset.

8 With the battery on the bench at the three monthly interval check, measure the specific gravity with a hydrometer to determine the state of charge and condition of the electrolyte. There should be very little variation between the different cells and, if a variation in excess of 0·025 is present, it will be due to either:

(a) Loss of electrolyte from the battery at some time caused by spillage or a leak, resulting in a drop in the specific gravity of the electrolyte when the deficiency was replaced with distilled water instead of fresh electrolyte

(b) An internal short circuit caused by buckling of the plates or a similar fault, pointing to the likelihood of total battery failure in the near future

9 The specific gravity of the electrolyte for fully charged conditions at the electrolyte temperature indicated, is listed in Table A. The specific gravity of a fully discharged battery at different temperatures of the electrolyte is given in Table B.

Table A

Specific gravity – battery fully charged

1·268 at 100°F or 38°C electrolyte temperature
1·272 at 90°F or 32°C electrolyte temperature
1·276 at 80°F or 27°C electrolyte temperature
1·280 at 70°F or 21°C electrolyte temperature
1·284 at 60°F or 16°C electrolyte temperature
1·288 at 50°F or 10°C electrolyte temperature
1·292 at 40°F or 4°C electrolyte temperature
1·296 at 30°F or – 1·5°C electrolyte temperature

Table B
Specific gravity – battery fully discharged
1·098 at 100°F or 38°C electrolyte temperature
1·102 at 90°F or 32°C electrolyte temperature
1·106 at 80°F or 27°C electrolyte temperature
1·110 at 70°F or 21°C electrolyte temperature
1·114 at 60°F or 16°C electrolyte temperature
1·118 at 50°F or 10°C electrolyte temperature
1·122 at 40°F or 4°C electrolyte temperature
1·126 at 30°F or – 1·5°C electrolyte temperature

4 Battery – electrolyte replenishment

1 If the battery is in a fully charged state and one of the cells maintains a specific gravity reading which is 0·025 or lower than the others, and a check of each cell has been made with a voltage meter to check for short circuits (a four to seven second test should give a steady reading of between 1·2 and 1·8 volts), then it is likely that electrolyte has been lost from the cell with the low reading at some time.
2 Top up the cell with a solution of 1 part sulphuric acid to 2·5 parts of water. If the cell is already fully topped up, draw some electrolyte out of it with a pipette. The total capacity of each cell is approximately ⅓ pint.
3 **Note**: *When mixing the sulphuric acid and water, never add water to sulphuric acid. Always pour the acid slowly onto the water in a glass container. If water is added to sulphuric acid it will explode.*
4 Continue to top up the cell with the freshly made electrolyte and to recharge the battery and check the hydrometer readings.

5 Battery – charging

1 In winter time when a heavy demand is placed on the battery, such as when starting from cold, and much electrical equipment is continually in use, it is a good idea to occasionally have the battery fully charged from an external source at a rate of 3·5 to 4 amps.
2 Continue to charge the battery at this rate until no further rise in specific gravity is noted over a four hour period.
3 Alternatively, a trickle charger charging at the rate of 1·5 amps, can be safely used overnight.
4 Special rapid 'boost' charges which are claimed to restore the power of the battery in 1 to 2 hours are most dangerous unless they are thermostatically controlled as they can cause serious damage to the battery plates through overheating.
5 Whilst charging the battery, note that the temperature of the electrolyte should never exceed 100°F.

6 Alternator – general description

A Lucas alternator is fitted as standard to the models covered by this manual. The main advantage of the alternator lies in its ability to provide a high charge at low revolutions. Driving slowly in heavy traffic with a dynamo invariably means no charge is reaching the battery. In similar conditions even with the wipers, heater, lights and perhaps radio switched on, the alternator will ensure a charge reaches the battery.

The system provides for direct connection of a charge light and eliminates the need for a field switching relay and warning light control unit, necessary with former systems.

The alternator is of the rotating field ventilated design and comprises pricipally a laminated stator on which is wound a star connected 3-phase output winding, a twelve pole rotor carrying the field windings – each end of the rotor shaft runs in ball race bearings which are lubricated for life, natural finish aluminium die-cast end brackets incorporating the mounting lugs, a rectifier pack for converting the AC output of the machine to DC for battery charging, and an output control regulator.

The rotor is belt driven from the engine through a pulley keyed to the rotor shaft. A pressed steel fan adjacent to the pulley draws cooling air through the machine. This fan forms an integral part of the alternator specification. It has been designed to provide adequate air flow with a minimum of noise and to withstand the high stresses associated with maximum speed. Rotation is clockwise viewed on the drive end. Maximum continuous rotor speed is 15 000 rpm.

Rectification of alternator ouput is achieved by six silicone diodes housed in a rectifier pack and connected as a 3-phase full wave bridge.The rectifier pack is attached to the outer face of the slip ring end bracket and contains also three 'field' diodes. At normal operating speeds rectified current from the stator output windings flows through these diodes to provide self excitation of the rotor field via brushes bearing on face type slip rings.

The slip rings are carried on a small diameter moulded drum attached to the rotor shaft outboard of the rotor shaft axle, whilst the outer ring has a mean diameter of ¾ in. By keeping the mean diameter of the slip rings to a minimum, relative speeds between brushes and rings, and hence wear, are also minimal. The slip rings are connected to the rotor field winding by wires carried in grooves in the rotor shaft.

The brush gear is housed in a moulding screwed to the outside of the slip ring end bracket. This moulding thus encloses the slip ring and brush gear assembly, and together with the shielded bearing, protects the assembly against the entry of dust and moisture.

The regulator is set during manufacture and requires no further attention.

7 Alternator – testing the charging circuit in situ

1 Initially ensure that the battery terminals are clean, the battery is fully charged, all cables and terminals are in good condition and that the drivebelt is correctly tensioned.

Battery voltage test

2 Remove the cable connector from the alternator then connect the negative terminal of a voltmeter to a chassis earth point. Switch on the ignition and connect the voltmeter positive lead to each of the alternator cable connectors in turn.
3 If there is no voltage at the 'IND' cable connector, check the charge indicator lamp and associated wiring.
4 If there is no voltage at the main charging cable connector, check for continuity between the battery and alternator.
5 If satisfactory at paragraph 3 and 4, proceed to the alternator test below.

Alternator test

6 Reconnect the alternator cable connector then disconnect the brown eyelet-ended cable at the starter motor solenoid. Connect an ammeter between the solenoid terminal and the brown cable, and a voltmeter across the battery terminals. Run the engine to obtain an alternator speed of 6000 rpm (approx 2500 rpm of the engine).
7 If there is no current reading, remove and overhaul the alternator.
8 If there is a negative ammeter reading and a voltmeter reading of 13·6 to 14·6 volts and if the battery is in a low state of charge a bench test of the alternator should be carried out to check for the specified current. This is not a do-it-yourself task and must be entrusted to an automotive electrical specialist.
9 If less than 10 amps is registered and a voltage below 13·6 is obtained, the alternator voltage regulator must be renewed.
10 If more than 10 amps is registered and a voltage above 14·6 volts is obtained, the alternator voltage regulator must be renewed.

8 Alternator – removal and refitting

1 Disconnect the battery negative terminal.
2 Loosen the pivot and adjustment bolts and swivel the alternator towards the engine to facilitate the removal of the drivebelt.
3 Disconnect the multi-connector from the alternator end cover.
4 Unscrew and remove the pivot and adjustment bolts and withdraw the alternator from the engine.
5 Refitting is a reversal of the removal procedure but adjust the alternator drivebelt as described in Section 9.

9 Alternator drivebelt – adjustment

1 It is important to keep the alternator drivebelt tension correctly adjusted. If the belt is loose it will slip, wear rapidly and result in malfunction of the alternator and water pump. If the belt is too tight, the alternator and water pump bearings will wear rapidly and cause

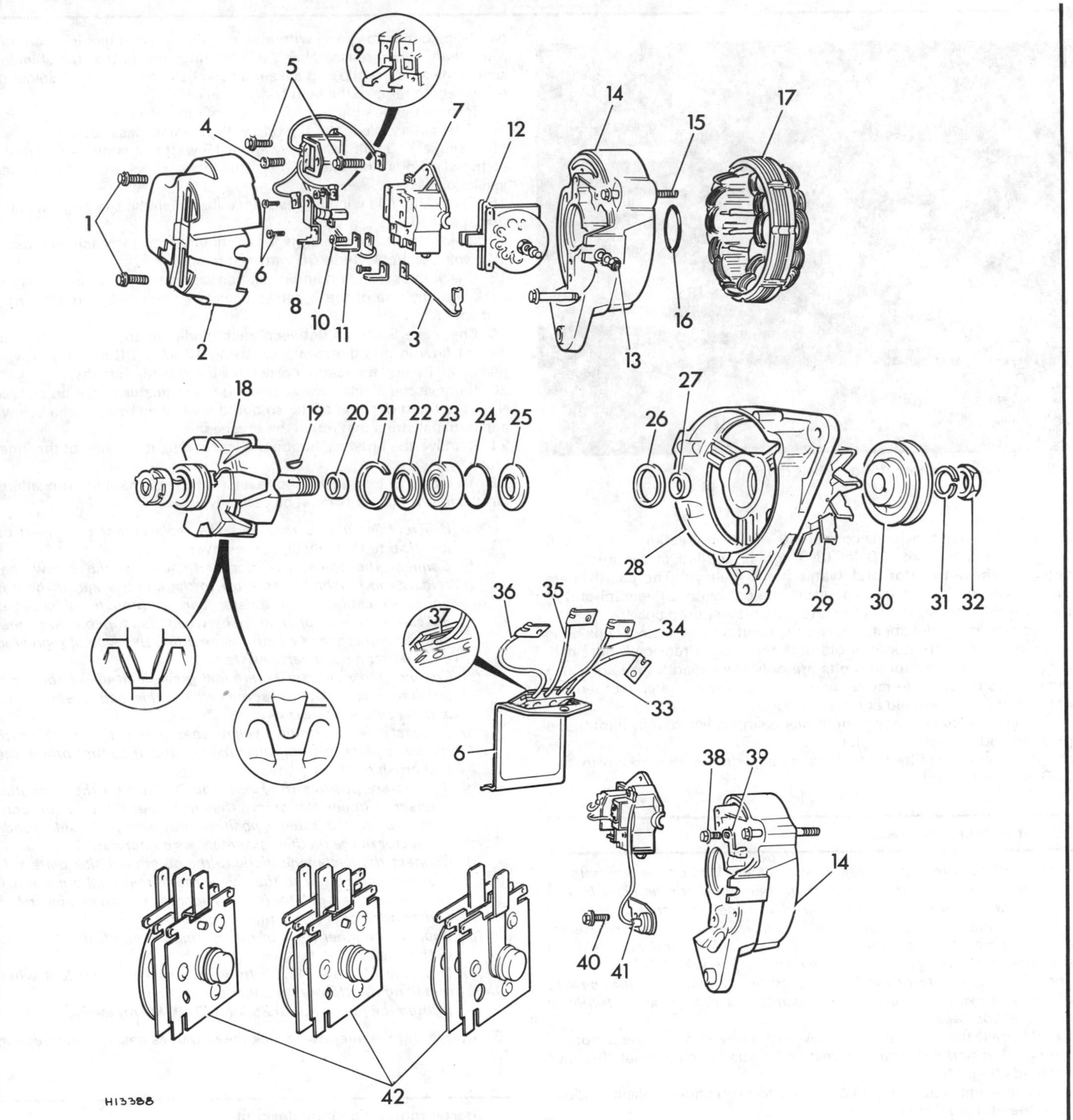

Fig. 10.1 Exploded view of alternator (Sec 10)

1 Screw
2 End cover
3 Lead (typical)
4 Screw
5 Screws
6 Regulator
7 Brush box mounting
8 Brush assembly
9 Brush spring
10 Screws
11 Terminals
12 Rectifier assembly
13 Nut and washers
14 Slip ring end bracket
15 Through-bolts
16 O-ring
17 Stator lamination pack
18 Rotor assembly with bearing and slip ring
19 Key
20 Distance piece
21 Circlip
22 Cover plate
23 Bearing
24 O-ring
25 Cover plate
26 Felt washer
27 Distance piece
28 Drive end bracket
29 Fan
30 Pulley
31 Spring washer
32 Nut
33 Black earth lead
34 F green lead
35 + yellow lead
36 B + red lead
37 Copper strip (alternative to item 33)
38 Screw
39 Earthing link
40 Screw
41 Avalanche diode
42 Rectifiers (alternative types to item 12)

9.2 Checking the alternator drivebelt tension

premature failure.

2 The drivebelt tension is correct if there is a deflection of 0·25 in (6 mm) when a load of 8·0 lbf (3·6 kgf) is applied at the mid-point between the alternator and water pump pulleys. The approximate tension can be judged by depressing the belt with a finger (photo) at the mid-point between the alternator and water pump pulley.

3 To adjust the drivebelt, slacken the pivot and adjustment link bolts and move the alternator in or out until the correct tension is obtained. It is easier if the alternator bolts are only slackened a little so that it requires some effort to move the unit. In this way the correct tension of the belt can be arrived at more quickly.

4 When the correct adjustment has been achieved, fully tighten the pivot and adjustment link bolts.

5 If a new belt is fitted, it will require adjustment after approximately 250 miles (400 km).

10 Alternator – overhaul

Note: *Before commencing repair work of any kind on the alternator, it must be appreciated that more harm than good can be done by any person inexperienced with the use of a soldering iron and electrical test equipment in connection with semi-conductor devices. Before commencing, read through this Section carefully and ascertain the availability of any spare parts which may be required. If in any doubt about the feasibility of the job, or your own capabilities, it is best to obtain a service exchange unit or contact a recognised automotive electrical specialist.*

1 Remove the alternator end cover (2 screws) and make a note of the position and colour of the rectifier spade terminal leads. Remove the leads (Fig. 10.1).

2 Where applicable, remove the surge protection avalanche diode from the end bracket (1 screw).

3 Remove the brush box moulding (2 screws) and the regulator (1 screw) from the end bracket.

4 Where applicable, remove the rectifier earthing link (1 screw).

5 Unsolder the three stator cables from the rectifier using the minimum practicable amount of heat.

6 Remove the rectifier after slackening the nut.

7 Mark the drive-end bracket, stator lamination pack and slip ring end bracket to aid reassembly.

8 Remove the three through-bolts and withdraw the end bracket and lamination pack. Remove the O-ring from inside the end bracket.

9 Remove the nut and washer then withdraw the pulley and fan from the rotor.

10 Press the rotor out of the drive-end bracket bearing and withdraw the distance piece from the rotor.

11 Remove the circlip, bearing, cover plates. O-ring and felt washer from the drive-end bracket.

12 If it is necessary to remove the slip ring end bearing, unsolder the two field connections and withdraw the slip ring and bearing from the rotor shaft. When reassembling this bearing, ensure that the shielded side is towards the slip ring assembly and use only Fry's HT3 solder or equivalent to remake the field connections.

13 Inspect the bearings for roughness of running wear.

14 If necessary, clean the slip ring with very fine glass paper.

15 Using a 110 volt AC supply and a 15 watt test lamp between one of the slip rings and one of the rotor lobes, check the field winding insulation.

16 Check the field winding resistance (between the slip rings) against that specified.

17 Check the stator windings for continuity between each lead using a 12 volt DC supply and a 36 watt test lamp.

18 Check the stator winding insulation between the stator lamination pack and any one of the 3 cables using 110 volts AC and a 15 watt test lamp.

19 Check the 9 diodes between each diode pin and its heat sink for current flow in one direction only using 12 volts DC and a 1·5 watt test lamp. Renew the rectifier assembly if any diode is faulty.

20 Remove the single screw retaining the regulator to the brush box. Note the fitted position of the coloured leads then remove the screws and terminal strips that retain the brushes.

21 Remove the brushes, noting the leaf spring at the side of the inner brush.

22 Assembly is essentially the reverse of the dismantling procedure, but the following points must be carefully noted:

(a) If a rectifier unit is renewed, the replacement unit must be identical to that which was removed

(b) Connect the black regulator earth lead to the screw that retains the brushbox assembly to the end bracket. If there is no earth cable, the regulator earths through its case as indicated by a copper strip between the top bracket inner surface and the lead clamp. When fitting this type of regulator, ensure that a good earth exists

(c) Connect the regulator F terminal green lead under the inner brush retaining plate (on some versions this lead is replaced by a metal connecting strip)

(d) Connect the + terminal yellow lead above the outer brush retaining plate ensuring that the conductor cannot bridge the two brush retaining plates

(e) If a battery positive (B+) red lead is fitted on the regulator, connect it under the screw that retains the B+ spade connector, or to the middle positive heat sink plate side spade connector of the rectifier assembly, as appropriate

(f) Connect the avalanche diode to the outer heat sink plate side spade connector, or to the screw that retains the outer brush and the yellow regulator lead, as appropriate; an extension link and an additional screw may be provided

(g) Support the inner track of the bearing when refitting the rotor to the end bracket

(h) Use only M grade 45/55 tin-lead solder or equivalent when remaking the stator to rectifier connections

(j) Tighten the pulley nut to 25 lbf ft (3·46 kgf m) torque

23 After fitting the alterator, check the output as described in Section 7.

11 Starter motor – general description

The type of starter motor fitted can be either the inertia or pre-engaged type.

Both starter motors are interchangeable and engage with a common flywheel starter ring gear. The relay for the inertia starter motor is mounted next to the ignition coil whereas the pre-engaged type has the solenoid switch on the top of the motor.

The principle of operation of the inertia type starter motor is as follows: When the ignition switch is turned, current flows from the battery to the starter motor solenoid switch which causes it to become energized. Its internal plunger moves inwards and closes an internal switch so allowing full starting current to flow from the battery to the starter motor. This creates a powerful magnetic field to be induced into the field coils which causes the armature to rotate.

Mounted on helical spines is the drive pinion which, because of the sudden rotation of the armature, is thrown forwards along the armature shaft and so into engagement with the flywheel ring gear.

The engine crankshaft will then be rotated until the engine starts to operate on its own and, at this point, the drive pinion is thrown out of mesh with the flywheel ring gear.

The pre-engaged starter motor operates by a slightly different method but still using end face commutator brushes instead of brushes located on the side of the commutator.

The method of engagement on the pre-engaged starter differs considerably in that the drive pinion is brought into mesh with the starter ring gear before the main starter current is applied.

When the ignition is switched on, current flows from the battery to the solenoid which is mounted on the top of the starter motor body. The plunger in the solenoid moves inwards so causing a centrally pivoted lever to move in such a manner that the forked end pushes the drive pinion into mesh with the starter ring gear. When the solenoid plunger reaches the end of its travel, it closes an internal contact and full starting current flows to the starter field coils. The armature is then able to rotate the crankshaft so starting the engine.

A special one-way clutch is fitted to the starter drive pinion, so that when the engine fires and starts to operate on its own it does not drive the starter motor.

12 Starter motor (M35J) – testing on engine

1 If the starter motor fails to operate, then check the condition of the battery by turning on the headlamps. If they glow brightly for several seconds and then gradually dim, the battery is in an uncharged condition.

2 If the headlamps glow brightly and it is obvious that the battery is in good condition, then check the tightness of the battery wiring connections (and in particular the earth lead from the battery terminal to its connection on the body frame). Check the tightness of the connections at the relay switch and at the starter motor. Check the wiring with a voltmeter for breaks or shorts due to failure of insulation.

3 If the wiring is in order, then check that the starter motor switch is operating. To do this, press the rubber covered button in the centre of the relay switch located next to the ignition coil, if it is working the starter motor will be heard to click as it tries to rotate. Alternatively, check it with a voltmeter.

4 If the battery is fully charged with wiring in order, and the switch working, but the starter motor fails to operate, then it will have to be removed from the car for examination. Before this is done however, ensure that the starter pinion has not jammed in mesh with the flywheel. Check by turning the square end of the armature shaft with a spanner. This will free the pinion if it is stuck in engagement with the flywheel teeth.

13 Starter motor (M35J) – removal and refitting

1 Disconnect the negative and then the positive terminals from the battery. Also disconnect the starter motor cable from the terminal on the starter motor end cover (Fig. 10.2).

2 Undo and remove the bolt and spring washer and the nut and spring washer that secure the starter motor to the engine backplate.

3 Lift the starter motor away by manipulating the drivegear out from the ring gear area and then from the engine compartment.

4 Refitting is the reverse sequence to removal. Make sure that the starter motor cable, when secured in position by its terminal retaining nut, does not touch any part of the body or power unit which could damage the insulation. Fit the engine earth strap to the lower bolt.

14 Starter motor (M35J) – dismantling and reassembly

1 With the starter motor on the bench, first mark the relative positions of the starter motor body to the two end brackets.

2 Undo and remove the two screws and spring washers that secure the drive end bracket to the body. The drive end bracket, complete with armature and drive, may now be drawn forwards from the starter motor body (Fig. 10.3).

3 Lift away the thrustwasher from the commutator end of the armature shaft.

4 Undo and remove the two screws that secure the commutator end bracket to the starter motor body. The commutator end bracket may now be drawn back about an inch, allowing sufficient access so as to

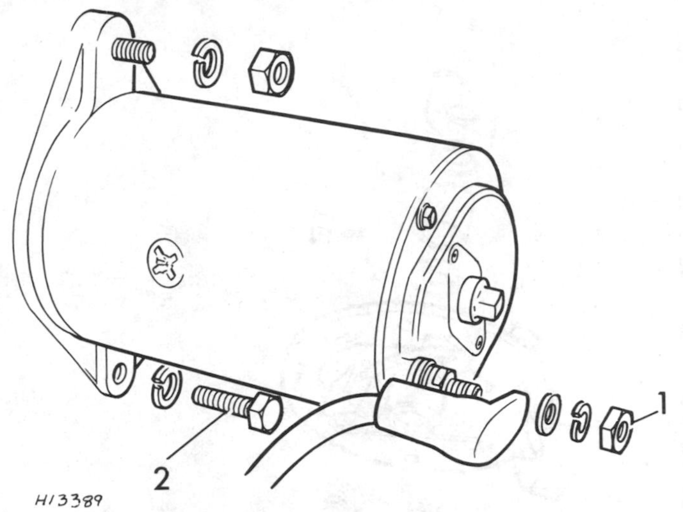

Fig. 10.2 M35J starter motor removal (Sec 13)

1 Cable terminal nut *2 Securing bolt*

disengage the field bushes from the bracket. Once these are free, the end bracket may now be completely removed.

5 With the motor stripped, the brushes and brush gear may be inspected. To check the brush spring tension, fit a new brush into each holder in turn and, using an accurate spring balance, push the brush on the balance tray until the brush protrudes approximately 0·063 in (1·6 mm) from the holder. Make a note of the reading which should be approximately 28 oz (0·8 kg). If the spring pressures vary considerably, the commutator end bracket must be renewed as a complete assembly.

6 Inspect the brushes for wear and fit new brushes if the ones fitted are nearing the minimum length of 0·375 in (9·53 mm). To renew the end bracket brushes, cut the brush cables from the terminal posts and, with a small file or hacksaw, slot the head of the terminal posts to a sufficient depth to accommodate the new leads. Solder the new brush leads to the posts.

7 To renew the field winding brushes, cut the brush leads approximately 0·25 in (6·35 mm) from the field winding junction and carefully solder the new brush leads to the remaining stumps, making sure that the insulation sleeves provide adequate cover.

8 If the commutator surface is dirty or blackened, clean it with a petrol dampened rag. Carefully examine the commutator for signs of excessive wear, burning or pitting. If evident, it may be reconditioned by having it skimmed at the local engineering works or BL dealer who possesses a centre lathe. The thickness of the commutator copper must not be less than 0·08 in (2·0 mm). For minor reconditioning, the commutator may be polished with glass paper. *Do not undercut the mica insulators between the commutator segments.*

9 With the starter motor dismantled, test the field coils for open circuit. Connect a 12 volt battery, with a 12 volt bulb in one of the leads, between each of the field brushes and a clean part of the body. The lamp will light if continuity is satisfactory between the brushes, windings and body connection.

10 Renewal of the field coils calls for the use of a wheel operated screwdriver, a soldering iron and caulking and riveting operations. This is beyond the scope of the majority of owners. The starter motor body should be taken to an automotive electrical engineering works for new field coils to be fitted. Alternatively, purchase an exchange Lucas starter motor.

11 Check the condition of the brushes, they should be renewed when they are sufficiently worn to allow visible side movement of the armature shaft.

12 To renew the commutator end bracket bush, drill out the rivets that secure the brush box moulding and remove the moulding, bearing seal retaining plate and felt washer seal.

13 Screw in a 0·5 in (12·7 mm) tap and withdraw the bush with the tap.

14 As the bush is of the phosphor bronze type, it is essential that it is allowed to stand in engine oil for at least 24 hours before fitment. Alternatively soak in oil at 100°C for 2 hours.

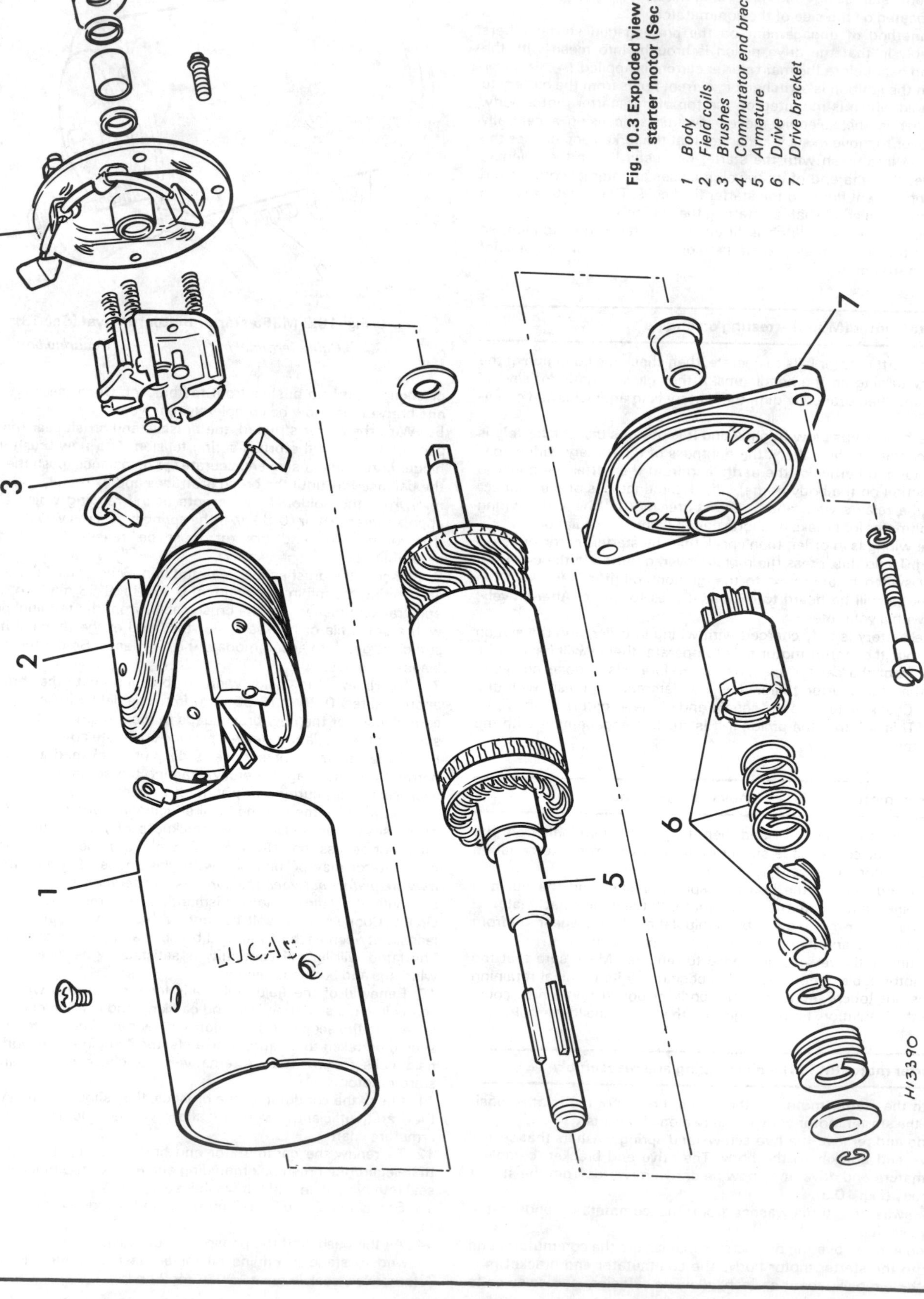

Fig. 10.3 Exploded view of M35J starter motor (Sec 14)

1 *Body*
2 *Field coils*
3 *Brushes*
4 *Commutator end bracket*
5 *Armature*
6 *Drive*
7 *Drive bracket*

15 Using a suitable diameter drift, drive the new bush into position. Do not ream the bush as its self lubricating properties will be impaired.
16 To remove the drive end bracket bush, it will be necessary to remove the drivegear as described in paragraphs 18 and 19.
17 Using a suitable diameter drift, remove the old bush and fit a new one as described in paragraphs 14 and 15.
18 To dismantle the starter motor drive, first use a press to push the retainer clear of the circlip which can then be removed. Lift away the retainer and main spring.
19 Slide off the remaining parts with a rotary action of the armature shaft.
20 It is most important that the drivegear is completely free from oil, grease and dirt. With the drivegear removed, clean all parts thoroughly in paraffin. *Under no circumstances oil the drive components.* Lubrication of the drive components could easily cause the pinion to stick.
21 Reassembly of the starter motor drive is the reverse sequence to dismantling. Use a press to compress the spring and retainer sufficiently to allow a new circlip to be fitted to its groove on the shaft. Remove the drive from the press.
22 Reassembly of the starter motor is the reverse sequence to dismantling.

15 Starter motor (M35JPE and 2M100) – testing on engine

1 The testing procedure is basically similar to the inertia engagement type as described in Section 12. However, note the following instructions before finally deciding to remove the starter motor.
2 Ensure that the pinion gear has not jammed in mesh with the flywheel due either to a broken solenoid spring or dirty pinion gear splines. To release the pinion, engage top gear and, with the ignition switched off, rock the car backwards and forwards. This should release the pinion from mesh with the ring gear. If the pinion still remains jammed, the starter motor must be removed for further examination.

16 Starter motor (M35JPE and 2M100) – removal and refitting

1 Disconnect the negative terminal from the battery.
2 Remove the securing nut and disconnect the top heavy duty cable (Fig. 10.4).
3 Disconnect the two leads from the Lucar terminals at the rear of the solenoid. There is no need to disconnect the lower heavy duty cable at the rear of the solenoid.
4 Undo and remove the two bolts that secure the starter motor to the flywheel housing. Note the earth strap fitted at the bottom bolt.
5 Remove the bolts that secure the starter motor bracket to the engine and lift out the starter motor.
6 Refitting is the reverse of the removal procedure. Do not forget the earth strap when refitting the bottom bolt.

17 Starter motor (M35J pre-engaged) – dismantling and reassembly

1 Detach the heavy duty cable that links the solenoid STA terminal to the starter motor terminal by undoing and removing the securing nuts and washers (Fig. 10.5).
2 Undo and remove the two nuts and spring washers that secure the solenoid to the drive end bracket.
3 Carefully withdraw the solenoid coil unit from the drive end bracket.
4 Lift off the solenoid plunger and return spring from the engagement lever.
5 Remove the rubber sealing block from the drive end bracket.
6 Remove the retaining ring (spire nut) from the engagement lever pivot pin and withdraw the pin.
7 Unscrew and remove the two drive end bracket securing nuts and spring washers and withdraw the bracket.
8 Lift away the engagement lever from the drive operating plate.
9 Extract the split-pin from the end of the armature and remove the shim washers and thrust plate from the commutator end of the armature shaft.
10 Remove the armature, together with its internal thrustwasher.
11 Withdraw the thrustwasher from the armature.

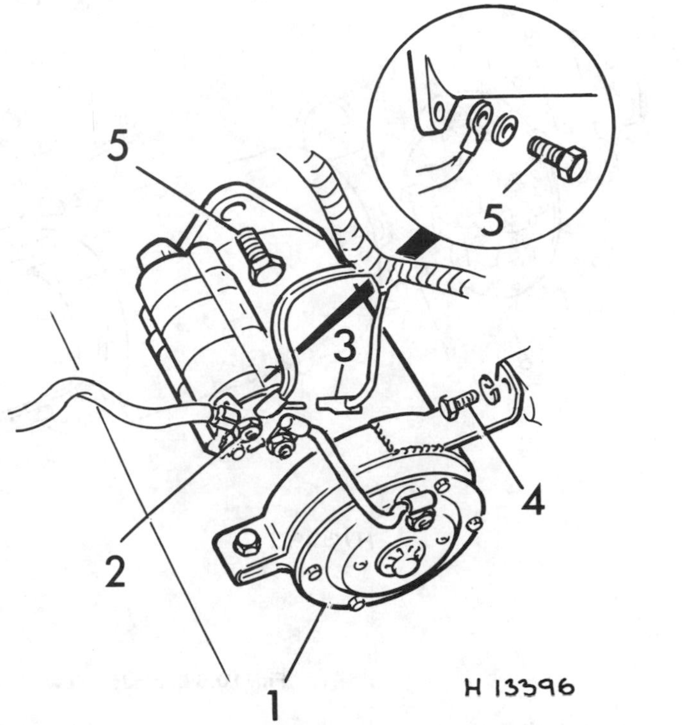

Fig. 10.4 M35JPE and 2M100 starter motor removal (Sec 16)

1 *Starter motor*
2 *Top heavy duty cable*
3 *Lucar connectors*
4 *Starter motor bracket securing bolt*
5 *Starter motor to flywheel housing bolt*

12 Undo and remove the two screws that secure the commutator end bracket to the starter motor body.
13 Carefully detach the end bracket from the yoke, at the same time disengaging the field brushes from the brush gear. Lift away the end bracket.
14 Move the thrust collar clear of the jump ring and then remove the jump ring. Withdraw the drive assembly from the armature shaft.
15 Inspection and renovation is basically the same as for the Lucas 2M100/PE starter motor and full information will be found in Section 18. The following necessitated by the fitting of the solenoid coil should be noted:
16 If a bush is worn, so allowing excessive side movement of the armature shaft, the bush must be renewed. Drift out the old bush with a piece of suitable diameter rod, preferably with a shoulder on it to stop the bush collapsing.
17 Soak a new bush in engine oil for 24 hours or if time does not permit, heat in an oil bath at 100°C for 2 hours prior to fitting.
18 As new bushes must not be reamed after fitting they must be pressed into position using a small mandrel of the same diameter as the bush and with a shoulder on it. Place the bush on the mandrel and press into position using a bench vice.
19 Use a test light and battery to test the continuity of the coil windings between terminal STA and a good earth point on the solenoid body. If the light fails to come on, the solenoid should be renewed.
20 To test the solenoid contacts for correct opening and closing, connect a 12 volt battery and a 60 watt test light between the main unmarked Lucar terminal and the STA terminal. The light should not come on.
21 Energise the solenoid with a separate 12 volt supply connected to the small unmarked Lucar terminal and a good earth on the solenoid body.
22 As the coil is energised, the solenoid should be heard to operate and the test lamp should light with full brilliance.
23 The contacts may only be renewed as a set, ie moving and fixed contacts. The fixed contacts are part of the moulded cover.

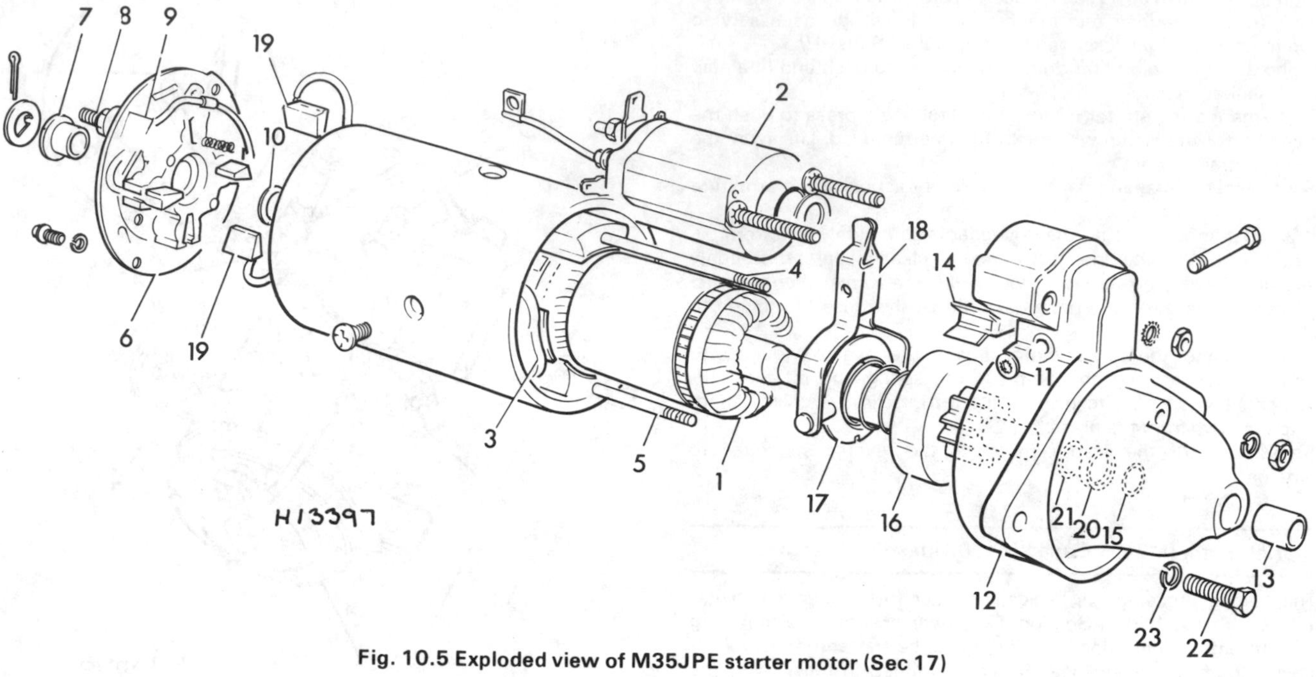

Fig. 10.5 Exploded view of M35JPE starter motor (Sec 17)

1	*Armature*	7	*Commutator end bracket bush*	12	*Drive end bracket*	18	*Lever and pivot assembly*
2	*Solenoid*	8	*Field terminal*	13	*End bracket bush*	19	*Brush*
3	*Field coil*	9	*Terminal insulating bush*	14	*Grommet*	20	*Thrust collar*
4	*Pole piece and long stud*	10	*Thrust plate*	15	*Jump ring*	21	*Shim*
5	*Pole piece and short stud*	11	*Pivot pin retaining clip*	16	*Roller clutch drive*	22	*Fixing bolt*
6	*Commutator end bracket*			17	*Bearing bush*	23	*Lock washer*

24 To fit a new set of contacts, first undo and remove the moulded cover securing screws.
25 Unsolder the coil connections from the cover terminals.
26 Lift away the cover and moving contact assembly.
27 Fit a new cover and moving contact assembly, soldering the connections to the cover terminals.
28 Refit the moulded cover securing screws.
29 Whilst the motor is apart, check the operation of the drive clutch. It must provide instantaneous take up of the drive in one direction and rotate easily and smoothly in the opposite direction.
30 Make sure that the drive moves smoothly on the armature shaft splines without binding or sticking.
31 Reassembly of the starter motor is the reverse sequence to dismantling. The following additional points should be noted:
32 When assembling the drive, always use a new retaining ring (spire nut) to secure the engagement lever pivot pin.
33 Make sure that the internal thrustwasher is fitted to the commutator end of the armature shaft before the armature is fitted.
34 Make sure that the thrust washers and plate are assembled in the correct order and are prevented from rotating separately by engaging the collar pin with the locking piece on the thrust plate.

18 Starter motor (2M100 pre-engaged) – dismantling and reassembly

1 Undo and remove the nut and spring washer that secure the connecting link between the solenoid and starter motor at the solenoid STA terminal. Carefully ease the connecting link out of engagement of the terminal post on the solenoid (Fig. 10.6).
2 Undo and remove the two nuts and spring washers that secure the solenoid to the drive end bracket.
3 Carefully ease the solenoid back from the drive end bracket, lift the solenoid plunger and return spring from the engagement lever and completely remove the solenoid.
4 Recover the shaped rubber block that is placed between the solenoid and starter motor body.
5 Carefully remove the end cap seal from the commutator end cover.
6 Ease the armature shaft retaining ring (spire nut) from the armature shaft. **Note:** *The retaining ring must not be reused but a new one obtained ready for fitting.*
7 Undo and remove the two long through-bolts and spring washers.
8 Detach the commutator end cover from the yoke, at the same time disengaging the filed brushes from the brush box moulding.
9 Lift away the thrustwasher from the armature shaft.
10 The starter motor body may now be lifted from the armature and drive end assembly.
11 Ease the retaining ring (spire nut) from the engagement lever pivot pin. **Note**: *The retaining ring must not be reused but a new one obtained ready for fitting.*
12 Using a parallel pin punch of suitable size, remove the pivot pin from the engagement lever and drive end bracket.
13 Carefully move the thrust collar clear of the jump ring and slide the jump ring from the armature shaft.
14 Slide off the thrust collar and finally remove the roller clutch drive and engagement lever assembly from the armature shaft.
15 With the motor stripped, the brushes and brush gear may be inspected. To check the brush spring tension, fit a new brush into each holder in turn and, using an accurate spring balance, push the brush on the balance tray until the brush protrudes approximately $\frac{1}{16}$ in (1·5 mm) from the holder. Make a note of the reading which should be approximately 36 oz (1·02 kg). If the spring pressures vary considerably, the commutator end bracket must be renewed as a complete assembly.
16 Inspect the brushes for wear and fit new brushes if the ones fitted are nearing the minimum length of $\frac{3}{8}$ in (10 mm). To renew the end bracket brushes, cut the brush cables from the terminal posts and, with a small file or hacksaw, slot the head of the terminal posts to a sufficient depth to accommodate the new leads. Solder the new brush leads to the posts.
17 To renew the field winding brushes, cut the brush leads approximately $\frac{1}{4}$ in (6·35 mm)) from the field winding junction and carefully solder the new brush leads to the remaining stumps, making sure that the insulation sleeves provide adequate cover.

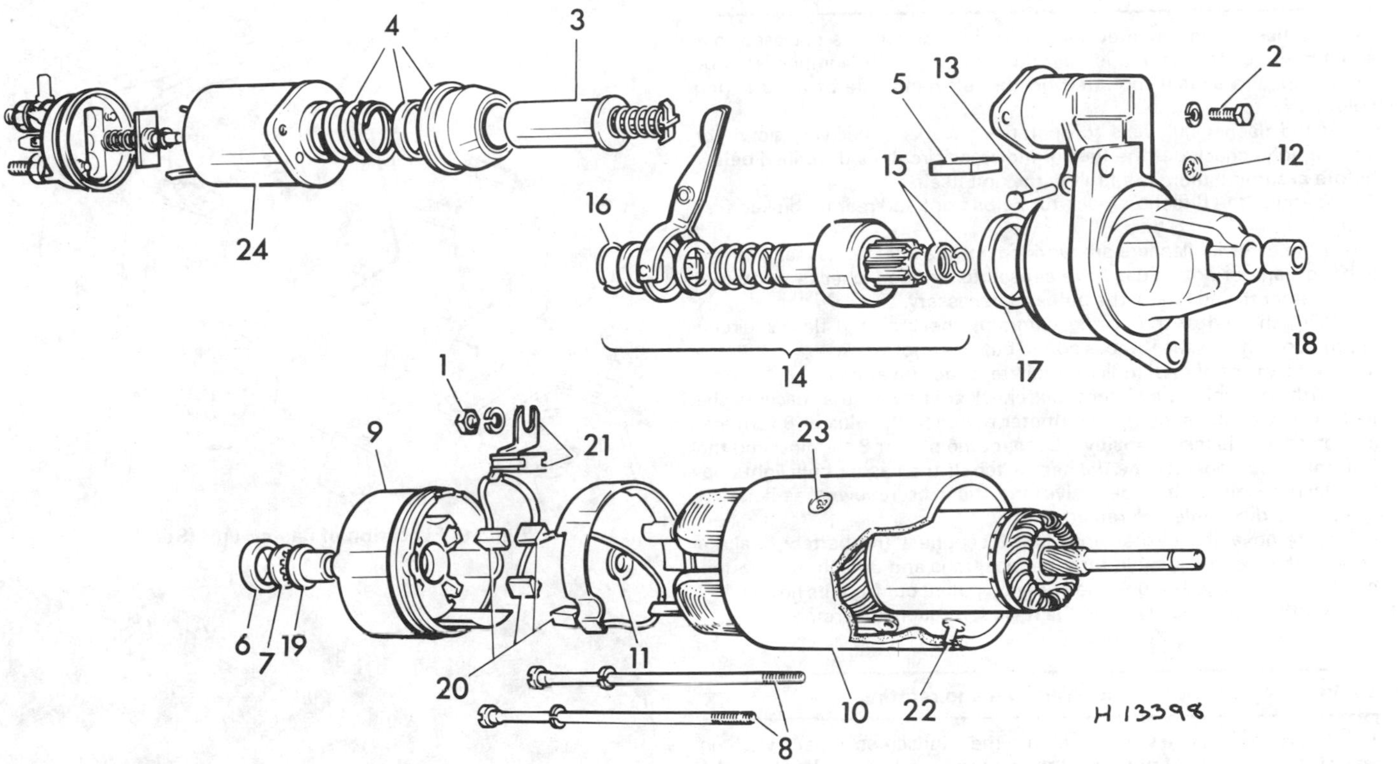

Fig. 10.6 Exploded view of 2M100 starter motor (Sec 18)

1 *Connecting link securing nut*
2 *Solenoid to drive end bracket securing set-screw*
3 *Solenoid plunger*
4 *Solenoid plunger return spring, spring seat and dust excluder*
5 *Rubber grommet*
6 *Armature end cap seal*
7 *Armature shaft retaining ring (spire nut)*
8 *Through-bolts*
9 *Commutator end cover*
10 *Yoke*
11 *Thrust washer*
12 *Retaining ring (spire nut)*
13 *Engagement lever pivot pin*
14 *Armature and roller clutch drive assembly*
15 *Thrust collar and jump ring*
16 *Spring ring*
17 *Dirt seal*
18 *Drive end bracket armature shaft bush*
19 *Commutator end cover armature shaft bush*
20 *Field coil brushes*
21 *Terminal and rubber grommet*
22 *Rivet*
23 *Pole shoe retaining screw*
24 *Solenoid*

18 If the commutator surface is dirty or blackened, clean it with a petrol dampened rag. Carefully examine the commutator for signs of excessive wear, burning or pitting. If evident, it may be reconditioned by having it skimmed at the local engineering works or BL dealer who possesses a centre lathe. The thickness of the commutator must not be less than 0·14 in (3·5 mm). For minor reconditioning, the commutator may be polished with glass paper. *Do not undercut the mica insulators between the commutator segments.*
19 With the starter motor dismantled, test the field coils for open circuit. Connect a 12 volt battery with a 12 volt bulb in one of the leads between each of the field brushes and a clean part of the body. The lamp will light if continuity is satisfactory between the brushes, windings and body connection.
20 Renewal of the field coils calls for the use of a wheel operated screwdriver, a soldering iron,and caulking and riveting operations. This is beyond the scope of the majority of owners. The starter motor body should be taken to an automobile electrical engineering works for new field coils to be fitted. Alternatively, purchase an exchange Lucas starter motor.
21 Check the condition of the bushes. They should be renewed when they are sufficiently worn to allow visible side movement of the armature shaft.
22 To renew the commutator end bracket bush, drill out the rivets that secure the brush box moulding and remove the moulding, bearing seal retaining plate and felt washer seal.
23 Screw in a ½ in tap and withdraw the bush with the tap.
24 As the bush is of the phosphor bronze type, it is essential that it is allowed to stand in engine oil for at least 24 hours before fitment. Alternatively, soak in oil at 100°C for 2 hours.
25 Using a suitable diameter drift, drive the new bush into position. Do not ream the bush as its self lubricating properties will be impaired.
26 To renew the drive end bracket bush, drive out the old bush with a suitable diameter drift and fit a new one as described in paragraphs 24 and 25.
27 Whilst the motor is apart, check the operation of the drive clutch. It must provide instantaneous take up of the drive in one direction and rotate easily and smoothly in the opposite direction.
28 Make sure that the drive moves smoothly on the armature shaft splines without binding or sticking.
29 Reassembling the starter motor is the reverse sequence to dismantling. The following additional points should be noted:
30 When assembling the drive end bracket, always use a new retaining ring (spire nut) to secure the engagement lever pivot pin.
31 Make sure that the internal thrustwasher is fitted to the commutator end of the armature shaft before the armature end cover is fitted.
32 Always use a new retaining ring (spire nut) on the armature shaft. There should be a maximum clearance of 0·010 in (0·25 mm) between the retaining ring and the bearing shoulder. This will be the armature endfloat.
33 Tighten the through-bolts to a torque wrench setting of 8 lbf ft (1·1 kgf m) and the nuts that secure the solenoid to the drive bracket to 4·5 lbf ft (0·6 kgf m).

19 Flasher unit and circuit – fault tracing and rectification

The flasher unit located as shown in Fig. 10.7, is enclosed in a small metal container and is operated only when the ignition is on by the composite switch mounted on the left-hand side of the steering column.

If the flasher unit fails to operate or works either very slowly or very rapidly, check out the flasher indicator circuit as described below before assuming there is a fault in the unit itself.

1 Examine the direction indicator bulbs front and rear for broken filaments.

2 If the external flashers are working but the internal flasher warning lights on one or both sides have ceased to function, check the internal bulb filaments and renew the bulbs as necessary.

3 With the aid of the wiring diagram, check all the flasher circuit connections if a flasher bulb is sound but does not work.

4 In the event of total indicator failure, check fuse No 1.

5 With the ignition switched on, check that current is reaching the flasher unit by connecting a voltmeter between the plus or B terminal and earth. If this test is positive, connect the plus or B terminal and the L terminal and operate the flasher switch. If the flasher bulb lights up, the flasher unit itself is defective and must be renewed as it is not possible to dismantle and repair it.

6 To remove the flasher unit, first disconnect the battery. Make a note of the electrical cable terminal positions and detach the two terminal connections. The unit may now be pulled out from its holder.

7 Refitting the flasher unit is the reverse sequence to removal.

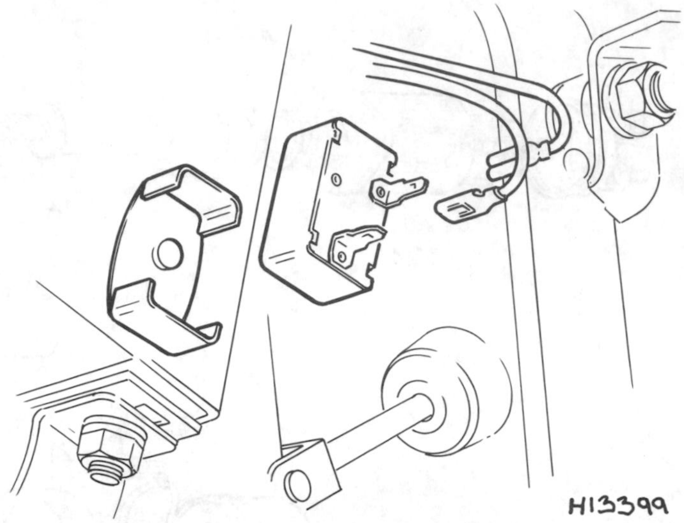

Fig. 10.7 Location of flasher unit (Sec 19)

20 Windscreen wiper arms – removal and refitting

1 Before removing a wiper arm, turn the windscreen wiper switch on and off to ensure the arms are in their normal parked position with the blades parallel with the bottom of the windscreen.

2 To remove the arm, pivot the arm back and pull the wiper arm head off the splined drive, at the same time easing back the clip with a screwdriver (photos).

3 When refitting an arm, place it so it is in the correct relative parked position and then press the arm head onto the splined drive until the retaining clip clicks into place.

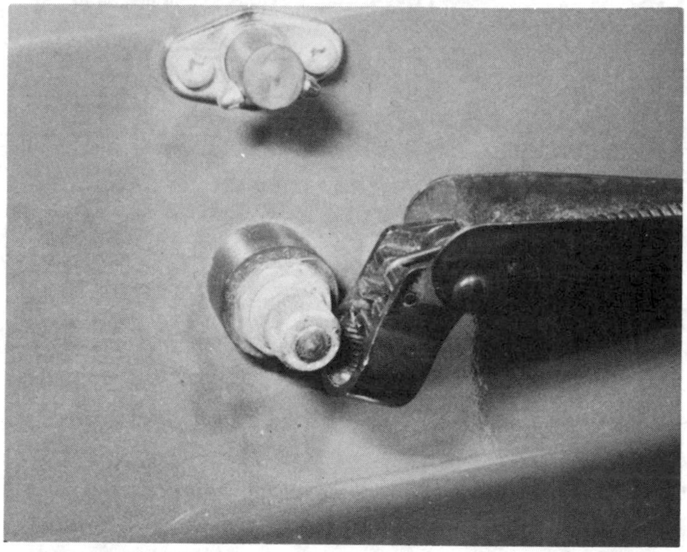

20.2a The tailgate wiper arm removed from the splined drive

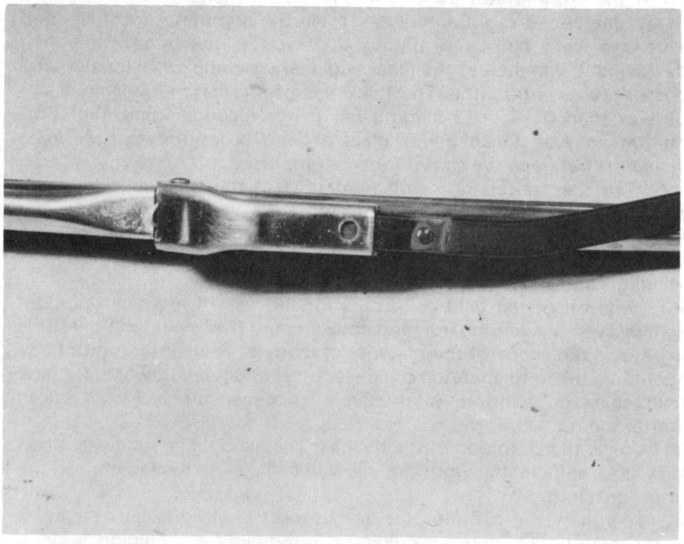

20.2b The windscreen wiper blade is removed by holding the retainer clip away from the arm and pulling the blade out

21 Windscreen wiper mechanism – fault diagnosis and rectification

1 Should the windscreen wipers fail or work very slowly, then check the terminals for loose connections and make sure the insulation of the external wiring is not broken or cracked. If this is in order, then check the current the motor is taking by connecting up an ammeter in the circuit and turning on the wiper switch. Consumption should be 1·5 amps for normal speed or 2·5 amps for high speeds.

2 If no current is passing, check the No 3 fuse. If the fuse has blown, renew it after having checked the wiring to the motor and other electrical circuits serviced by this fuse for short circuits. Further information will be found in Section 49. If the fuse is in good condition, check the wiper switch. Should the wiper take a very high current, check the wiper blades for freedom of movement. If this is satisfactory, check the wiper motor and drive cable for signs of damage. Measure the endfloat which should be between 0·002 to 0·008 in (0·051 to 0·203 mm). The endfloat is set by the thrust screw. Check that excessive friction in the cable connecting tubes (caused by too small a curvature) is not the cause of the high current consumption.

3 If the motor takes a very low current, ensure that the battery is fully charged. Check the brush gear after removing the commutator yoke assembly, and ensure that the brushes are free to move. If necessary, renew the tension springs. If the brushes are very worn they should be replaced with new ones. The armature may be checked by substitution.

22 Windscreen wiper blades – changing wiper arc

If it is wished to change the area through which the wiper blades move, this is simply done by removing each arm in turn from each splined drive and then refitting it on the drive in a slightly different position.

23 Windscreen wiper motor – removal and refitting

1 Refer to Section 20 and remove the wiper arms and blades.
2 Undo and remove the screw and plain washer that secure the wiper motor clamp to the body valance (photo). Release the clamp and rubber moulding, by pressing the clamp band into the release slot (Fig. 10.8).
3 Undo the wiper drive tube securing nut and slide the nut down the tube.
4 Next disconnect the electrical cable plug from the motor socket.
5 Lift the motor clear of the body valance, whilst at the same time pulling the inner cable from the tube.
6 Refitting the wiper motor and inner cable is the reverse sequence to removal. Take care in feeding the inner cable through the outer tube and engaging the inner cable with each wiper wheelbox spindle. Lubricate the inner cable with general purpose grease.

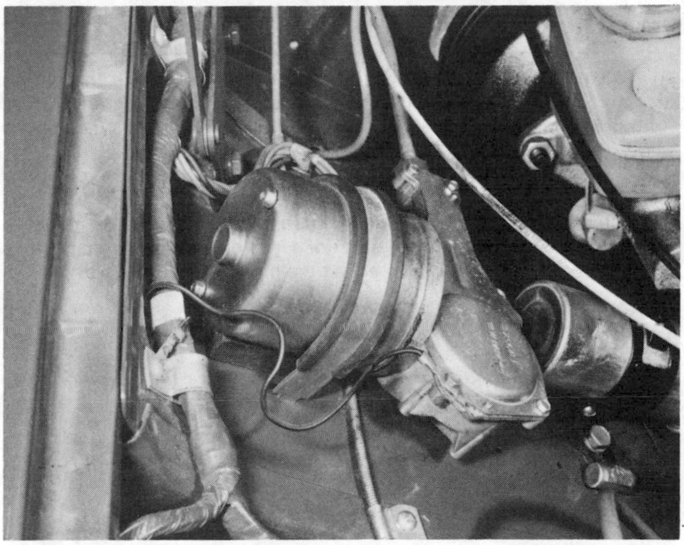

23.2 The windscreen wiper motor is mounted on the right-hand wing

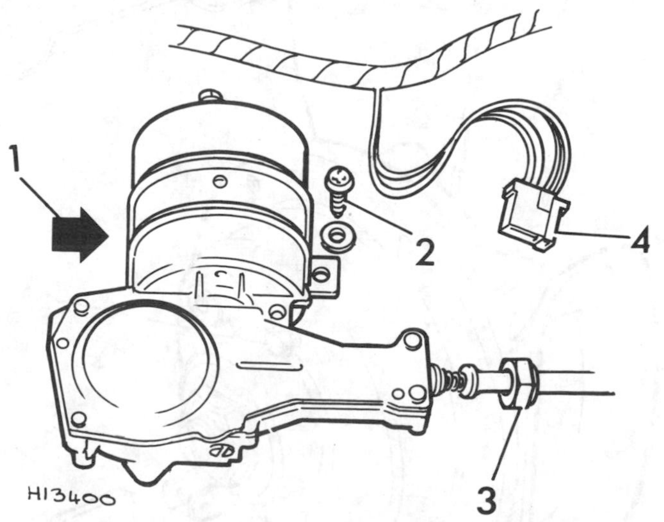

Fig. 10.8 Windscreen wiper motor removal (Sec 23)

1 Clamp band
2 Clamp securing screw
3 Drive tube securing nut
4 Electrical connector

24 Windscreen wiper motor – dismantling, inspection and reassembly

The only repair which can be effectively undertaken by the do-it-yourself mechanic to a wiper motor is brush renewal. Anything more serious than this will mean either exchanging the complete motor or having a repair done by an auto electrician. Spare part availability is really the problem. Brush renewal is described as follows:
1 Refer to Fig. 10.9, remove the four gearbox cover retaining screws and lift away the cover. Release the circlip and flat washer that secure the connecting rod to the crankpin on the shaft and gear. Lift away the connecting rod followed by the second flat washer.
2 Release the circlip and flat washer that secure the shaft and gear to the gearbox body.
3 De-burr the gear shaft and lift away the gear, making careful note of the location of the dished washer.
4 Scribe a mark on the yoke assembly and gearbox to ensure correct reassembly and unscrew the two yoke bolts from the motor yoke assembly. Part the yoke assembly, including the armature, from the gearbox body. As the yoke assembly has residual magnetism, ensure that the yoke is kept well away from metallic dust.
5 Unscrew the two screws that secure the brush gear and the terminal and switch assembly and remove both the assemblies.
6 Inspect the brushes for excessive wear. If the main brushes are worn to a limit of $\frac{3}{16}$ in (4·76 mm) or the narrow section of the third brush is worn to the full width of the brush, fit a new brush gear assembly. Ensure that the three brushes move freely in their boxes.
7 Reassembly at this stage is a straight reversal of disassembly.

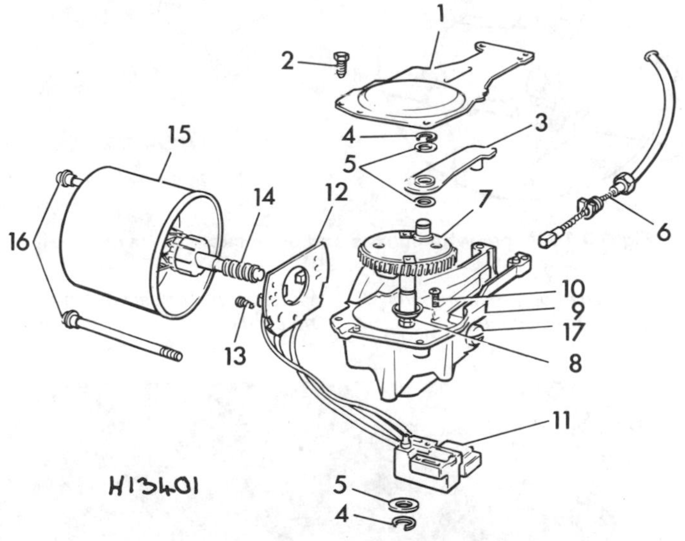

Fig. 10.9 Exploded view of windscreen wiper motor (Sec 24)

1 Gearbox cover
2 Screw for cover
3 Connecting rod
4 Circlip
5 Plain washers
6 Cross head
7 Shaft and gear
8 Dished washer
9 Gearbox
10 Screw for limit switch
11 Limit switch assembly
12 Brush gear
13 Screw for brush gear
14 Armature
15 Yoke assembly
16 Yoke bolts
17 Armature thrust screw

25 Windscreen wiper wheelboxes and drive cable tubes – removal and refitting

1 Refer to Section 22 and remove the windscreen wiper motor.
2 Refer to Chapter 12 and remove the instrument panel.
3 Refer to Chapter 12 and remove the glovebox.
4 Refer to Section 19 and remove the windscreen wiper arms.
5 Undo and remove the nuts that secure the wheelboxes to the body. Lift away the shaped spacer from each wheelbox (Fig. 10.10).
6 Slacken the two nuts that clamp the two wheelbox plates on the glovebox side. Carefully pull out the drive tube from the wheelbox.
7 Carefully remove the free drive tube and its grommet through the glovebox opening.

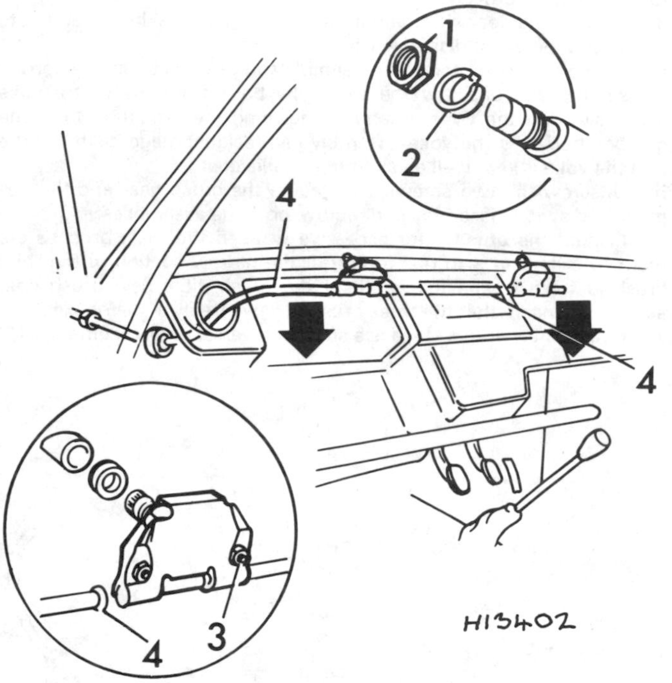

Fig. 10.10 Wiper wheelboxes and drive cable tubes removal (Sec 25)

1 Wheelbox securing nut
2 Shaped spacer
3 Wheelbox plates securing nut
4 Drive tube

26.2 The tailgate window wiper motor is mounted behind the trim panel

8 Lift away the two wheelbox units through the instrument panel opening.
9 Recover the spacer and washer from each wheelbox unit.
10 Refitting the wheelboxes and drive cable tubes is the reverse sequence to removal.

26 Tailgate wiper motor – removal and refitting

1 Remove the wiper arm and blades as described in Section 20.
2 Remove the tailgate trim pad (photo).
3 Disconnect the electrical cable plug from the motor socket.
4 Undo the wiper rack tube securing nut and slide the nut down the tube.
5 Undo and remove the six screws that secure the mounting plate to the tailgate and withdraw the cable rack from the tube.
6 Undo the motor clamp securing screw and lift away the motor assembly.
7 Refitting is the reverse of the removal procedure.

27 Horns – fault tracing and rectification

1 If a horn works badly or fails completely, first check the wiring leading to it for short circuits and loose connections. Also check that the horn is firmly secured and that there is nothing lying on the horn body.
2 The horn is protected by the No 5 fuse and if this has blown, the circuit should be checked for short circuits.
3 The horn should never be dismantled, but it is possible to adjust it. This adjustment is to compensate for wear of the moving parts only and will not affect the tone. To adjust the horn proceed as follows:

(a) There is a small adjustment screw on the broad rim of the horn, nearly opposite the two terminals (See Fig. 10.11). Do not confuse this with the large screw in the centre
(b) Turn the adjustment screw anti-clockwise until the horn just fails to sound. Then turn the screw a quarter of a turn clockwise which is the optimum setting

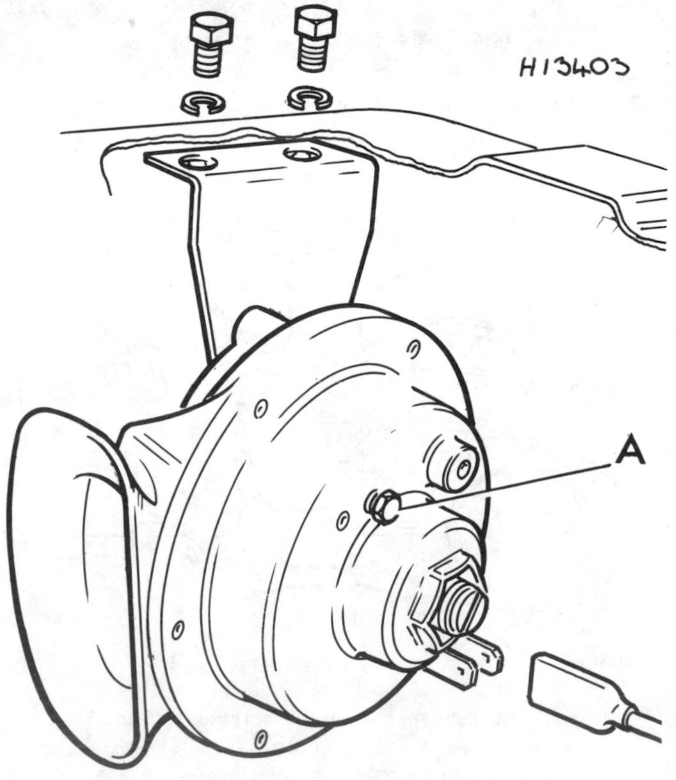

Fig. 10.11 The horn (Sec 27)
A Adjustment screw

(c) *It is recommended that if the horn has to be reset in the car, the No 5 fuse should be removed and temporarily replaced with a piece of wire, otherwise the fuse will continually blow due to the high current required for the horn in continual operation*

(d) *Should twin horns be fitted, the horn which is not being adjusted should be disconnected while adjustments of the other takes place*

28 Headlight units – removal and refitting

1 Sealed beam (or renewable bulb) light units are fitted.
2 The method of gaining access to the light unit for renewal is identical for all types of light units and bulbs.
3 Undo and remove the four screws that secure the top of the front grille to the body panel.
4 Carefully lift the grille outwards and upwards so releasing it from its locating holes in the body.
5 Undo and remove the three securing screws and lift away the headlamp rim (Fig. 10.12).
6 *Sealed beam unit:* Disconnect the plug from the rear of the light unit, detach the sidelamp holder assembly and then pull back the rubber cover and lift out the light unit.
7 *Spring clip bulb holder:* Disconnect the plug from the bulb holder, pull back the rubber cover then release the spring clip and lift out the bulb.
8 Refitting is the reverse of the removal procedure. Where a bulb is fitted, make sure that the locating clip or slot on the bulb is correctly located in the reflector.

29 Headlight beam – adjustment

The headlights may be adjusted for both vertical and horizontal beam positions by the two screws, these being shown in Fig. 10.12.

They should be set on a full or high beam, the beams are set slightly below parallel with a level road surface. Do not forget that the beam position is affected by how the car is normally loaded for night driving.

Although this adjustment can be set approximately at home, it is recommended that this be left to a local garage who will have the necessary equipment to do the job more accurately.

30 Front flasher bulb – renewal

1 Unscrew and remove the lens securing screws and lift away the lens.
2 The bulb has a bayonet fitting and is removed by pressing in and turning anti-clockwise.
3 Refitting the bulb and lens is the reverse of the removal procedure.

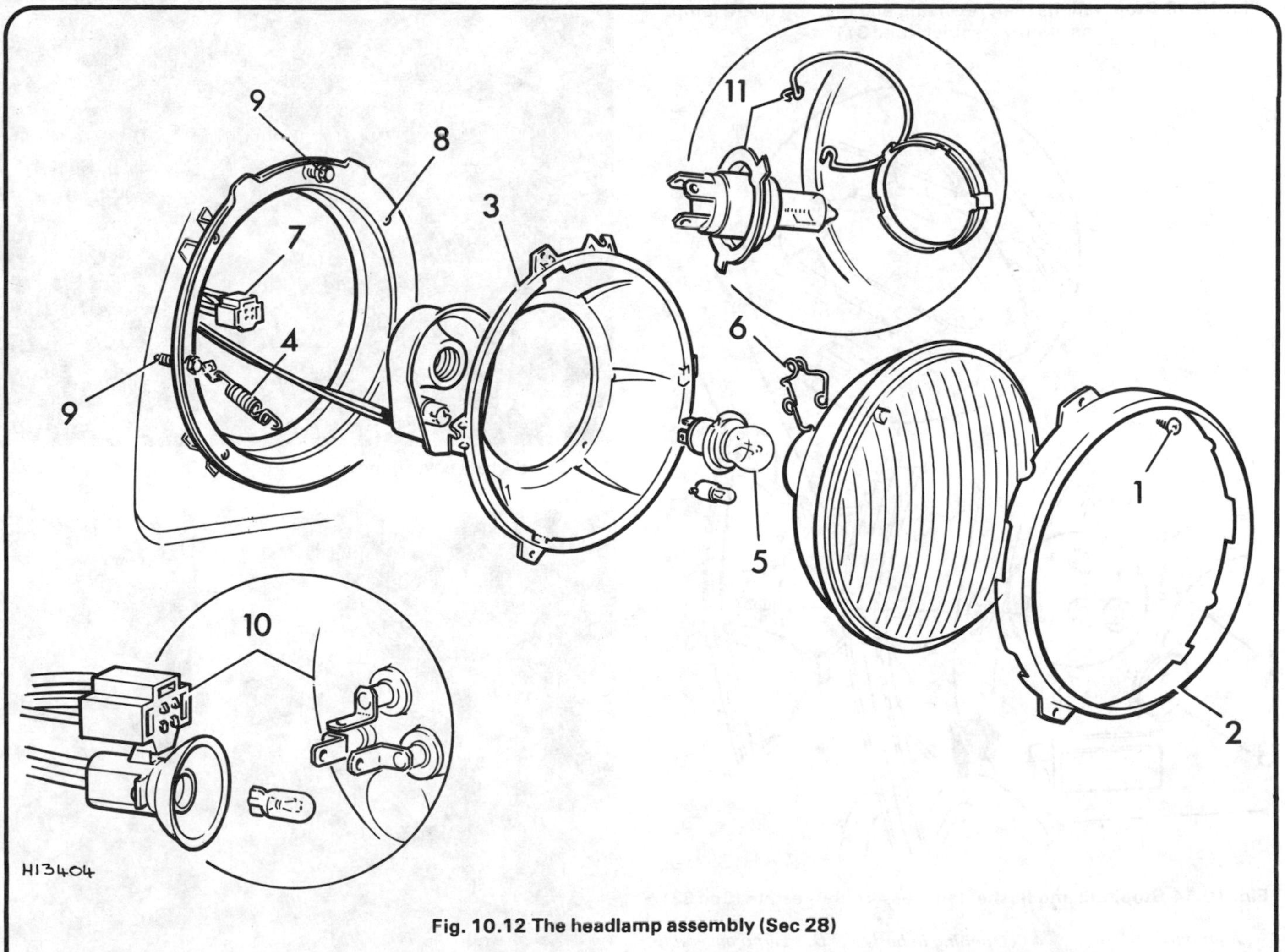

Fig. 10.12 The headlamp assembly (Sec 28)

1 *Rim securing screw*
2 *Rim*
3 *Inner shell*
4 *Inner shell tensioning spring*
5 *Bulb*
6 *Bulb clip*
7 *Electrical connector*
8 *Snap rivet*
9 *Vertical and horizontal adjustment screws*
10 *Sealed beam type fitting*
11 *Renewable bulb type fitting*

Fig. 10.13 Stop, tail, flasher, reversing and rear fog guard lamp assembly – saloon (Sec 31)

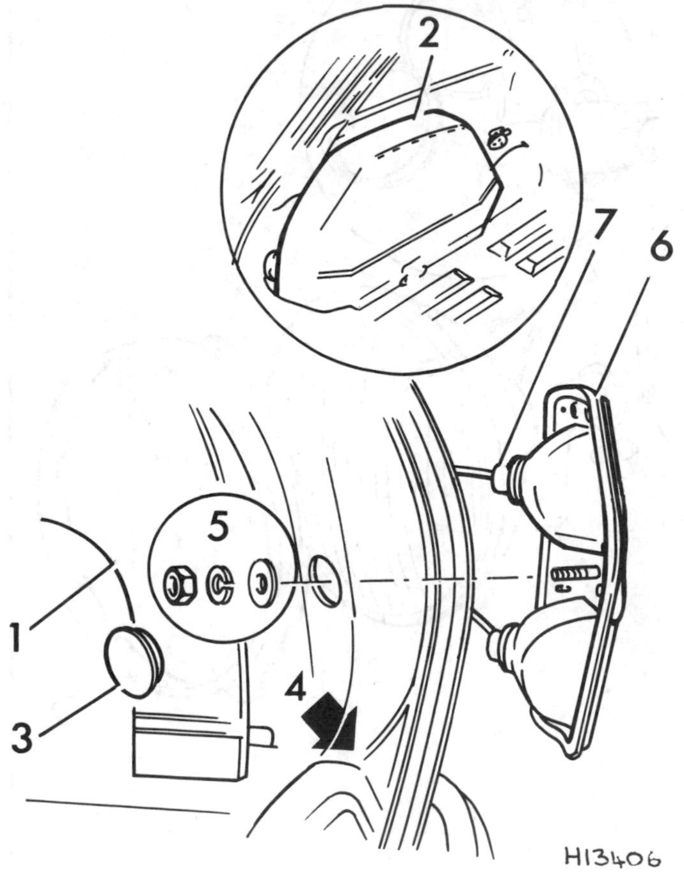

Fig. 10.14 Stop, tail and flasher lamp assembly – estate (Sec 32)

1 Trim pad
2 Spare wheel cover
3 Rubber plug
4 Opening in body
5 Nut and washers
6 Lamp assembly
7 Bulb holder

31 Stop, tail, flasher, reversing and rear fog guard lamp bulbs (Saloon) – renewal

1 Remove the six screws that secure the lamp cluster lens and lift away the lens (Fig. 10.13).
2 The bulb has a bayonet fitting and is removed by depressing and turning anti-clockwise.
3 Refitting is the reverse of the removal procedure but note that the stop/tail lamp bulb has offset pins on the bayonet cap.

32 Stop, tail and flasher lamp (Estate) – removal and refitting

1 *Right-hand side:* Remove the rear compartment trim pad.
2 *Left-hand side:* Remove the spare wheel cover and spare wheel (photo).
3 Remove the rubber plug located at the side of the light housing in the body.
4 Cover up the opening in the body directly below the rear of the light assembly (Fig. 10.14).
5 Undo and remove the nut, spring and plain washers that secure the light assembly to the body.
6 Withdraw the light assembly from the body.
7 Pull out the bulb holders from the rear of the light (photo).

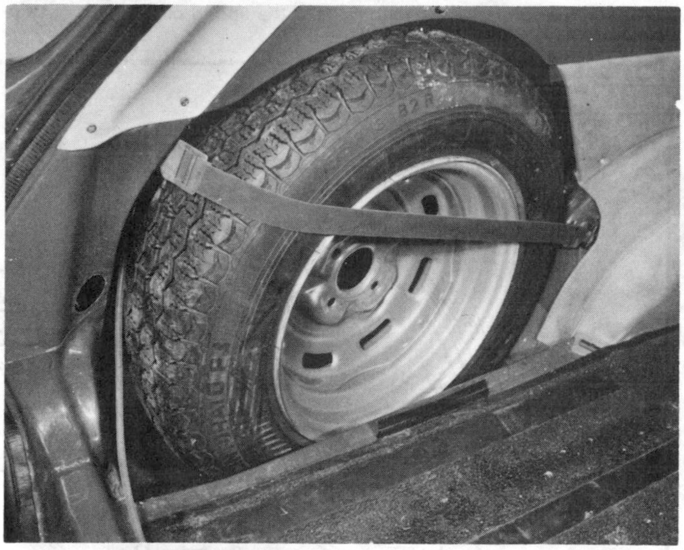

32.2 Spare wheel cover removed

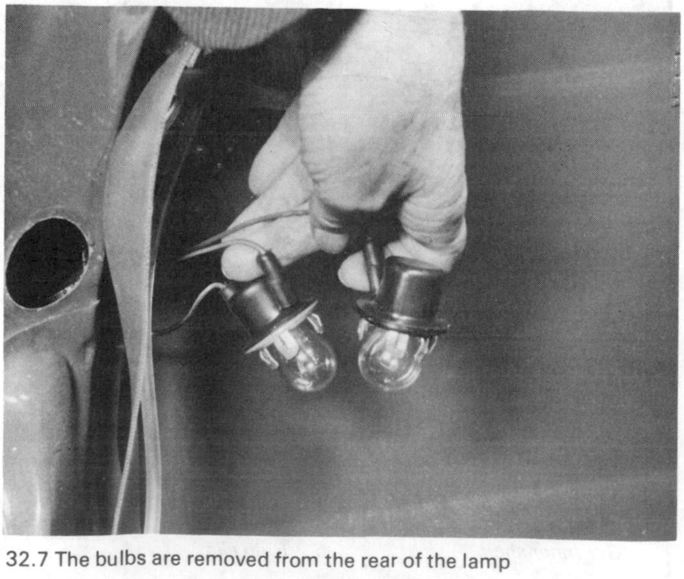

32.7 The bulbs are removed from the rear of the lamp

8 To remove the lens, undo and remove the six screws that secure the lens to the light body.
9 Lift away the lens, noting that the flasher lens laps over the stop and tail light lens and must be removed first.
10 Reassembling and refitting the light assembly is the reverse sequence to removal.

33 Rear number plate bulb – renewal

1 The lamp is retained in the rear bumper by two clips; release the clips by pressing them in and then pull the lamp out of the bumper (Fig. 10.15).
2 Withdraw the bulb holder, taking care nut to disturb the lens. Remove the bulb by pressing in and turning anti-clockwise.
3 Refitting is the reverse of the removal procedure. Ensure that the lamp retaining lugs are engaged correctly, then press the lamp into position on the bumper (photo).

34 Panel illumination lamp bulb – renewal

1 Disconnect the battery.
2 Refer to Chapter 12 and remove the instrument panel.
3 Pull out the bulb holder from the rear of the panel giving access to the bulb (photo).
4 Carefully pull on the bulb and it will be released from its holder. It will be noticed that capless bulbs are used.
5 Refitting is the reverse sequence to removal.

35 Interior lamp – removal and refitting

1 Disconnect the battery negative terminal.
2 Squeeze the lens together and detach it from the lamp body (Fig. 10.16).

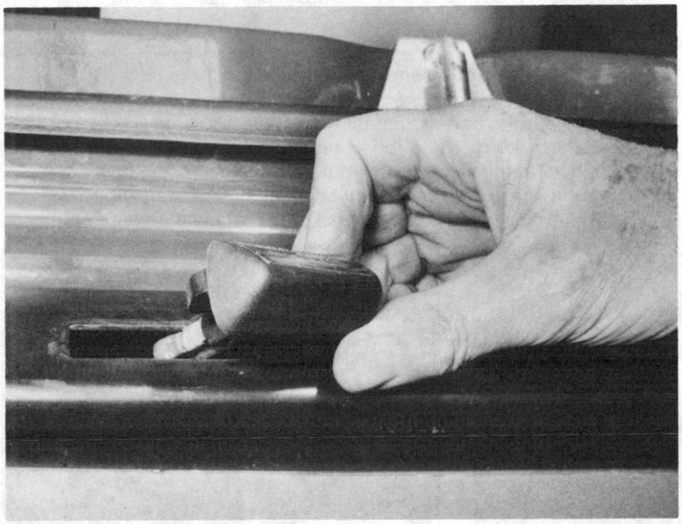

33.3 Press the lamp into the bumper

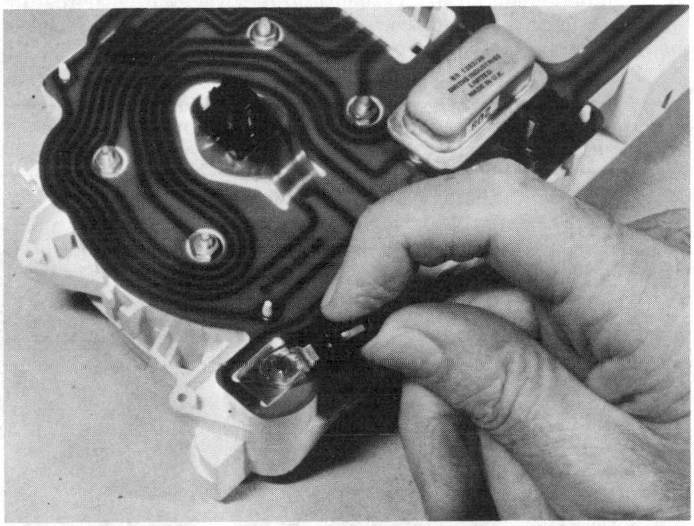

34.3 Panel light bulb holders pull out of the rear of the instrument cluster

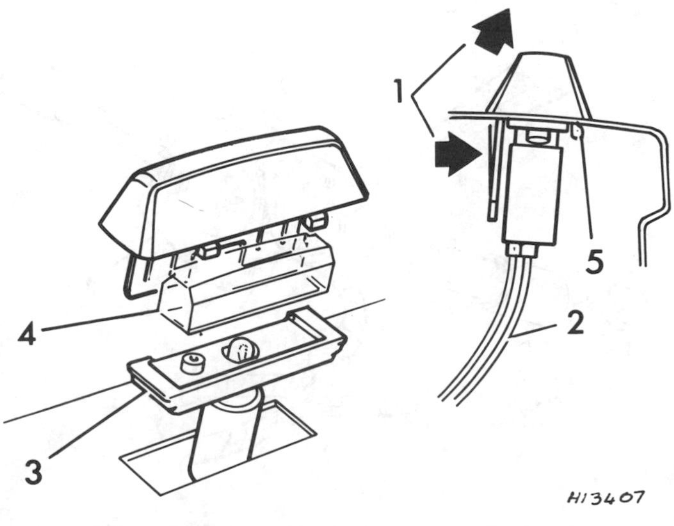

Fig. 10.15 Rear number plate lamp (Sec 33)

1 Retaining clip
2 Supply lead
3 Lamp body
4 Lens
5 Retaining lug

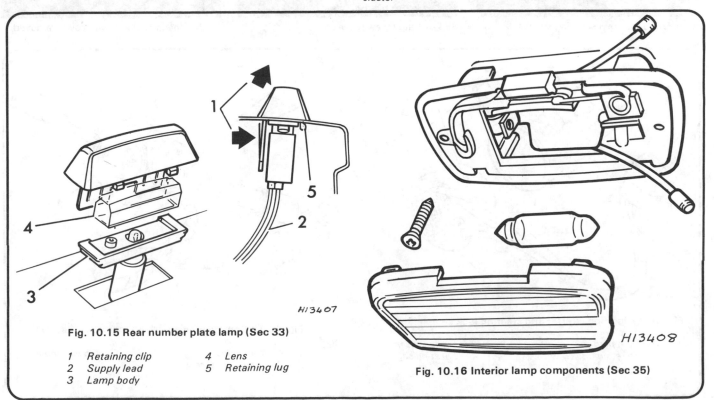

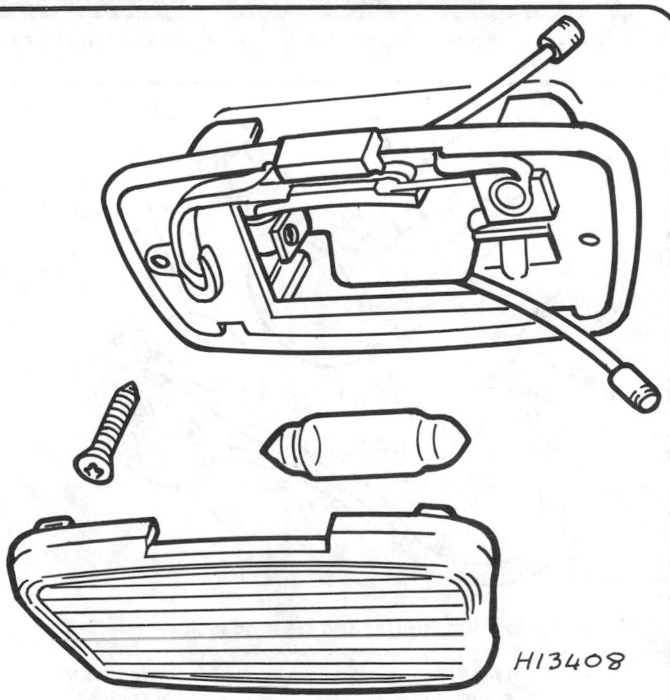

Fig. 10.16 Interior lamp components (Sec 35)

3 Extract the festoon type bulb and then remove the two lamp body retaining screws.
4 Carefully pull the lamp from the roof, at the same time disconnecting the two supply leads.
5 Refitting is a reversal of the removal procedure.

36 Ignition, starter and steering lock switch – removal and refitting

1 Disconnect the battery.
2 Undo and remove the four screws that secure the steering column cowls. Lift the cowls from over the switch stalks (photos).
3 Detach the multi-pin plug from the electrical leads to the switch assembly at the harness socket (Fig. 10.17).
4 Undo and remove the one screw that retains the switch assembly in the lock housing.
5 Slide the switch assembly from the lock housing.
6 Refitting is the reverse sequence to removal. Make sure that the locating peg on the switch correctly registers in the groove in the lock housing.

37 Headlight dip/flasher, horn, direction indicator switch – removal and refitting

1 Disconnect the battery and remove the steering column cowl.
2 Refer to Chapter 11 and remove the steering wheel.
3 Detach the multi-pin plug from the electrical leads to the switch assembly at the harness socket located under the facia (Fig. 10.18).
4 Slacken the switch clamp tightening screw located on the underside of the switch and ease the switch assembly from the steering column.
5 Refitting the switch assembly is the reverse sequence to removal. Make sure that the lug on the inner diameter of the switch locates in the slot in the outer steering column as shown in B, Fig. 10.19 and also that the striker dog on the nylon switch centre is in line with and towards the switch stalk A.

38 Lighting switch – removal and refitting

1 Compress the spring clips that retain the switch in the panel and push the switch out of the panel (Fig. 10.20).
2 Disconnect the plug connector from the rear of the switch and lift away the switch.
3 Refitting is the reverse of the removal procedure (photo).

39 Heater fan switch – removal and refitting

1 Disconnect the battery negative terminal.
2 Depress the upper and lower retaining springs on the rear of the switch body and press the switch out from the panel.
3 Disconnect the plug connector from the rear of the switch.
4 Refitting the switch is the reverse of the removal procedure.

36.2a Undo the cowl securing screws ...

36.2b ... and lift away the cowls

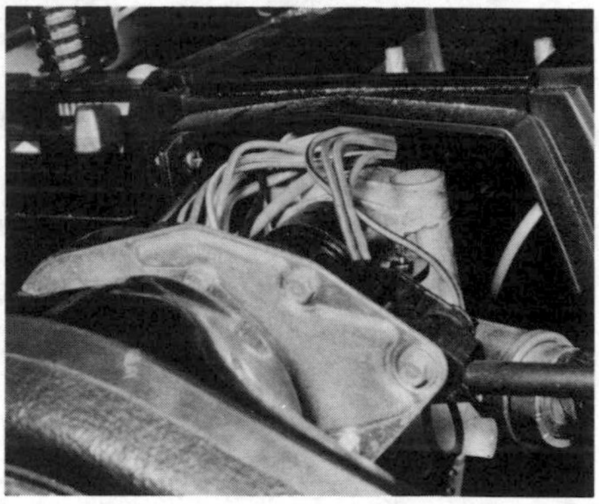
36.2c Steering column with cowls removed

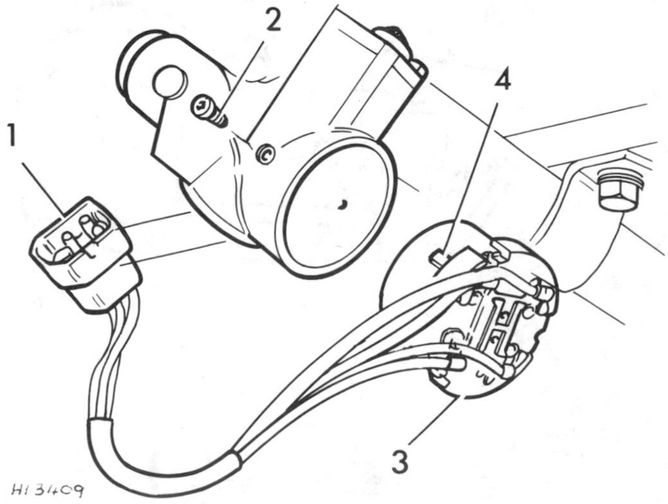

Fig. 10.17 Ignition, starter and steering lock switch (Sec 36)

1 Multi-pin plug
2 Switch retaining screw
3 Switch assembly
4 Locating peg

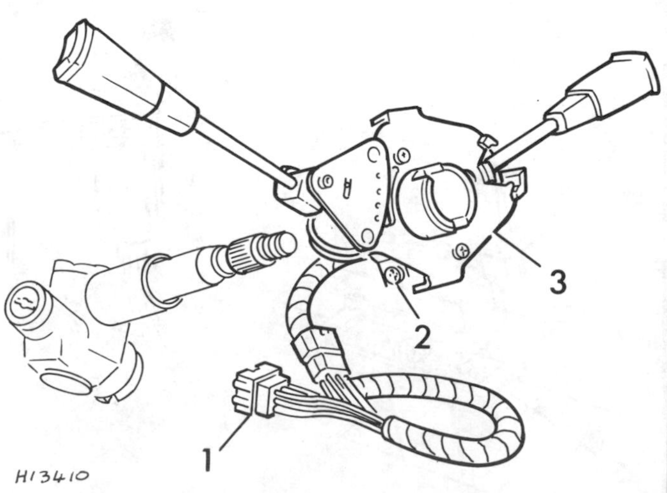

Fig. 10.18 Combination switch attachments (Sec 37)

1 Multi-pin plug
2 Switch clamp screw
3 Switch assembly

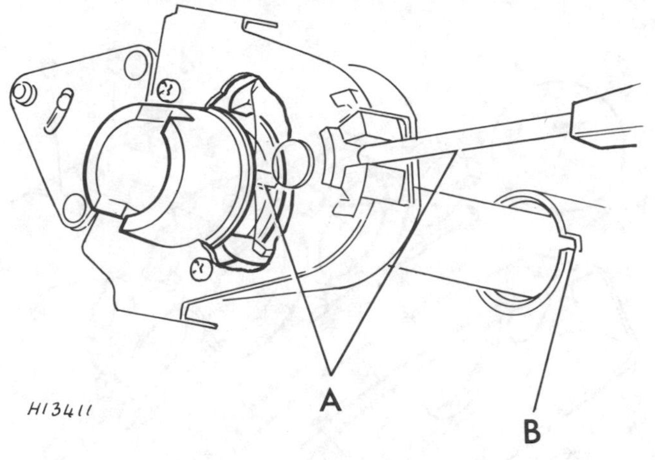

Fig. 10.19 Combination switch alignment (Sec 37)

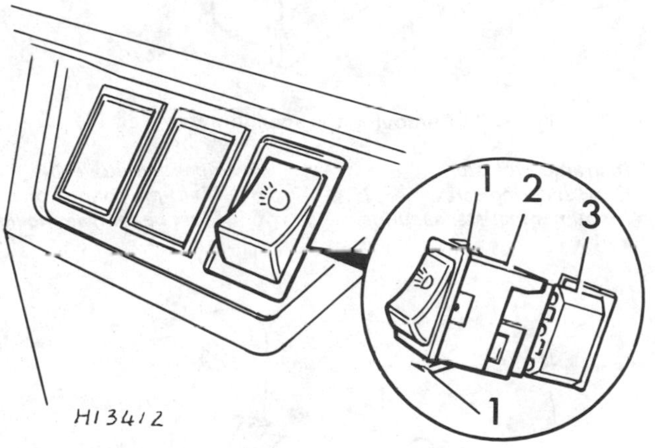

Fig. 10.20 Lighting switch removal (Sec 38)

1 *Spring clips*
2 *Lighting switch*
3 *Electrical connector*

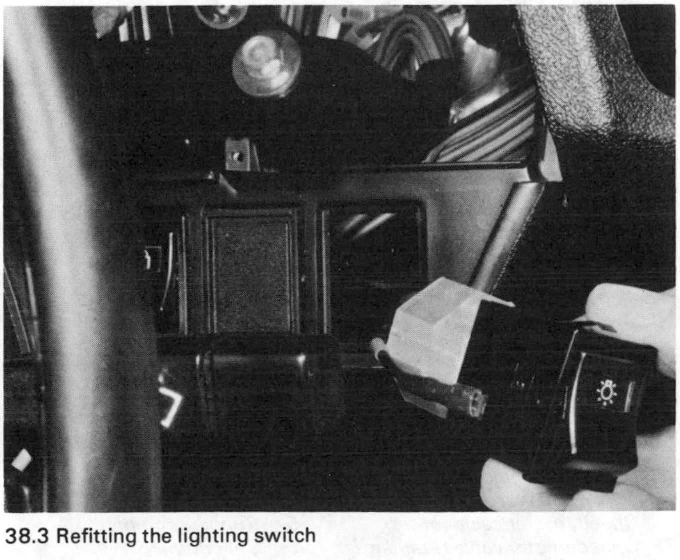

38.3 Refitting the lighting switch

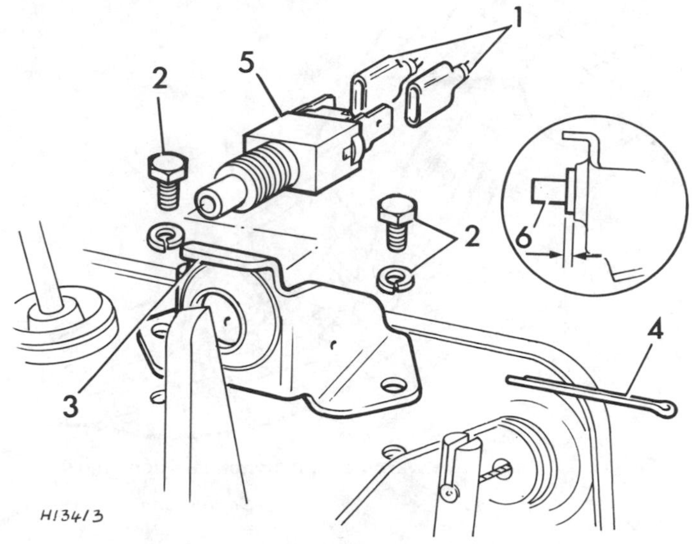

Fig. 10.21 Stop light switch removal (Sec 40)

1 *Switch terminal connections*
2 *Switch bracket securing screw and spring washer*
3 *Bracket*
4 *Split-pin*
5 *Switch*
6 *Switch adjustment*

40 Stop light switch – removal and refitting

1 Make a note of the two cable connections at the rear of the switch located on the top of the brake pedal mounting bracket. Detach the two terminals (Fig. 10.21).
2 Undo and remove the two bolts and spring washers that secure the switch mounting bracket to the brake pedal mounting bracket. Lift away the switch and bracket.
3 Straighten the legs of the switch locking split-pin and withdraw the split-pin.
4 The switch may now be unscrewed from its mounting bracket.
5 Refitting the stop light switch is the reverse sequence to removal. It is however, necessary to adjust the position of the switch when refitting to its mounting bracket.
6 Screw the switch into its mounting bracket until one complete thread of the switch housing is visible on the pedal side of the bracket. Lock with a new split-pin.

41 Handbrake warning light switch – removal and refitting

1 Remove the centre console as described in Chapter 12.
2 Remove the gaiter retainer securing screws and pull the gaiter up the handbrake lever. To provide access to the retainer securing screws, it may be necessary to make a small cut in the carpet at each corner of the handbrake assembly aperture in the floor.
3 Disconnect the switch supply lead, unscrew the switch securing screw and remove the switch (Fig. 10.22).
4 Refitting is the reverse of the removal procedure. To prevent entry of water from beneath the car, it is advisable to coat the gaiter securing screw holes with a suitable sealing compound.

42 Instrument panel printed circuit – removal and refitting

1 Refer to Chapter 12 and remove the instrument panel.
2 Withdraw the voltage stabilizer from the rear of the instrument panel printed circuit.
3 Withdraw the warning light and panel light bulb holders from the speedometer and gauges.
4 Remove the main beam warning light connections.
5 Remove the plastic retaining pegs.
6 Undo and remove the nuts that secure the gauges and printed

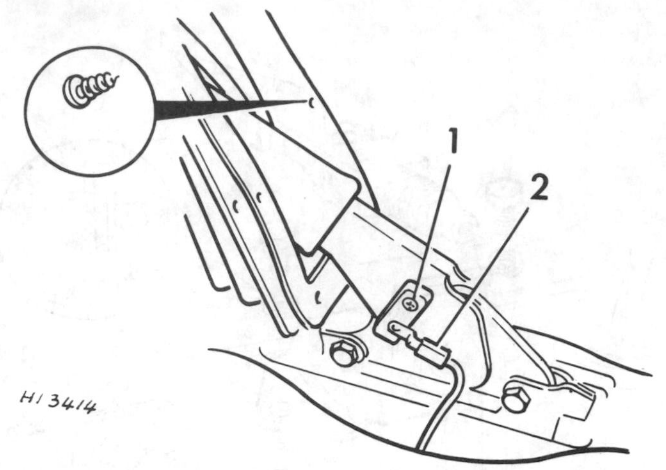

Fig. 10.22 Handbrake warning light switch location (Sec 41)

1 Switch *2 Supply lead*

circuit to the back of the instrument cluster.
7 Lift away the printed circuit.
8 Refitting the printed circuit is the reverse of the removal procedure.

43 Gauge units – removal and refitting

1 Remove the instrument panel as described in Chapter 12.
2 Remove the bulb holders.
3 Remove the two nuts and plastic peg that secure the temperature gauge and printed circuit to the instrument cluster housing, then undo and remove the two screws that secure the temperature gauge to the instrument cluster and lift away the gauge unit.
4 Undo and remove the two fuel gauge securing screws and lift out the fuel gauge.
5 Refitting is the reverse of the removal procedure.

44 Speedometer – removal and refitting

1 Disconnect the battery negative terminal.
2 Unscrew and remove the two instrument cowl retaining screws and withdraw the cowl (Fig. 10.23).
3 Depress the speedometer cable locking clip and release the cable from the speedometer head.
4 Unscrew and remove the instrument cluster retaining screws and pull the assembly forwards.
5 Note the location of the various leads and bulb holders, then disconnect them and withdraw the instrument cluster.
6 Remove the speedometer and instrument lens from its retaining clips followed by the face plate (where applicable).
7 Unscrew and remove the two speedometer retaining screws and carefully withdraw the speedometer from the cluster.
8 Refitting is the reverse of the removal procedure.

45 Speedometer cable assembly – removal and refitting

1 Working under the car, disconnect the speedometer cable from the gearbox by removing the screw that secures the cable locking flange. Then withdraw the cable from the drive pinion in the gearbox (Fig. 10.24).
2 Working from behind the instrument panel, release the speedometer cable from the speedometer head by depressing the cable locking clip and withdrawing the cable from the speedometer head.
3 The inner cable can now be withdrawn from the outer cable.
4 To remove the outer cable, pull it through the brake pedal housing.
5 Refitting is the reverse of the removal procedure. When refitting the outer cable, ensure that the rubber grommet is properly located in the brake pedal housing.

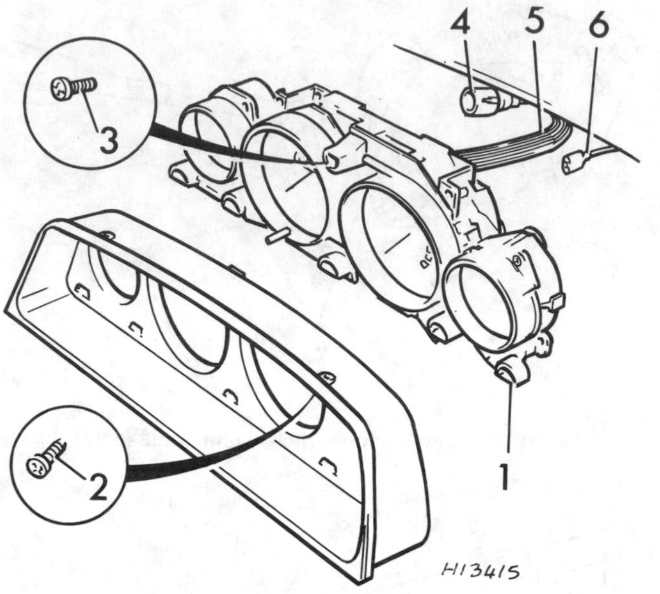

Fig. 10.23 Removing the speedometer (Sec 44)

1 Instrument cluster
2 Cowl securing screw
3 Instrument cluster securing screw
4 Speedometer cable
5 Multi-pin connector
6 Main beam warning light lead

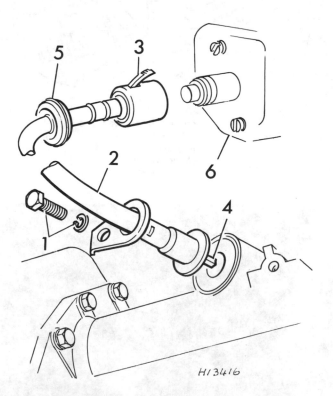

Fig. 10.24 Speedometer cable removal (Sec 45)

1 Locking flange securing screw and washer
2 Speedometer cable (outer)
3 Speedometer cable retaining clip
4 Speedometer cable (inner)
5 Rubber grommet
6 Speedometer

46 Tachometer – removal and refitting

1 Remove the instrument cluster as described in Chapter 12.
2 Unscrew and remove the three tachometer and printed circuit retaining nuts and withdraw the printed circuit.
3 Unscrew and remove the three tachometer retaining screws and withdraw the unit from the cluster.
4 Refitting is the reverse of the removal procedure.

47 Voltage stabilizer – removal and refitting

1 The voltage stabilizer is a push fit into the rear of the instrument panel printed circuit board.
2 Before removal, as a safety precaution disconnect the battery.
3 Carefully pull the voltage stabilizer from the rear of the printed circuit.
4 Refitting is the reverse sequence to removal. Note that the terminals of the stabilizer are offset so that it cannot be fitted the wrong way round.

48 Instrument operation – testing

1 The bi-metal resistance equipment for the fuel and thermal type temperature gauges comprises an indicator head and transmitter with the unit connected to a common voltage stabilizer. This item is fitted because the method of operation of the equipment is voltage sensitive and a voltage stabilizer is necessary to ensure a constant voltage supply at all times.
2 Special test equipment is necessary when checking the operation of the stabilizer, but should it be found that both the fuel and temperature gauges are reading inaccurately, it is worthwhile removing the stabilizer and tapping it firmly onto a hard surface; in many cases this will provide at the very least a temporary cure.
3 The gauges can be checked by applying 8 volts dc directly to their terminals; this can be done in-situ by removing the voltage stabilizer and connecting 8 volts (+) to the 1 socket on the rear of the printed circuit panel and 8 volts (–) to earth. The gauges should give a full scale deflection. If the leads to the thermal transmitter or fuel tank unit are disconnected, the gauges should not give any reading when the ignition is switched on. If these leads are then earthed, both gauges should give a full scale deflection.

49 Fuses

1 The fuse block is located inside the car at the back of the parcel shelf. Five fuses are used and protect the circuits as listed in the Specifications at the beginning of this Chapter.
2 If any of the fuses should blow, check the circuits protected by that fuse then trace and rectify the fault before renewing the fuse. Always fit a fuse of the correct rating.

50 Radios and tape players – fitting (general)

A radio or tape player is an expensive item to buy, and will only give its best performance if fitted properly. It is useless to expect concert hall performance from a unit that is suspended from the dashpanel by string with its speaker resting on the back seat or parcel shelf! If you do not wish to do the fitting yourself, there are many in-car entertainment specialists who will do the fitting for you.

Make sure the unit purchased is of the same polarity as the vehicle. Ensure that units with adjustable polarity are correctly set before commencing the fitting operations.

It is difficult to give specific information with regard to fitting, as final positioning of the radio/tape player, speakers and aerial is entirely a matter of personal preference. However, the following paragraphs give guidelines to follow which are relevant to all fittings:

Radios

Most radios are a standardised size of 7 in wide by 2in deep. This ensures that they will fit into the radio aperture provided in most cars. If your car does not have such an aperture, then the radio must be fitted in a suitable position either in or beneath the dashpanel. Alternatively, a special console can be purchased which will fit between the dashpanel and the floor or on the transmission tunnel. These consoles can also be used for additional switches and instrumentation if required. Where no radio aperture is provided, the following points should be borne in mind before deciding exactly where to fit the unit:

(a) The unit must be within easy reach of the driver wearing a seat belt
(b) The unit must not be mounted in close proximity to an electronic tachometer, the ignition switch and its wiring, or the flasher unit and associated wiring
(c) The unit must be mounted within reach of the aerial lead, and in such a place that the aerial lead will not have to be routed near the components detailed in the preceding paragraph (b)
(d) The unit should not be positioned in a place where it might cause injury to the car occupants in an accident; for instance, under the dashpanel above the driver's or passenger's legs
(e) The unit must be fitted securely

Some radios will have mounting brackets provided, together with instructions: others will need to be fitted using drilled and slotted metal strips, bent to form mounting brackets. These strips are available from most accessory shops. The unit must be properly earthed by fitting a separate earthing lead between the casing of the radio and the vehicle frame.

Use the radio manufacturers' instructions when wiring the radio into the vehicle's electrical system. If no instructions are available, refer to the relevant wiring diagram to find the location of the radio 'feed' connection in the vehicle's wiring circuit. A 1 to 2 amp 'in-line' fuse must be fitted in the radio's feed wire; a choke may also be necessary (see the following Section).

The type of aerial used and its fitted position, is a matter of personal preference. In general, the taller the aerial the better the reception. It is best to fit a fully retractable aerial, especially if a mechanical car-wash is used or if you live in an area where cars tend to be vandalised. In this respect, electric aerials which are raised and lowered automatically when switching the radio on or off are convenient, but are more likely to give trouble than the manual type.

When choosing a position for the aerial, the following points should be considered:

(a) The aerial lead should be as short as possible; this means that the aerial should be mounted at the front of the car
(b) The aerial must be mounted as far away from the distributor and HT leads as possible
(c) The part of the aerial which protrudes beneath the mounting point must not foul the roadwheels, or anything else
(d) If possible, the aerial should be positioned so that the coaxial lead does not have to be routed through the engine compartment
(e) The plane of the panel on which the aerial is mounted should not be so steeply angled that the aerial cannot be mounted vertically (in relation to the 'end-on' aspect of the car). Most aerials have a small amount of adjustment available

Having decided on a mounting position, a relatively large hole will have to be made in the panel. The exact size of the hole will depend upon the specific aerial being fitted, although generally, the hole required is of $\frac{3}{4}$ in (19 mm) diameter. On metal bodied cars, a 'tank-cutter' of the relevant diameter is the best tool to use for making the hole. This tool needs a small diameter pilot hole drilled through the panel, through which the tool clamping bolt is inserted. On GRP bodied cars, a 'hole-saw' is the best tool to use. Again, this tool will require the drilling of a small pilot hole. When the hole has been made, the raw edges should be de-burred with a file and then painted to prevent corrosion.

Fit the aerial according to the manufacturer's instructions. If the aerial is very tall, or if it protrudes beneath the mounting panel for a considerable distance, it is a good idea to fit a stay between the aerial and the vehicle frame. This stay can be manufactured from the slotted and drilled metal strips previously mentioned. The stay should be securely screwed or bolted in place. For best reception, it is advisable to fit an earth lead between the aerial and the vehicle frame; this is essential for GRP bodied cars.

It will probably be necessary to drill one or two holes through bodywork panels in order to feed the aerial lead into the interior of the car. Where this is the case, ensure that the holes are fitted with rubber grommets to protect the cable and to stop possible entry of water.

Positioning and fitting of the speaker depends mainly on its type. Generally, the speaker is designed to fit directly into the aperture already provided in the car. Where this is the case, fitting the speaker is just a matter of removing the protective grille from the aperture and screwing or bolting the speaker in place. Take great care not to damage the speaker diaphragm whilst doing this. It is a good idea to fit a 'gasket' between the speaker frame and the mounting panel. In order to prevent vibration, some speakers will already have such a gasket fitted.

If a 'pod' type speaker was supplied with the radio, this can be secured to the mounting panel with self-tapping screws.

When connecting a rear mounted speaker to the radio, the wires should be routed through the vehicle beneath the carpets or floor mats, preferably along the side of the floorpan where they will not be trodden on by passengers. Make the relevant connections as directed by the radio manufacturer.

By now you will have several yards of additional wiring in the car, use PVC tape to secure this wiring out of harm's way. Do not leave electrical leads dangling. Ensure that all new electrical connections are properly made (wires twisted together will not do) and completely secure.

The radio should now be working, but before you pack away your tools it will be necessary to 'trim' the radio to the aerial. If specific instructions are not provided by the radio manufacturer, proceed as follows: Find a station with a low signal strength on the medium-wave band, slowly turn the trim screw of the radio in or out until the loudest reception of the selected station is obtained. The set is then trimmed to the aerial.

Tape players

Fitting instructions for both cartridge and cassette stereo tape players are the same, and in general the same rules apply as when fitting a radio. Tape players are not usually prone to electrical interference like radio, although it can occur, so positioning is not so critical. If possible, the player should be mounted on an 'even-keel'. Also it must be possible for a driver wearing a seat belt to reach the unit in order to change or turn over tapes.

For the best results from speakers designed to be recessed into a panel, mount them so that the back of the speaker protrudes into an enclosed chamber within the car (eg door interiors or the boot cavity).

To fit recessed type speakers in the front doors, first check that there is sufficient room to mount the speakers in each door without it fouling the latch or window winding mechanism. Hold the speaker against the skin of the door and draw a line around the periphery of the speaker. With the speaker removed, draw a second 'cutting' line within the first to allow enough room for the entry of the speaker back, but at the same time providing a broad seat for the speaker flange. When you are sure that the 'cutting-line' is correct, drill a series of holes around its periphery. Pass a hacksaw blade through one of the holes and then cut through the metal between the holes until the centre section of the panel falls out.

De-burr the edges of the hole and then paint the raw metal to prevent corrosion. Cut a corresponding hole in the door trim panel, ensuring that it will be completely covered by the speaker grille. Now drill a hole in the door edge and a corresponding hole in the door surround. These holes are to feed the speaker leads through, so fit grommets. Pass the speaker leads through the door trim, door skin and out through the holes in the side of the door and door surround. Refit the door trim panel and then secure the speaker to the door using self-tapping screws. **Note:** *If the speaker is fitted with a shield to prevent water dripping on it, ensure that this shield is at the top.*

Pod type speakers can be fastened anywhere offering a corresponding mounting point on each side of the car. Pod speakers sometimes offer a better reproduction quality if they face the rear window which then acts as a reflector, so it is worthwhile to do a little experimenting before finally fixing the speaker.

51 Radios and tape players – suppression of interference (general)

To eliminate buzzes and other unwanted noises costs very little and is not as difficult as sometimes thought. With a modicum of common sense and patience, and following the instructions in the following paragraphs, interference can be virtually eliminated.

The first cause for concern is the generator. The noise this makes over the radio is like an electric mixer and the noise speeds up when the engine is revved. (To prove the point, remove the fanbelt and try it). The remedy for this is simple; connect a 1.0 to 3.0 mfd capacitor between earth (probably the bolt that holds down the generator base) and the positive (+) terminal on the alternator. This is most important, for if it is connected to the small terminal, the generator will probably be damaged permanently (see Fig. 10.25).

A second common cause of electrical interference is the ignition system. Here a 1.0 mfd capacitor must be connected between earth and the SW or + terminal on the coil (see Fig.10.26). This may stop the tick-tick sound that comes over the speaker. Next comes the spark itself.

There are several ways of curing interference from the ignition HT system. One is the use of carbon-cored HT leads as original equipment. Where copper cable is substituted, then resistive spark plug caps must be used (see Fig. 10.27). These should be of about 10 000 to 15 000 ohm resistance. If due to lack of room these cannot be used, an alternative is to use 'in-line' suppressors. If the interference is not too bad, it may be possible to get away with only one suppressor in the coil to distributor line. If the interference does continue (a 'clacking' noise), then modify all HT leads.

At this stage it is advisable to check that the radio is well earthed, also the aerial and to see that the aerial plug is pushed well into the set and that the radio is properly trimmed (see preceding Section). In addition, check that the wire which supplies the power to the set is as short as possible. At this stage it is a good idea to check that the fuse is of the correct rating. For most sets this will be about 1 to 2 amps.

At this point, the more usual causes of interference have been suppressed. If the problem still exists, a look at the cause of interference may help to pinpoint the component generating the stray electrical discharges.

The radio picks up electromagnetic waves in the air. Some are made by regular broadcasters and some, which we do not want, are made by the car itself. The home made signals are produced by stray electrical discharges floating around in the car. Common producers of these signals are electrical motors, ie the windscreen wipers, electric screen washers, electric window winders, heater fan or an electric aerial if fitted. Other sources of interference are flashing turn signals and instruments. The remedy for these cases is shown in Fig. 10.28, for an electric motor whose interference is not too bad and Fig. 10.29 for instrument suppression. Turn signals are not normally suppressed. In recent years, radio manufacturers have included in the live line of the radio, in addition to the fuse, an 'in-line' choke. If your circuit lacks one of these, put one in as shown in Fig. 10.30.

All the foregoing components are available from radio stores or accessory stores. If you have an electric clock fitted, this should be suppressed by connecting a 0.5 mfd capacitor directly across it as shown for a motor in Fig. 10.28.

If after all this you are still experiencing radio interference, first assess how bad it is, for the human ear can filter out unobtrusive unwanted noises quite easily. But if you are still adamant about eradicating the noise, then continue.

As a first step, a few 'experts' seem to favour a screen between the radio and the engine. This is OK as far as it goes, literally! The whole set is screened anyway and if interference can get past that then a small piece of aluminium is not going to stop it.

A more sensible way of screening is to discover if interference is coming down the wires. First, take the live lead; interference can get between the set and the choke (hence the reason for keeping the wires short). One remedy here is to screen the wire and this is done by buying screened wire and fitting that. The loudspeaker lead could be screened also to prevent 'pick-up' getting back to the radio although this is unlikely.

Without doubt, the worse source of radio interference comes from the ignition HT leads, even if they have been suppressed. The ideal way of suppressing these is to slide screening tubes over the leads themselves. As this is impractical, we can place an aluminium shield over the majority of the lead areas. In a vee or twin-cam engine this is relatively easy but for a straight engine, the results are not particularly good.

Now for the really impossible cases, here are a few tips to try out. Where metal comes into contact with metal, an electrical disturbance is caused which is why good clean connections are essential. To remove interference due to overlapping or butting panels, you must

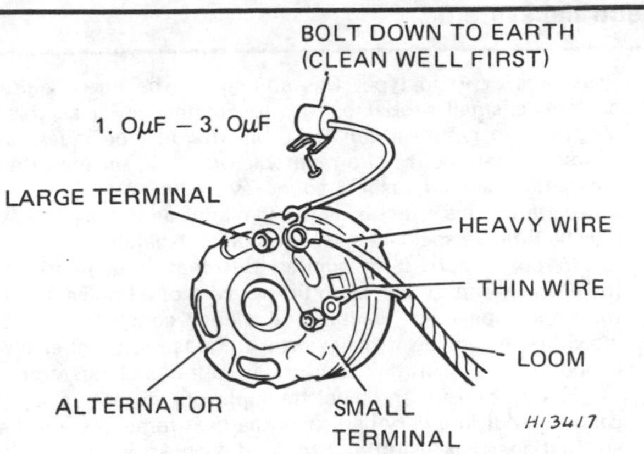

Fig. 10.25 The correct way to connect a capacitor to the alternator (Sec 51)

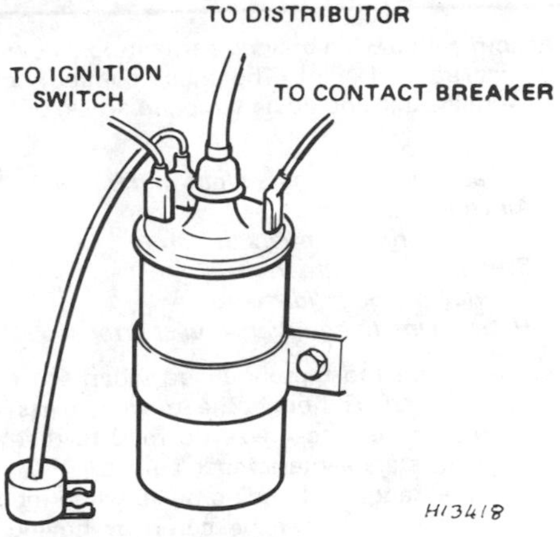

Fig. 10.26 The capacitor must be connected to the ignition side of the coil (Sec 51)

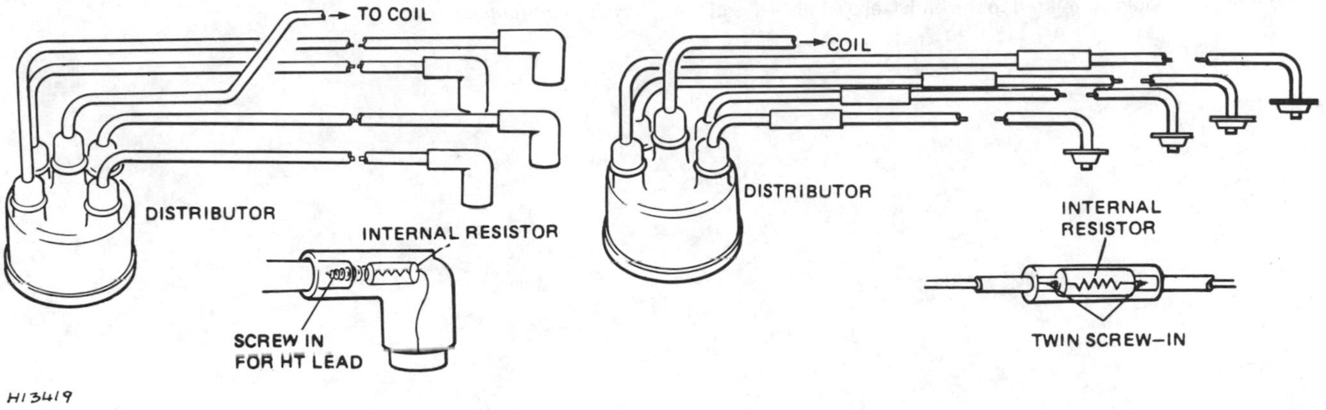

Fig. 10.27 Ignition HT lead suppressors (Sec 51)

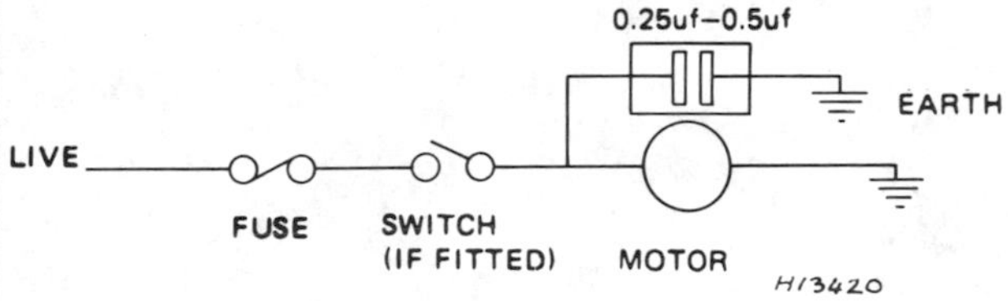

Fig. 10.28 Correct method of suppressing electric motors (Sec 51)

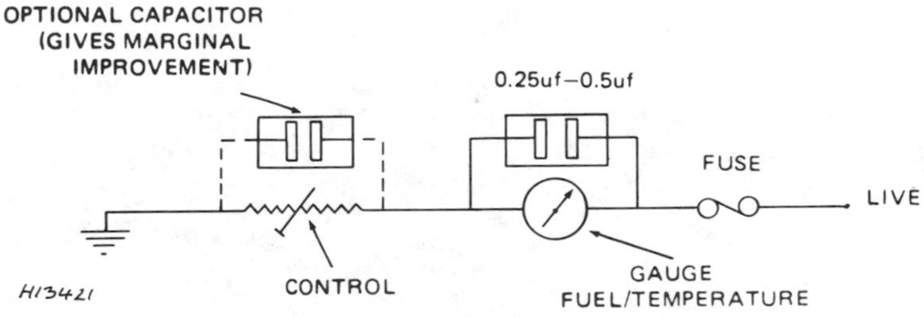

Fig. 10.29 Method of suppressing gauges and their control units (Sec 51)

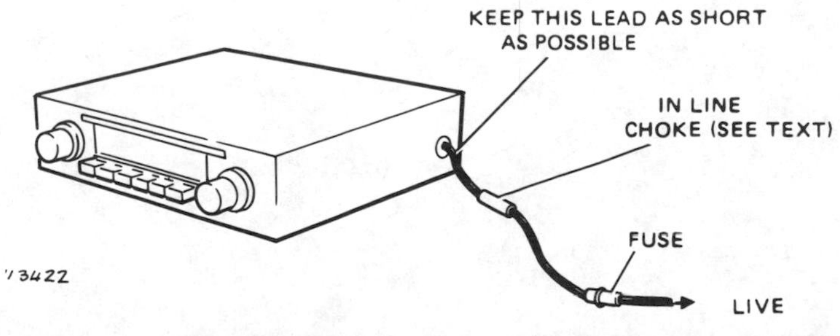

Fig. 10.30 An 'in-line' choke should be fitted with the line supply lead as close to the unit as possible (Sec 51)

bridge the join with a wide braided earth strap (like that from the frame to the engine/transmission). The most common moving parts that could create noise and should be strapped are, in order of importance:

(a) *Silencer to frame*
(b) *Exhaust pipe to engine block and frame*
(c) *Air cleaner to frame*
(d) *Front and rear bumpers to frame*
(e) *Steering column to frame*
(f) *Bonnet and boot lids to frame*
(g) *Hood frame to bodyframe on soft tops*

These faults are most pronounced when the engine is idling or labouring under load. Although the moving parts are already connected with nuts bolts, etc, these do tend to rust and corrode, this creating a high resistance interference source.

If you have a 'ragged' sounding pulse when mobile, this could be wheel or tyre static. This can be cured by buying some anti-static powder and sprinkling inside the tyres.

If the inteference takes the shape of a high pitched screeching noise that changes its note when the car is in motion and only comes now and then, this could be related to the aerial, especially if it is of the telescopic or whip type. This source can be cured quite simply by pushing a small rubber ball on top of the aerial as this breaks the electric field before it can form; but it would be much better to buy yourself a new aerial of a reputable brand. If, on the other hand, you are getting a loud rushing sound every time you brake, then this is brake static. This effect is most prominent on hot dry days and is cured only by fitting a special kit, which is quite expensive.

In conclusion, it is pointed out that it is relatively easy and therefore cheap, to eliminate 95 per cent of all noise, but to eliminate the final 5 per cent is time and money consuming. It is up to the individual to decide if it is worth it. Please remember also, that you cannot get a concert hall performance out of a cheap radio.

Finally, players and eight track players are not usually affected by car noise but in a very bad case, the best remedies are the first three suggestions plus using a 3 to 5 amp choke in the 'live' line and in incurable cases, screening the live and speaker wires.

Note: *If your car is fitted with electronic ignition, then it is not recommended that either the spark plug resistors or the ignition coil capacitor be fitted as these may damage the system. Most electronic ignition units have built in suppression and should, therefore, not cause interference.*

52 Fault diagnosis – electrical system

Symptom	Reason/s
Starter motor fails to turn engine	Battery discharged Battery defective internally Battery terminal leads loose or earth lead not securely attached to body Loose or broken connections in starter motor circuit Starter motor switch or solenoid faulty Starter motor pinion jammed in mesh with ring gear Starter brushes badly worn, sticking, or brush wires loose Commutator dirty, worn, or burnt Starter motor armature faulty Field coils earthed
Starter motor turns engine very slowly	Battery in discharged condition Starter brushes badly worn, sticking, or brush wires loose Loose wires in starter motor circuit
Starter motor operates without turning engine	Starter motor pinion sticking on the screwed sleeve Pinion or ring gear teeth broken or worn
Starter motor noisy or excessively rough engagement	Pinion or ring gear teeth broken or worn Starter drive main spring broken Starter motor retaining bolts loose
Battery will not hold charge for more than a few days	Battery defective internally Electrolyte level too low or electrolyte too weak due to leakage Plate separators no longer fully effective Battery plates severely sulphated Drivebelt slipping Battery terminal connections loose or corroded Short in lighting circuit causing continual battery drain
Ignition light fails to go out, battery runs flat in a few days	Alternator drivebelt loose and slipping, or broken Alternator brushes worn, sticking, broken or dirty Alternator brush springs weak or broken Commutator dirty, greasy, worn, or burnt Armature badly worn or armature shaft bent
Wiper motor fails to work	Blown fuse Wire connections loose, disconnected, or broken Brushes badly worn Armature worn or faulty Field coils faulty
Wiper motor works very slowly and takes excessive current	Commutator dirty, greasy, or burnt Drive to wheelboxes too bent or unlubricated Wheelbox spindle binding or damaged Armature bearings dry or unaligned Armature badly worn or faulty
Wiper motor works slowly and takes little current	Brushes badly worn Commutator dirty, greasy, or burnt Armature badly worn or faulty
Wiper motor works but wiper blades remain static	Driving cable rack disengaged or faulty Wheelbox gear and spindle damaged or worn Wiper motor gearbox parts badly worn

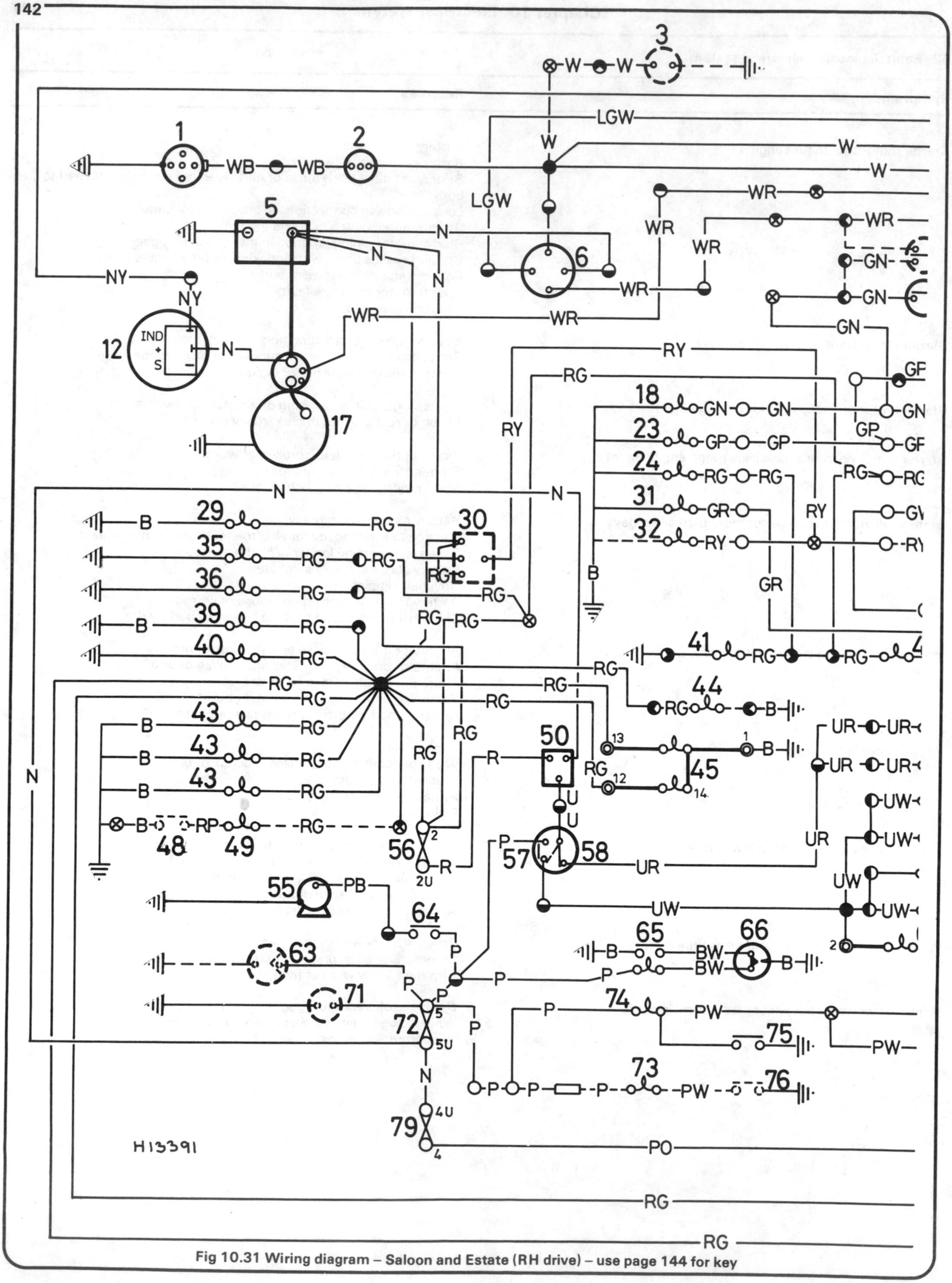

Fig 10.31 Wiring diagram – Saloon and Estate (RH drive) – use page 144 for key

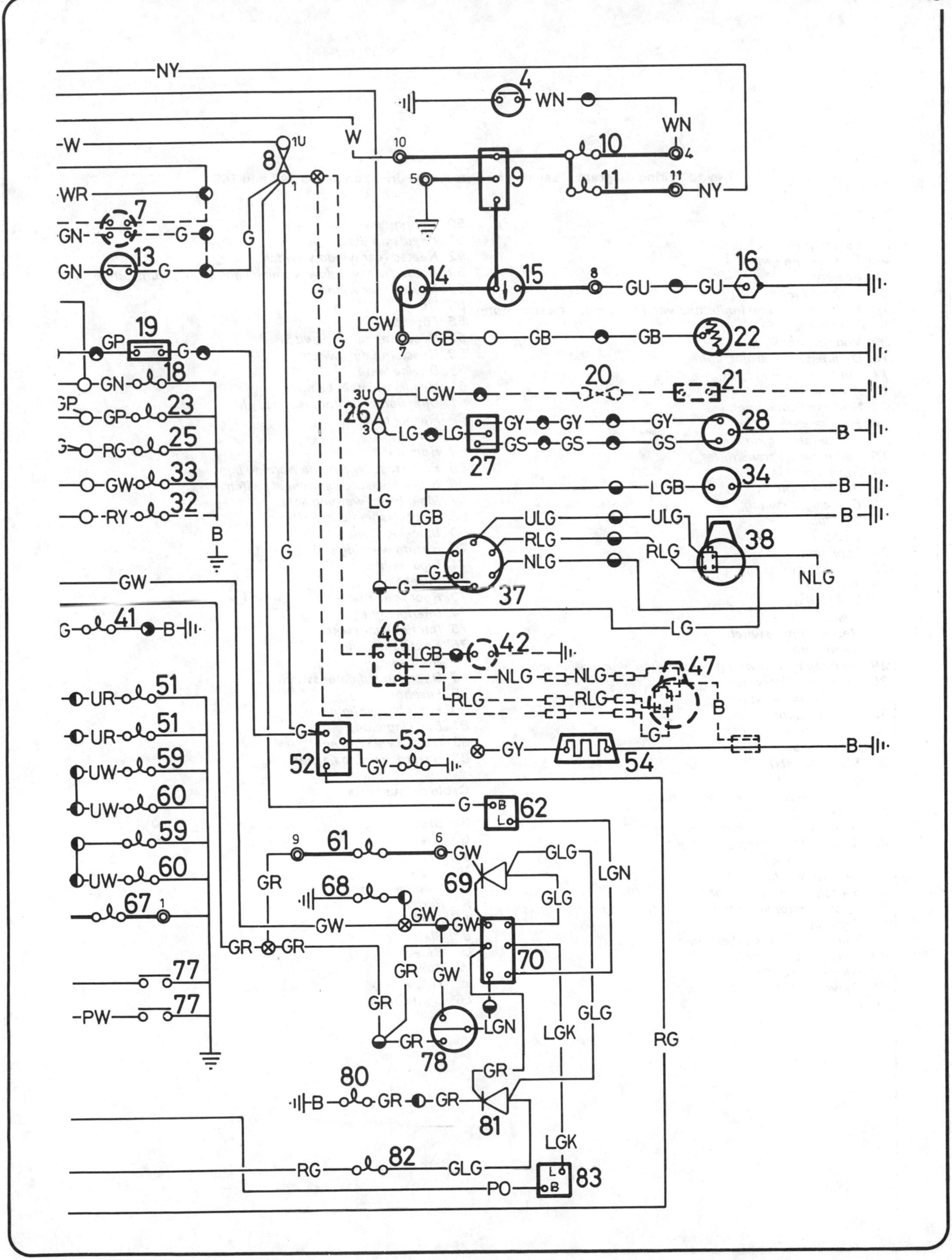
NY
4
WN
WN
W
1U
8
1
10
10
4
5
9
11
11
NY
WR
7
GN
G
G
13
GN
G
G
14
15
8
GU
GU
16
LGW
7
GB
GB
GB
22
19
GP
G
18
GN
G
20
21
23
GP
GP
3U
LGW
26
3
LG
LG
27
GY
GY
GY
28
GS
GS
GS
B
25
RG
33
GW
32
RY
LG
LGB
34
LGB
B
ULG
ULG
B
38
RLG
RLG
RLG
NLG
NLG
NLG
37
B
G
G
GW
LG
41
G
B
46
42
LGB
47
NLG
NLG
RLG
RLG
B
LG
51
UR
51
UR
G
53
GY
54
B
59
UW
52
GY
60
UW
G
62
B
L
59
61
9
6
GW
GLG
60
UW
LGN
GR
68
69
GLG
67
1
GW
GW
GW
GR
GR
70
GR
GW
77
GR
GLG
77
PW
LGN
LGK
GR
RG
78
80
GR
B
GR
GR
81
LGK
82
RG
GLG
L
83
PO
B

Key to Wiring diagram – Saloon and Estate (RH drive) on pages 142 and 143

1 *Distributor*
2 *Coil*
3 *Fuel pump*
4 *Oil pressure switch*
5 *Battery*
6 *Ignition switch*
7 *Reverse light and inhibitor switch (automatic transmission)*
8 *Fusebox*
9 *Voltage stabilizer*
10 *Oil pressure warning light*
11 *Ignition warning light*
12 *Alternator*
13 *Reverse light switch*
14 *Fuel gauge*
15 *Temperature gauge*
16 *Temperature transmitter*
17 *Starter motor*
18 *Reverse light*
19 *Brake light switch*
20 *Line fuse*
21 *Radio*
22 *Tank unit*
23 *Brake light*
24 *Tail light (LH)*
25 *Tail light (RH)*
26 *Fusebox*
27 *Heater motor switch*
28 *Heater motor*
29 *Rear fog light warning light/switch illumination light*
30 *Rear fog light switch*
31 *Rear flasher light (LH)*
32 *Rear fog light*
33 *Rear flasher light (RH)*
34 *Screen washer motor*
35 *Sidelight (RH)*
36 *Sidelight (LH)*
37 *Screen wiper and washer switch*
38 *Screen wiper motor*
39 *Heater control illumination light*
40 *Cigar lighter illumination light*
41 *Number plate light*
42 *Tailgate washer motor*
43 *Switch illumination light*
44 *Gear selector illumination light (automatic transmission)*
45 *Panel light*
46 *Tailgate wiper/washer switch*
47 *Tailgate wiper motor*
48 *Glovebox lid switch*
49 *Glovebox light*
50 *Lighting switch*
51 *Headlamp dip beam*
52 *Heated rear window switch*
53 *Heated rear window warning light/switch illumination light*
54 *Heated rear window*
55 *Horn*
56 *Fusebox*
57 *Headlamp flasher switch*
58 *Headlamp dip switch*
59 *Driving lamp*
60 *Headlamp main beam*
61 *Direction indicator warning light*
62 *Flasher unit*
63 *Clock*
64 *Horn push*
65 *Brake test switch and warning light*
66 *Brake pressure differential switch*
67 *Main beam warning light*
68 *Front flasher light (RH)*
69 *Blocking diode*
70 *Hazard warning switch*
71 *Cigar lighter*
72 *Fusebox*
73 *Boot light*
74 *Interior light*
75 *Interior light switch*
76 *Boot light switch*
77 *Door switch*
78 *Direction indicator switch*
79 *Fusebox*
80 *Front flasher light (LH)*
81 *Blocking diode*
82 *Hazard warning light*
83 *Hazard warning flasher unit*

Cable colour code

B *Black*
G *Green*
K *Pink*
N *Brown*
O *Orange*
P *Purple*
R *Red*
S *Slate*
U *Blue*
W *White*
Y *Yellow*
LG *Light Green*

Key to Wiring diagram – HL models (RH drive) on pages 146 and 147

1 Distributor
2 Ignition coil
3 Ballast resistor coil
4 Fuel pump
5 Battery
6 Ignition switch
7 Reverse light and inhibitor switch (automatic transmission)
8 Fusebox
9 Oil pressure switch
10 Voltage stabilizer
11 Oil pressure warning light
12 Ignition warning light
13 Tachometer
14 Alternator
15 Starter motor
16 Reverse light
17 Brake light
18 Brake light switch
19 Fuel gauge
20 Temperature gauge
21 Handbrake switch and warning light
22 Temperature transmitter
23 Fuel tank
24 Rear fog light warning light/switch illumination lamp
25 Rear fog light switch
26 Tail light (LH)
27 Tail light (RH)
28 Rear flasher light (LH)
29 Rear flasher light (RH)
30 Rear fog light
31 Fusebox
32 Heater motor switch
33 Line fuse
34 Radio
35 Heater motor
36 Sidelight (RH)
37 Sidelight (LH)
38 Heater control illumination light
39 Cigar lighter illumination light
40 Panel light switch
41 Number plate light
42 Windscreen wiper and washer switch
43 Windscreen wiper delay unit
44 Wiper motor
45 Washer motor
46 Switch illumination lights
47 Panel light resistor
48 Panel lights
49 Gear selector illumination light (automatic transmission)
50 Choke switch
51 Choke warning light
52 Glovebox lid switch
53 Glovebox light
54 Fusebox
55 Headlamp flasher switch
56 Lighting switch
57 Headlamp dip switch
58 Headlamp dip beam
59 Headlamp main beam
60 Driving lamp
61 Main beam warning light
62 Heated rear window switch
63 Heated rear window warning light/switch illumination light
64 Heated rear window
65 Horn
66 Horn push
67 Clock
68 Cigar lighter
69 Fusebox
70 Interior light
71 Interior light switch
72 Door switch
73 Boot light
74 Boot light switch
75 Brake test switch and warning light
76 Brake pressure differential switch
77 Direction indicator warning light
78 Flasher unit
79 Front flasher light (RH)
80 Blocking diode
81 Hazard warning switch
82 Direction indicator switch
83 Front flasher light (LH)
84 Hazard warning light
85 Hazard flasher unit

Cable colour code

B Black
G Green
K Pink
N Brown
O Orange
P Purple
R Red
S Slate
U Blue
W White
Y Yellow
LG Light Green

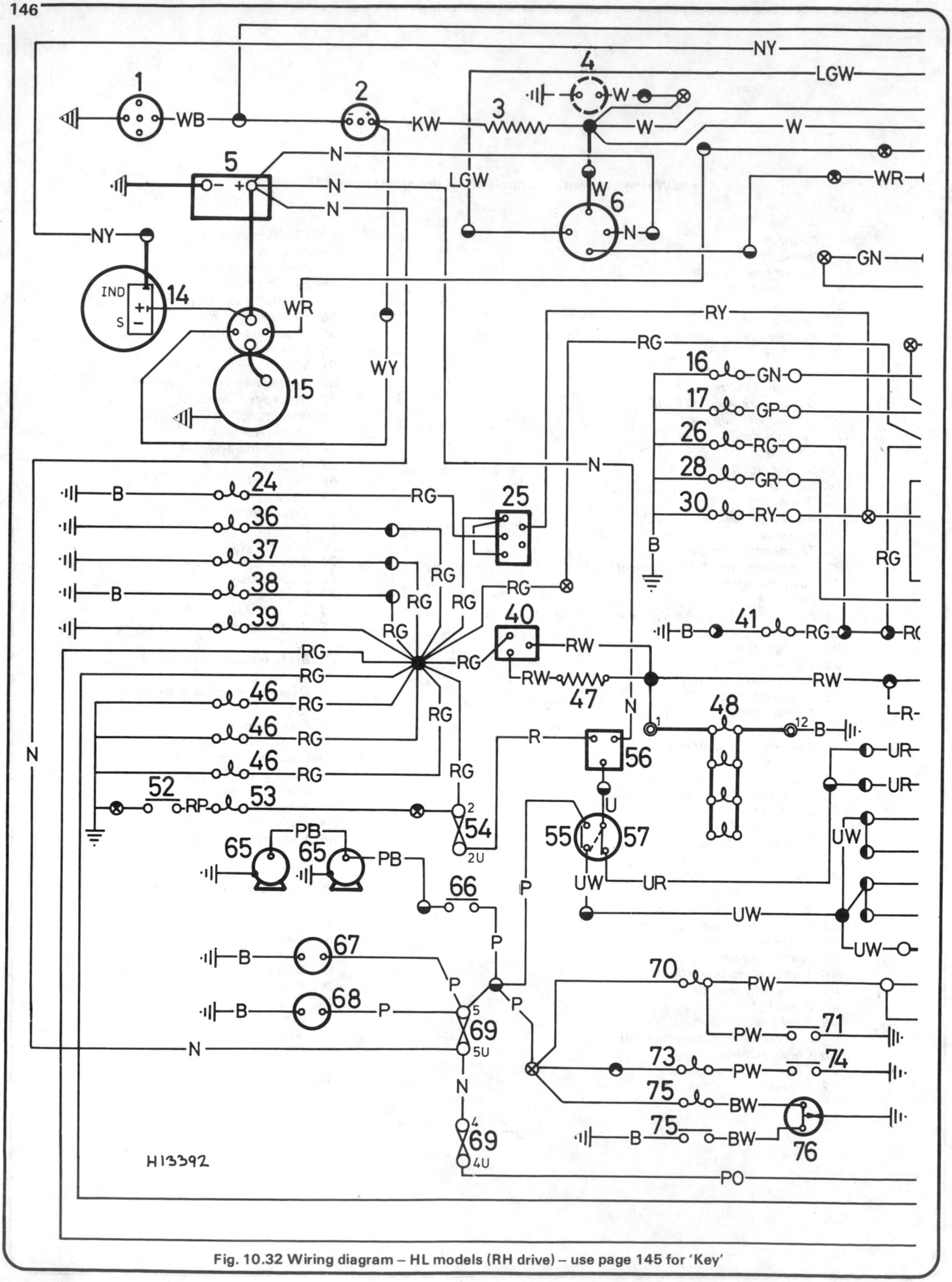

Fig. 10.32 Wiring diagram – HL models (RH drive) – use page 145 for 'Key'

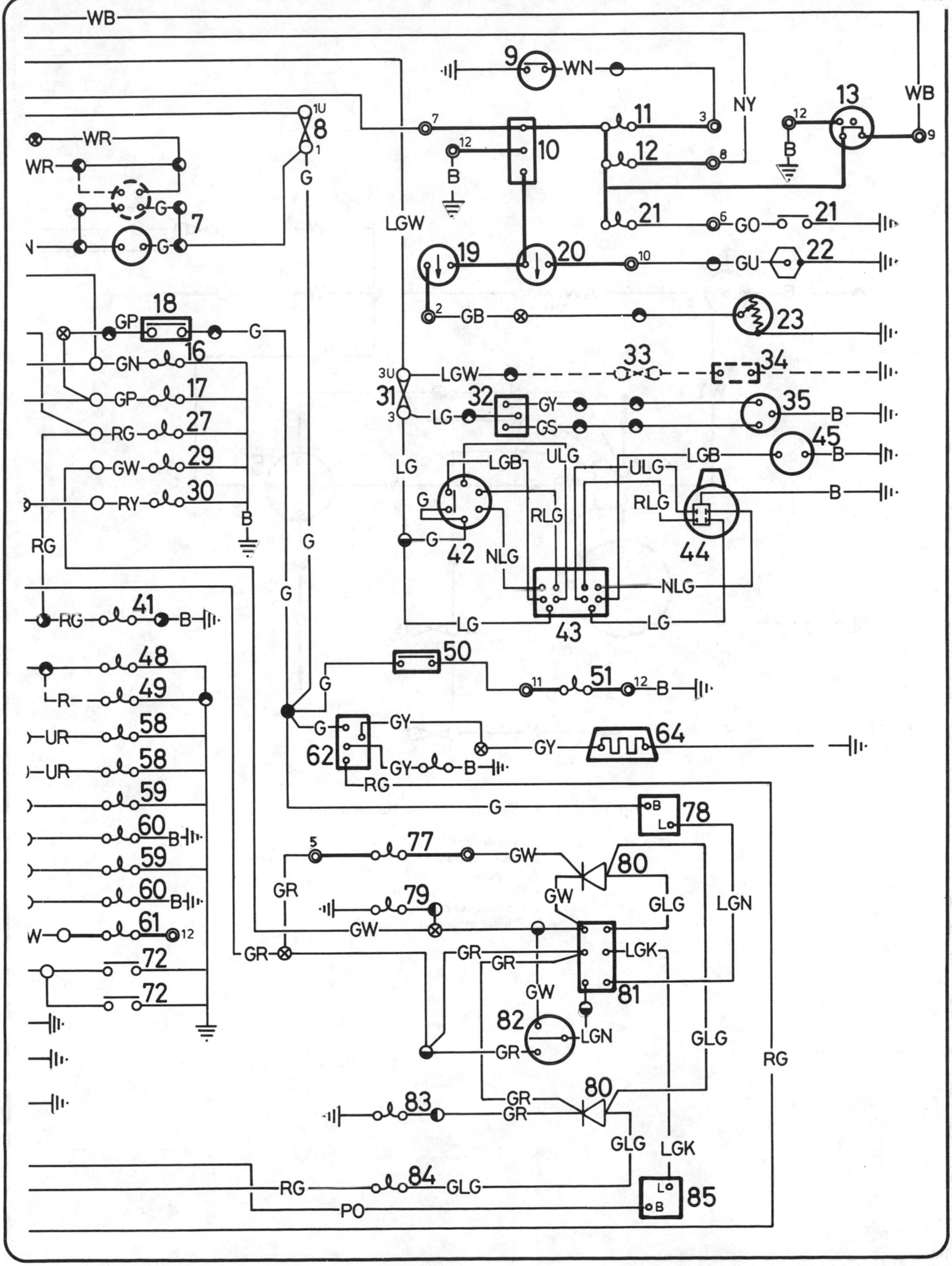
WB
9
WN
NY
13
WB
8
1U
1
WR
WR
G
7
G
G
7
10
11
12
3
8
12
B
12
B
9
LGW
21
21
6
GO
19
20
10
GU
22
18
GP
G
2
GB
23
GN
16
GP
17
RG
27
GW
29
RY
30
B
RG
G
G
3U
LGW
33
34
31
3
LG
32
GY
GS
35
B
45
LGB
B
LG
LGB
ULG
ULG
B
G
RLG
RLG
44
G
42
NLG
NLG
43
LG
LG
G
RG
41
B
48
R
49
UR
58
UR
58
59
60
B
59
60
B
W
61
12
72
72
50
G
11
51
12
B
G
GY
62
64
GY
GY
B
RG
G
B
78
L
5
77
GW
80
GR
79
GW
GLG
LGN
GW
GR
LGK
GR
GR
81
GW
82
LGN
GR
GLG
RG
80
GR
83
GR
GLG
LGK
RG
84
GLG
L
85
B
PO

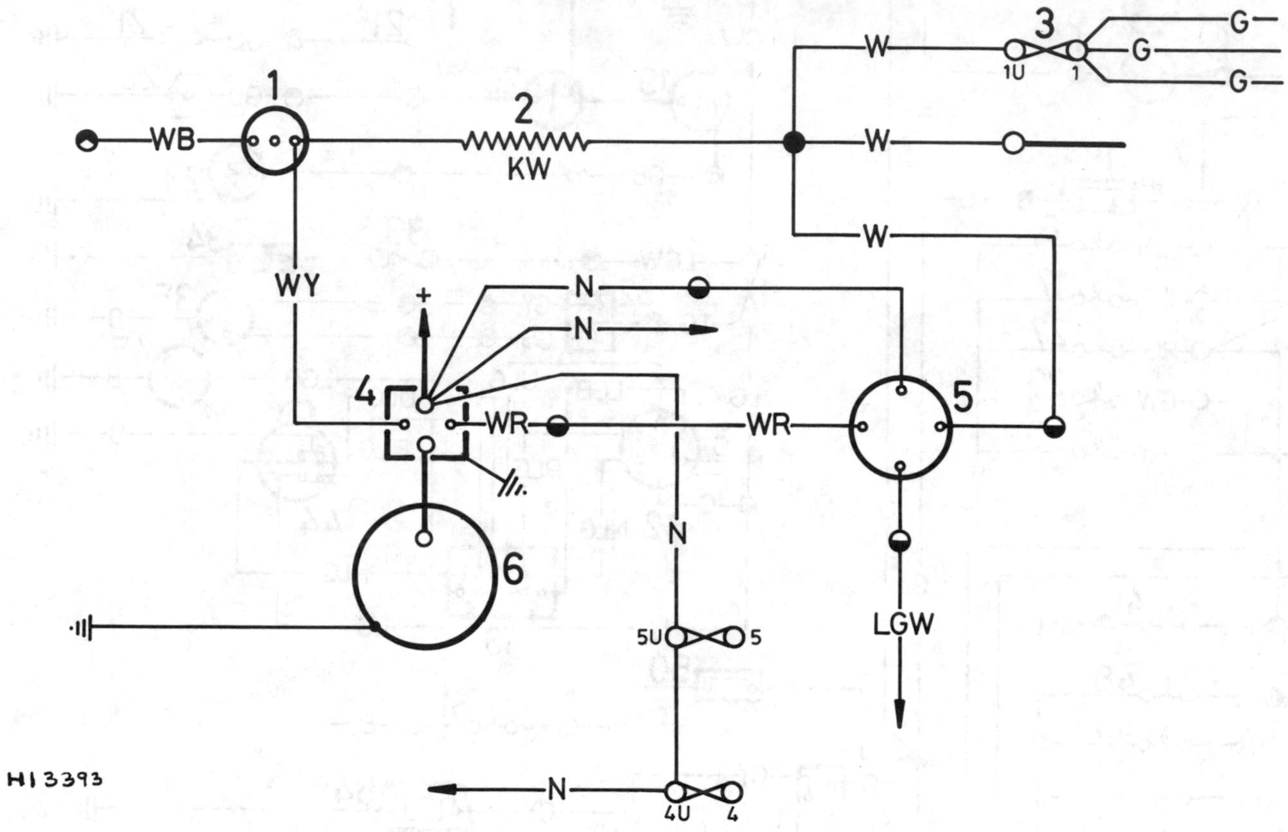

Fig. 10.33 Wiring diagram – inertia starter motor

1 *Ignition coil*
2 *Ballast resistor*
3 *Fusebox*
4 *Starter solenoid*
5 *Ignition switch*
6 *Starter motor*

Key to wiring diagram – Saloon and Estate (LH drive) on pages 150 and 151

1 *Distributor*
2 *Coil*
3 *Fuel pump*
4 *Line fuse*
5 *Induction heater and thermostat*
6 *Suction chamber heater*
7 *Fusebox*
8 *Oil pressure switch*
9 *Voltage stabilizer*
10 *Oil pressure warning light*
11 *Ignition warning light*
12 *Battery*
13 *Ignition switch*
14 *Reverse light and inhibitor switch (automatic transmission)*
15 *Reverse light switch*
16 *Fuel gauge*
17 *Temperature gauge*
18 *Temperature transmitter*
19 *Alternator*
20 *Starter motor*
21 *Reverse light*
22 *Brake light*
23 *Brake light switch*
24 *Heater motor switch*
25 *Tank unit*
26 *Radio*
27 *Heater motor*
28 *Rear fog light warning light/switch illumination light*
29 *Rear fog light switch*
30 *Tail light (LH)*
31 *Tail light (RH)*
32 *Rear flasher light (LH)*
33 *Rear flasher light (RH)*
34 *Rear fog light*
35 *Sidelight (RH)*
36 *Sidelight (LH)*
37 *Screen wiper and washer switch*
38 *Screen washer motor*
39 *Screen wiper motor*
40 *Heater control illumination*
41 *Cigar lighter illumination light*
42 *Number plate light*
43 *Tailgate wiper/washer switch*
44 *Tailgate washer motor*
45 *Tailgate wiper motor*
46 *Switch illumination light*
47 *Glovebox lid switch*
48 *Glovebox light*
49 *Fusebox*
50 *Lighting switch*
51 *Panel light*
52 *Gear selector illumination light (automatic transmission)*
53 *Headlamp dip beam*
54 *Heated rear window switch*
55 *Heated rear window warning light/switch illumination light*
56 *Heated rear window*
57 *Horn*
58 *Horn push*
59 *Headlamp flasher switch*
60 *Headlamp dip switch*
61 *Driving lamp*
62 *Headlamp main beam*
63 *Main beam warning light*
64 *Front flasher light (RH)*
65 *Repeater flasher light*
66 *Direction indicator warning light*
67 *Blocking diode*
68 *Flasher unit*
69 *Hazard warning switch*
70 *Clock*
71 *Cigar lighter*
72 *Fusebox*
73 *Brake test switch and warning light*
74 *Boot light*
75 *Interior light*
76 *Interior light switch*
77 *Boot light switch*
78 *Brake pressure differential switch*
79 *Door switch*
80 *Front flasher light (LH)*
81 *Repeater flasher*
82 *Hazard warning light*
83 *Direction indicator switch*
84 *Blocking diode*
85 *Hazard warning flasher unit*

Cable colour code

B *Black*
G *Green*
K *Pink*
N *Brown*
O *Orange*
P *Purple*
R *Red*
S *Slate*
U *Blue*
W *White*
Y *Yellow*
LG *Light Green*

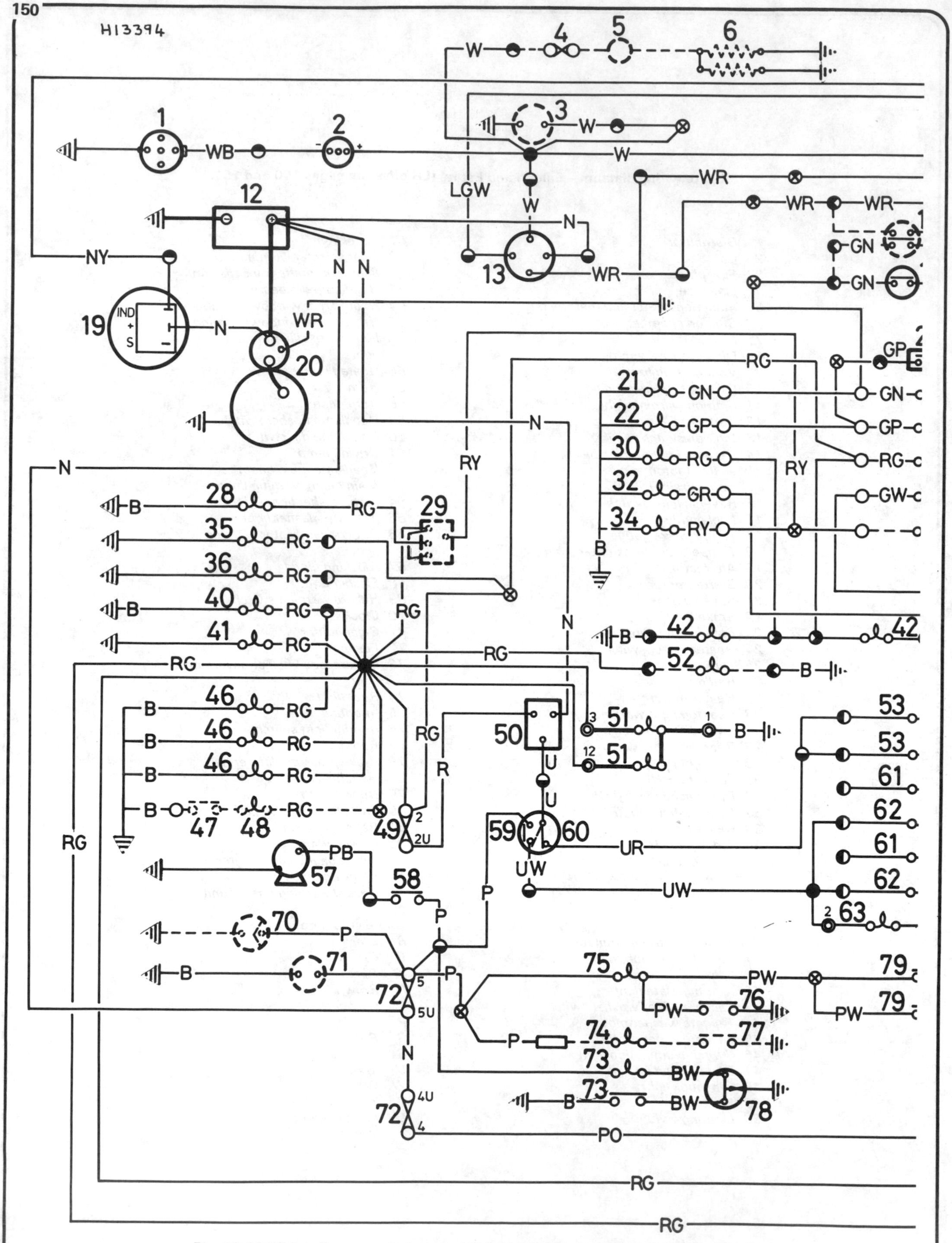

Fig. 10.34 Wiring diagram – Saloon and Estate (LH drive) – see page 149 for 'Key'

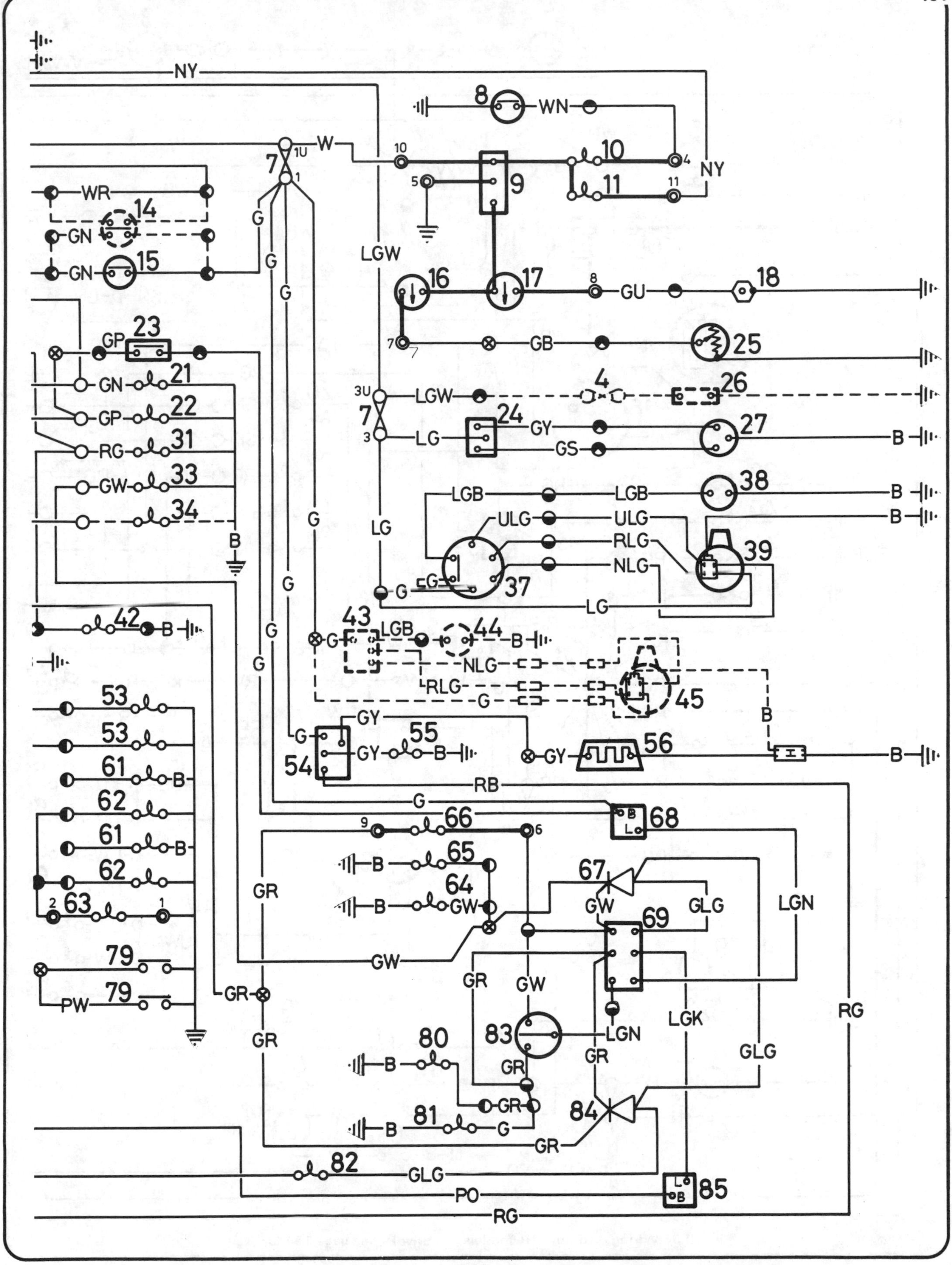
NY
8
WN
W
10
1U
7
1
10
4
NY
5
9
11
11
WR
14
GN
GN
15
G
LGW
16
17
8
GU
18
G
7
GB
25
23
GP
GN
21
GP
22
RG
31
GW
33
34
B
3U
LGW
4
26
7
3
LG
24
GY
GS
27
B
LGB
LGB
38
B
ULG
ULG
B
RLG
39
NLG
G
37
LG
G
LG
42
B
43
LGB
44
B
G
NLG
RLG
45
G
B
GY
53
53
55
G
GY
B
GY
56
B
61
B
54
RB
62
G
66
68
9
6
B
L
61
B
65
B
62
67
GR
64
B
GW
GW
GLG
LGN
63
2
1
69
GW
79
GR
GW
79
PW
GR
83
LGN
LGK
RG
GR
80
GR
GLG
B
GR
GR
84
81
B
G
GR
82
GLG
PO
85
L
B
RG

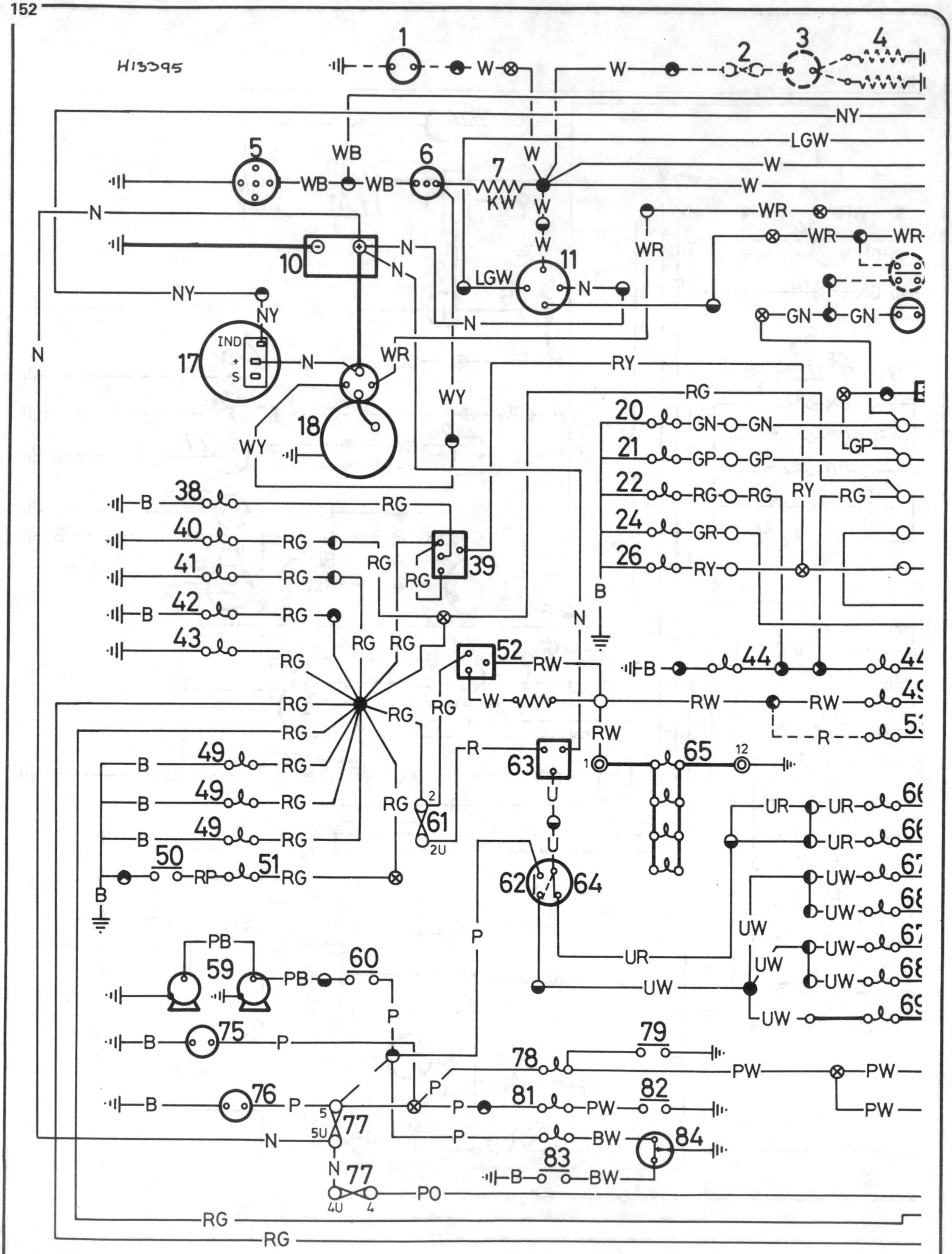

Fig. 10.35 Wiring diagram – HL models (LH drive) – see page 154 for 'Key'

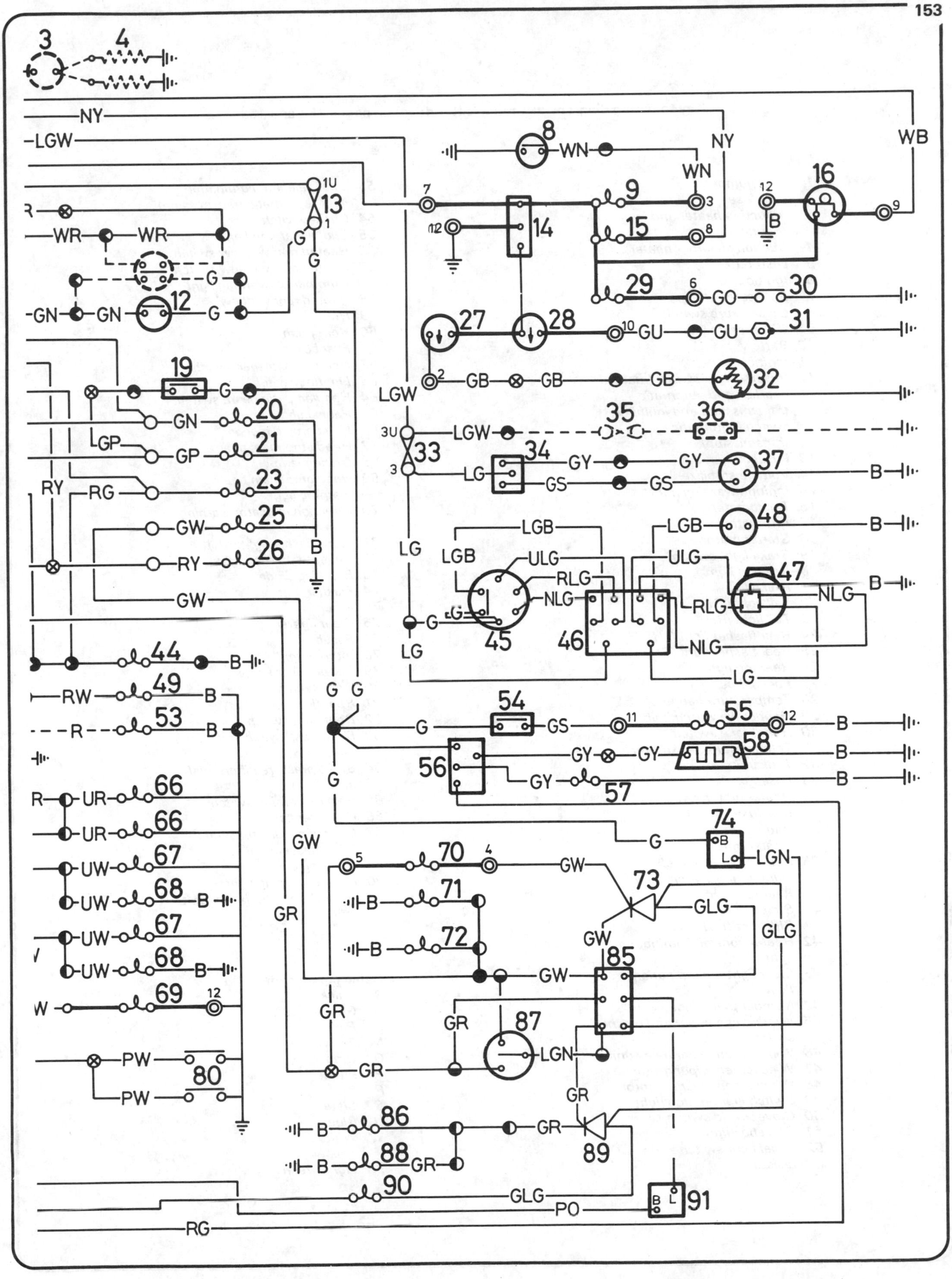

3
4
NY
LGW
WB
8
WN
NY
WN
1U
13
1
R
WR
WR
G
G
G
GN
GN
12
G
7
9
3
12
14
15
8
16
12
B
9
29
6
GO
30
27
28
10
GU
GU
31
19
G
2
GB
GB
GB
32
GN
20
LGW
GP
GP
21
35
36
3U
LGW
33
34
3
LG
GY
GY
37
GS
GS
B
RY
RG
23
GW
25
B
26
RY
LGB
48
B
LG
LGB
ULG
LGB
ULG
47
B
RLG
NLG
NLG
RLG
G
45
46
LG
NLG
GW
44
B
LG
RW
49
B
G
G
R
53
B
54
G
GS
11
55
12
B
56
GY
GY
58
B
G
GY
57
B
66
R
UR
66
UR
74
G
B
L
LGN
67
UW
GW
5
70
4
GW
68
UW
B
71
B
73
GLG
GR
67
UW
72
B
GLG
GW
68
UW
B
85
GW
69
W
12
GR
GR
87
LGN
PW
80
GR
PW
GR
86
B
GR
89
88
B
GR
90
GLG
PO
91
RG

Key to wiring diagram – HL models (LH drive) on pages 152 and 153

1 *Fuel pump*
2 *Line fuse*
3 *Induction heater and thermostat*
4 *Suction chamber heater*
5 *Distributor*
6 *Ignition coil*
7 *Ballast resistor*
8 *Oil pressure switch*
9 *Oil pressure warning light*
10 *Battery*
11 *Ignition switch*
12 *Reverse light switch (manual gearbox and automatic transmission) and inhibitor switch (automatic transmission)*
13 *Fusebox*
14 *Voltage stabilizer*
15 *Ignition warning light*
16 *Tachometer*
17 *Alternator*
18 *Starter motor*
19 *Brake light switch*
20 *Reverse light*
21 *Brake light*
22 *Tail light (LH)*
23 *Tail light (RH)*
24 *Rear flasher light (LH)*
25 *Rear flasher light (RH)*
26 *Rear fog light*
27 *Fuel gauge*
28 *Temperature gauge*
29 *Handbrake warning light*
30 *Handbrake switch*
31 *Temperature transmitter*
32 *Tank unit*
33 *Fusebox*
34 *Heater motor switch*
35 *Line fuse*
36 *Radio*
37 *Heater motor*
38 *Rear fog light switch illumination/warning light*
39 *Rear fog light*
40 *Sidelight (RH)*
41 *Sidelight (LH)*
42 *Heater control illumination light*
43 *Cigar lighter illumination light*
44 *Number plate light*
45 *Windscreen wiper and washer switch*
46 *Windscreen wiper delay unit*
47 *Windscreen wiper motor*
48 *Windscreen washer motor*
49 *Switch illumination light*
50 *Glovebox lid switch*
51 *Glovebox light*
52 *Panel light switch and resistor*
53 *Gear selector illumination light (automatic transmission)*
54 *Choke switch*
55 *Choke warning light*
56 *Heated rear window switch*
57 *Heated rear window switch illumination/warning light*
58 *Heated rear window*
59 *Horn*
60 *Horn push*
61 *Fusebox*
62 *Headlamp flasher switch*
63 *Lighting switch*
64 *Headlamp dip beam switch*
65 *Panel lights*
66 *Headlamp dip beam*
67 *Headlamp main beam*
68 *Driving lamp*
69 *Headlamp main beam warning light*
70 *Direction indicator warning light*
71 *Front flasher light (RH)*
72 *Repeater flasher light*
73 *Blocking diode*
74 *Flasher unit*
75 *Clock*
76 *Cigar lighter*
77 *Fusebox*
78 *Interior light*
79 *Interior light switch*
80 *Door switch*
81 *Boot light*
82 *Boot light switch*
83 *Brake test switch and warning light*
84 *Brake pressure differential switch*
85 *Hazard warning switch*
86 *Repeater flasher*
87 *Direction indicator switch*
88 *Front flasher light (LH)*
89 *Blocking diode*
90 *Hazard warning light*
91 *Hazard warning flasher unit*

Cable colour code

B *Black*
G *Green*
K *Pink*
N *Brown*
O *Orange*
P *Purple*
R *Red*
S *Slate*
U *Blue*
W *White*
Y *Yellow*
LG *Light Green*

Chapter 11 Suspension and Steering

Contents

Specifications

Front suspension

Type	Independent, by torsion bar with lever type shock absorbers and anti-roll bar
King pin inclination	7½° positive
Camber angle	0° 50′ positive
Castor angle:	
Saloon	2° positive
Estate	1° 18′ positive
Swivel pin link lower bush finished diameter	0.688 ± 0.0005 in (17.48 ± 0.013 mm)
Trim height	14.63 ± 0.25 in (371.6 ± 6.35 mm)

Rear suspension

Type	Semi-elliptic leaf spring with telescopic shock absorbers and anti-roll bar
Number of spring leaves:	
Saloon	2
Estate	4 or 3
Width of leaves	2 in (50.8 mm)
Gauge of leaves:	
Saloon	0.30 to 0.17 in (7.62 to 4.31 mm)
Estate:	
4 leaf	0.26 to 0.13 in (6.60 to 3.30 mm)
3 leaf	0.284 to 0.154 in (7.21 to 3.91 mm)
Working load:	
Saloon	270 lb (122 kg)
Estate (3 and 4 leaf)	320 lb (145 kg)

Steering

Type	Rack and pinion
Steering wheel turns lock-to-lock	3.7
Front wheel alignment	0 to ⅛ in (0 to 3.175 mm) toe-in
Rack travel	6.04 in (15.34 cm)
Rack travel either side of centre	3.02 in (7.67 cm)
Pinion bearing preload	0.001 to 0.003 in (0.025 to 0.076 mm)
Shims available	0.002 In (0.050 mm)
	0.005 in (0.127 mm)
	0.010 in (0.254 mm)
	0.030 in (0.762 mm)
	0.060 in (1.524 mm)
Cover gasket thickness	0.010 in (0.254 mm)

Yoke clearance	0.002 to 0.005 in (0.050 to 0.127 mm)	
Shims available	0.002 in (0.050 mm)	
	0.005 in (0.127 mm)	
	0.010 in (0.254 mm)	
Cover gasket thickness	0.010 in (0.254 mm)	
Steering rack oil capacity	$\frac{1}{3}$ pint (190 cm^3)	
Ball-pin centre dimension (ball-pins screwed to tie-rod equal amount)	43.7 in (111.0 cm)	

Wheels

Type	4-stud, pressed steel 4$\frac{1}{4}$C x 13

Tyres

Saloon and estate	155-13 radial ply	
Tyre pressures:	**Front**	**Rear**
Saloon	26 lbf/in² (1.8 kgf/cm²)	28 lbf/in² (2.0 kgf/cm²)
Estate:		
Light load	26 lbf/in² (1.8 kgf/cm²)	28 lbf/in² (2.0 kgf/cm²)
Heavy load	26 lbf/in² (1.8 kgf/cm²)	32 lbf/in² (2.25 kgf/cm²)

Torque wrench settings

	lbf ft	kgf m
Front suspension:		
Ball-pin retainer locknut	70 to 80	9.6 to 11.0
Eyebolt nut	50 to 54	6.9 to 7.4
Torsion bar reaction lever lockbolt	22	3.0
Reaction pad nut	35 to 40	4.8 to 5.5
Shock absorber retaining nuts	26 to 28	3.5 to 3.8
Tie-rod fork nut	48 to 55	6.6 to 7.6
Tie-rod to fork	22	3.0
Caliper bracket or dust shield bolts	35 to 42	4.8 to 5.8
Road wheels	38 to 41	5.2 to 5.8
Anti-roll bar link to bar and lower arm	48 to 55	6.6 to 7.6
Anti-roll bar to body	20 to 22	2.7 to 3.0
Rear suspension:		
Upper shackle pin nuts	28	3.9
Spring eye bolt nuts	40	5.5
Spring U-bolt nuts	15 to 18	2.0 to 2.4
Shock absorber to spring bracket	28	3.9
Shock absorber to body bracket	45	6.2
Anti-roll bar pivot bolt nut	48 to 55	6.6 to 7.6
Anti-roll bar clamp	26 to 28	3.5 to 3.8
Steering:		
Rack clamp brackets nuts	15 to 18	2.0 to 2.4
Tie-rod ball-pin nuts	20 to 24	2.77 to 3.3
Flexible joint pinch-bolt nut	17 to 20	2.35 to 2.77
Pinion end cover retaining bolts	12 to 15	1.6 to 2.0
Pinion preload	15 (lbf in)	0.17
Rack yoke cover bolts	12 to 15	1.6 to 2.0
Tie-rod housing locknut	33 to 37	4.6 to 5.6
Tie-rod ball spheres preload	32 to 52 (lbf in)	0.37 to 0.6
Steering column mounting bolts	14 to 18	1.94 to 2.49
Flexible joint coupling bolts	20 to 22	2.77 to 3.04
Steering column lock shear-screw	14	1.94
Steering wheel nut	33 to 37	4.6 to 5.6
Tie-rod locknuts	35 to 40	4.8 to 5.5
Steering lever to swivel pin nut	70 to 80	9.6 to 11.0
Lower suspension arm to torsion bar nut	50 to 54	6.9 to 7.4
Lower arm front and rear pivot and front to rear arm bolts	26 to 28	3.5 to 3.8
Tie-rod to body bracket nut	26 to 28	3.5 to 3.8

1 General description

The component parts of the right-hand side front suspension unit are shown in Fig. 11.1. Although the left-hand side front suspension is identical in principle, some parts are handed and therefore not interchangeable.

Attached to the hub is the road wheel as is also the brake disc, these being retained by bolts. The hub rotates on two opposed tapered roller bearings mounted on the swivel pin stub axle and is retained on the stub axle by a nut. Also attached to the swivel pin is the disc brake dust shield.

The shock absorber is attached to the body and its arm carries, at the outer end, the balljoint for the swivel pin top attachment. Its arm therefore acts as an upper suspension wishbone. The bottom end of the steering swivel screws into the lower link which is mounted between the outer ends of the lower arms. This link is mounted on a pivot pin so that the suspension is able to move in a vertical manner. Horizontal movement of the suspension is controlled by a tie-rod assembly. The inner ends of the lower arms are free to pivot about an eye bolt and the rear arm is spline attached to the torsion bar. The rear of the torsion bar is attached to the body so that both the body weight and road shocks are taken by the torsion bar.

The anti-roll bar is attached to the lower suspension arm by ball-joint swivel links.

Rear suspension is by semi-elliptic leaf springs, the springs being mounted on rubber bushed shackle pins. Double acting telescopic hydraulic shock absorbers are fitted to absorb road shocks and damp spring oscillations. An anti-roll bar is also fitted.

A rack-and-pinion steering is used. The steering wheel is splined to the upper inner column which in turn is connected to the lower column by a flexible coupling. A second flexible coupling connects the lower column to the steering gear pinion. The pinion teeth mesh with those machined in the rack so that rotation of the pinion moves the rack from one side of the housing to the other. Located at either end of the rack are tie-rods and balljoints which are attached to the suspension swivel pin steering arms.

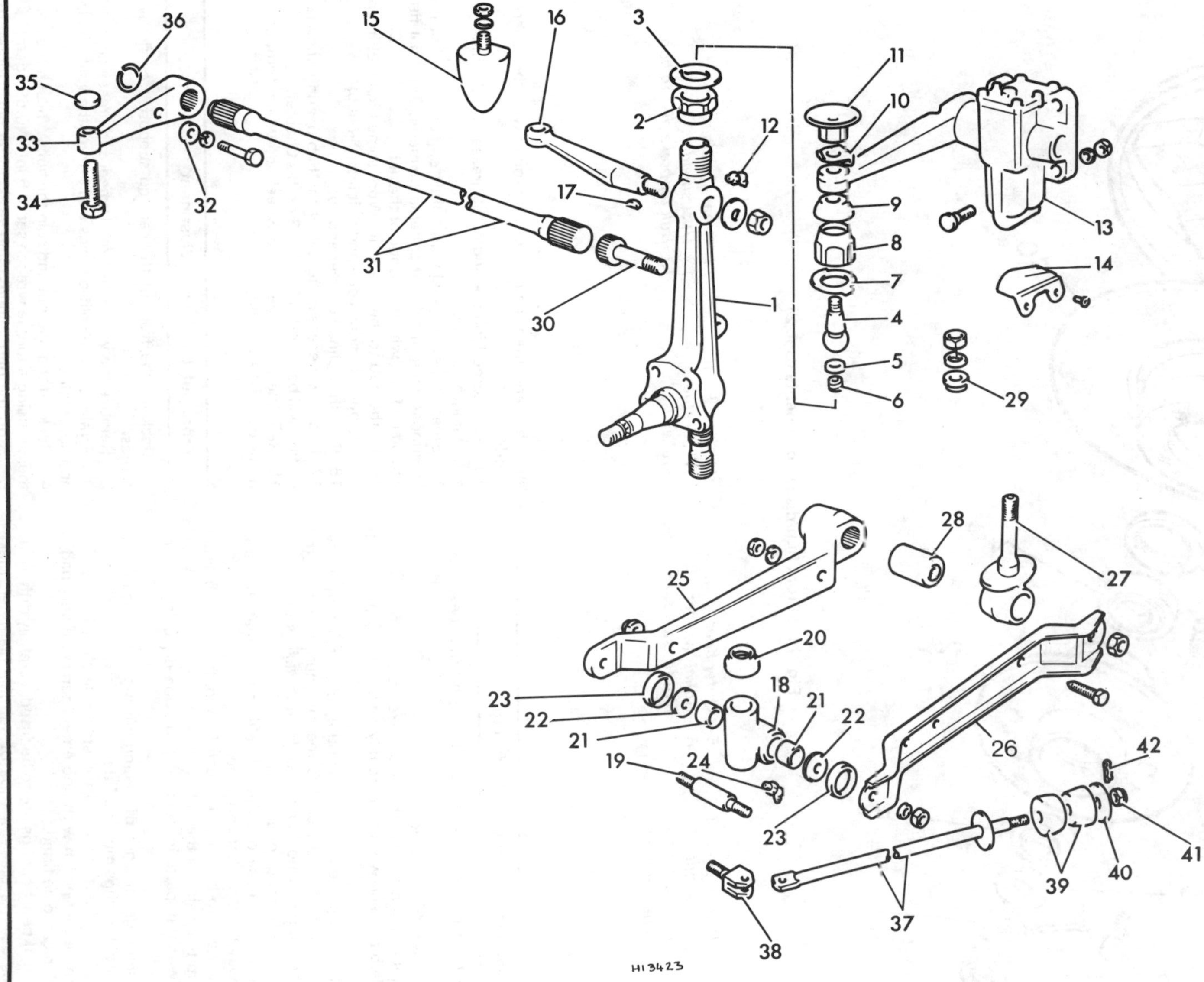

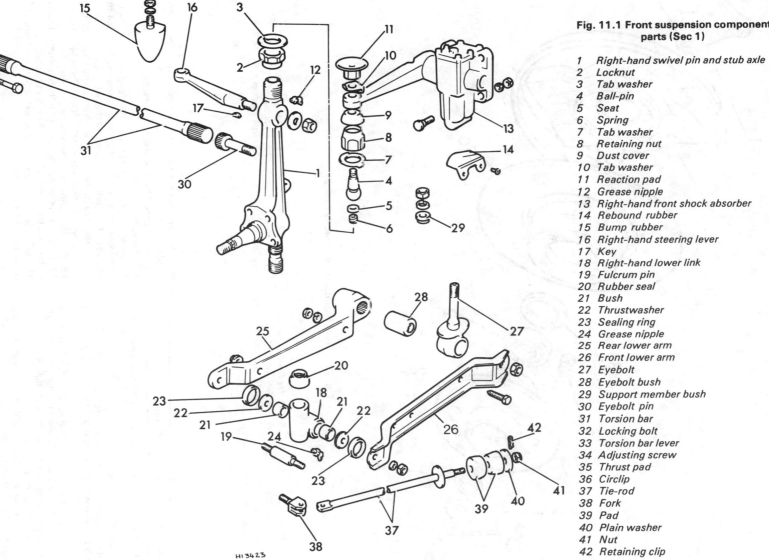

Fig. 11.1 Front suspension component parts (Sec 1)

1 *Right-hand swivel pin and stub axle*
2 *Locknut*
3 *Tab washer*
4 *Ball-pin*
5 *Seat*
6 *Spring*
7 *Tab washer*
8 *Retaining nut*
9 *Dust cover*
10 *Tab washer*
11 *Reaction pad*
12 *Grease nipple*
13 *Right-hand front shock absorber*
14 *Rebound rubber*
15 *Bump rubber*
16 *Right-hand steering lever*
17 *Key*
18 *Right-hand lower link*
19 *Fulcrum pin*
20 *Rubber seal*
21 *Bush*
22 *Thrustwasher*
23 *Sealing ring*
24 *Grease nipple*
25 *Rear lower arm*
26 *Front lower arm*
27 *Eyebolt*
28 *Eyebolt bush*
29 *Support member bush*
30 *Eyebolt pin*
31 *Torsion bar*
32 *Locking bolt*
33 *Torsion bar lever*
34 *Adjusting screw*
35 *Thrust pad*
36 *Circlip*
37 *Tie-rod*
38 *Fork*
39 *Pad*
40 *Plain washer*
41 *Nut*
42 *Retaining clip*

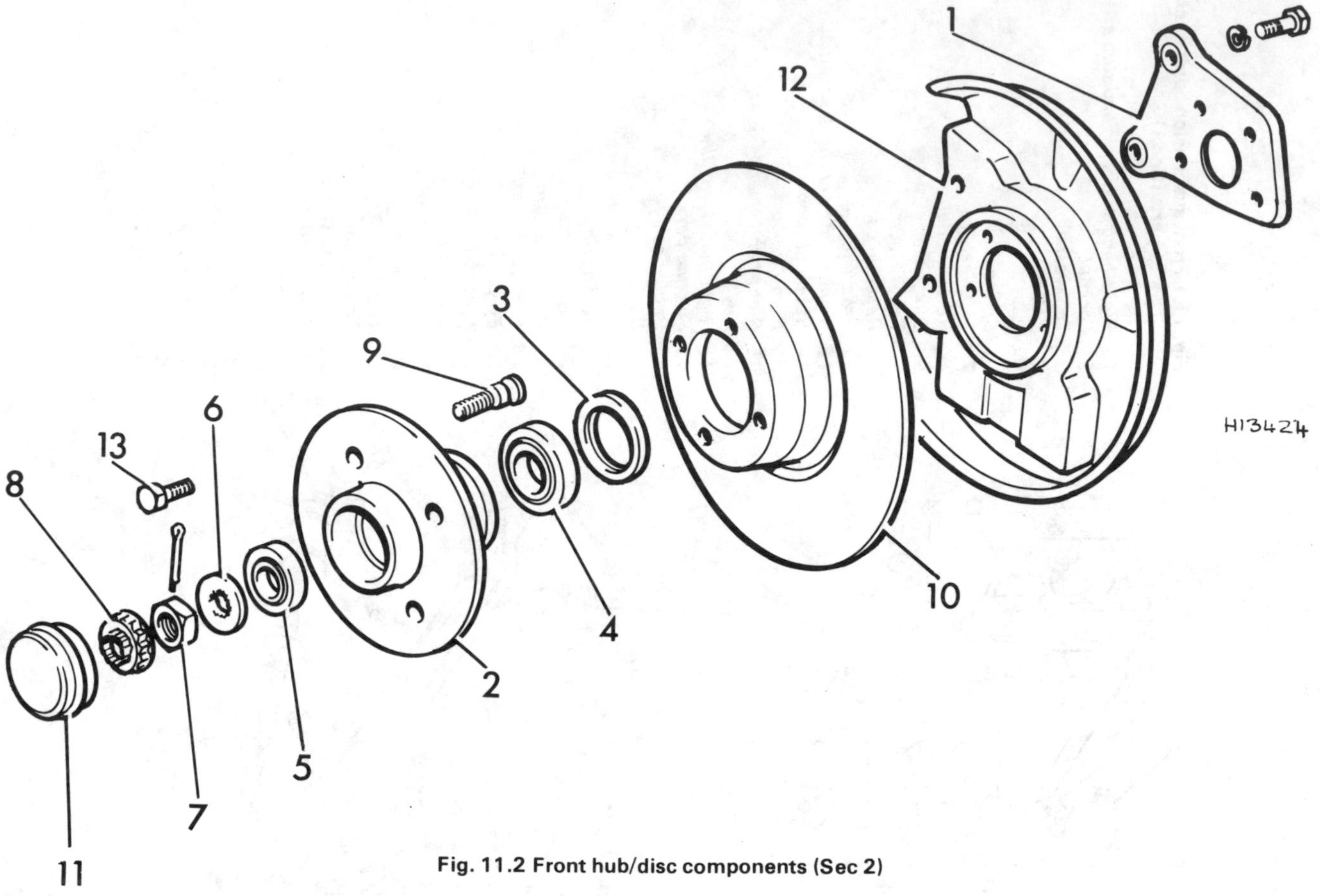

Fig. 11.2 Front hub/disc components (Sec 2)

1 Caliper bracket
2 Wheel hub
3 Oil seal for hub
4 Inner bearing
5 Outer bearing
6 Spline washer
7 Nut for stub axle
8 Retainer for nut
9 Wheel stud
10 Brake disc
11 Grease retaining cap
12 Dust shield
13 Bolt for brake disc to hub

2 Front hub bearings – removal and refitting

1 Jack up the front of the car and support it on firmly based axle-stands.
2 Remove the wheel trim and the road wheels.
3 Refer to Chapter 9 and remove the disc brake caliper.
4 Using a wide blade screwdriver, carefully ease off the grease cap (Fig. 11.2).
5 Straighten the split-pin legs and extract the split-pin. Lift away the nut retainer and then undo and remove the hub nut. Withdraw the splined washer. The hub may now be drawn from the axle stub.
6 Remove the outer bearing cone.
7 Using a screwdriver, ease out the oil seal, noting that the lip is innermost. Lift away the inner bearing cone.
8 If the bearings are to be renewed, carefully drift out the bearing cups working from the inside of the hub.
9 Thoroughly wash all parts in paraffin and wipe dry using a non-fluffy rag.
10 Inspect the bearings for signs of rusting, pitting or overheating. If evident, a new set of bearings must be fitted.
11 Inspect the oil seal journal face of the stub axle shaft for signs of damage. If evident, either polish with fine emery tape or if very bad, a new stub axle will have to be fitted.
12 To reassemble, if new bearings are to be fitted, carefully drift in the new bearing cups using a piece of tube of suitable diameter. Make sure they are fitted the correct way round with the tapers facing outwards.
13 Work some high melting point grease into the inner bearing cone and fit it into the hub.
14 Smear a new oil seal with a little engine oil and fit it with the lip innermost using a tube of suitable diameter. The final fitted position should be flush with the flange of the hub.
15 Fit the hub to the axle stub. Work some high melting point grease into the outer bearing cone and fit it into the hub.
16 Refit the splined washer and nut.
17 It is now necessary to adjust the hub bearing endfloat. Full information will be found in Section 3 of this Chapter.
18 Refit the grease cap, road wheel and wheel trim. Lower the front of the car to the ground.

3 Front hub bearings – adjustment

1 Jack up the front of the car and support it on firmly based axle stands.
2 Remove the wheel trim, road wheel and grease cap.
3 Straighten the split-pin legs and extract the split-pin. Lift away the hub nut retainer.
4 Back off the hub nut and spin the hub. Whilst it is spinning, tighten the nut using a torque wrench set to 5 lbf ft (0.69 kgf m).
5 Stop the hub spinning and slacken the nut. Tighten the nut again

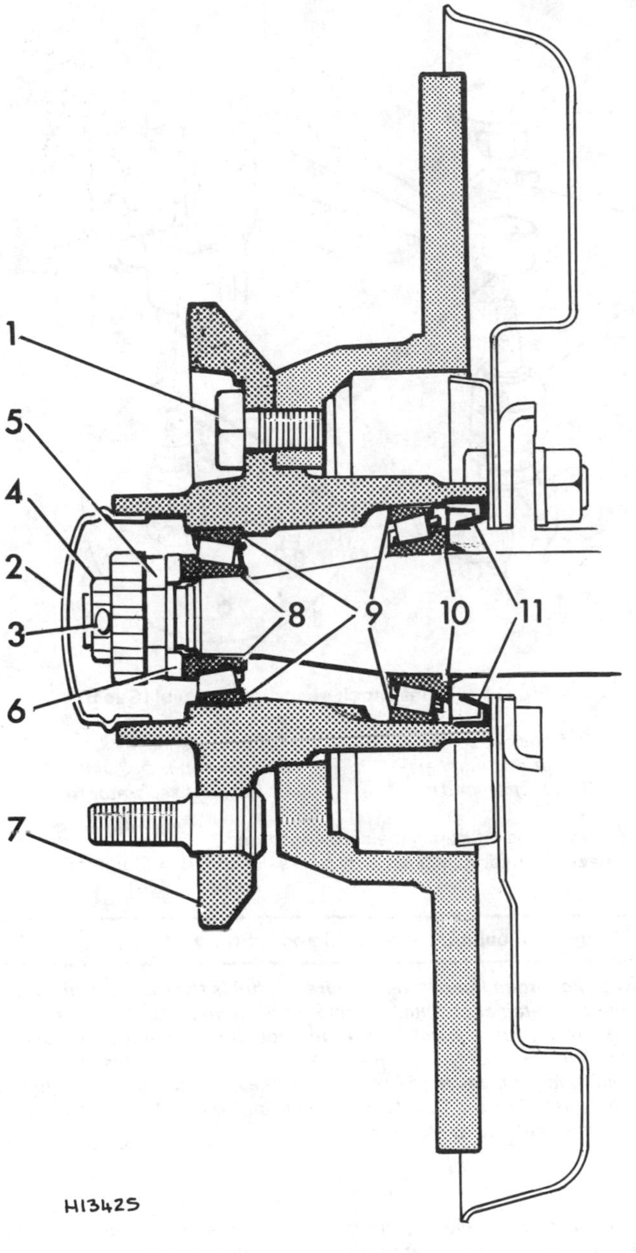

Fig. 11.3 Front hub assembly (Sec 2)

1 Disc to hub bolt
2 Grease cap
3 Split pin
4 Nut retainer
5 Nut
6 Splined washer
7 Hub
8 Outer bearing cone
9 Bearing cups
10 Inner bearing cone
11 Oil seal

but this time finger-tight only.

6 Position the nut retainer so that the left-hand half of the split-pin hole is covered by one of the arms of the retainer.

7 Slacken the nut and retainer until the split-pin hole is fully uncovered.

8 Fit a new split-pin and lock by opening the ears of the split-pin and bending circumferentially around the nut retainer.

9 Fit the grease cap and refit the road wheel and wheel trim.

10 It will be observed that the endfloat setting achieved can cause a considerable amount of movement when the tyre is rocked. Do not reduce the endfloat any further provided it has been set correctly as described. The bearings must not on any account be pre-loaded.

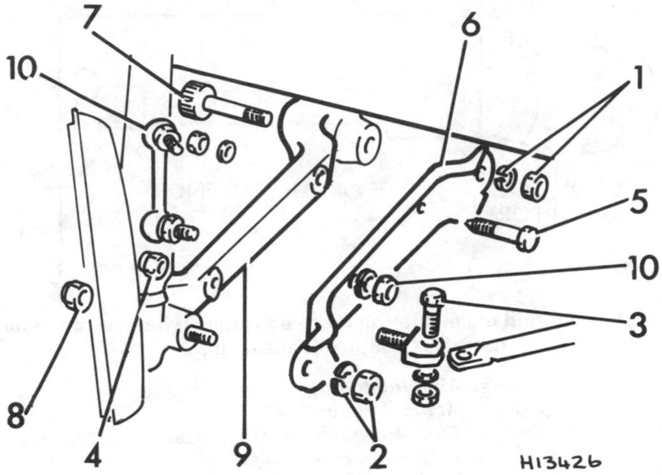

Fig. 11.4 Lower suspension arm removal (Sec 4)

1 Eyebolt pin nut and washer
2 Lower link pin securing nut and washer
3 Tie-rod securing bolt
4 Tie-rod fork retaining nut
5 Lower arm clamp bolt
6 Front lower suspension arm
7 Eyebolt pin
8 Swivel lower link pin nut
9 Lower suspension arm
10 Anti-roll bar link and securing nut

4 Lower suspension arm – removal and refitting

1 Jack up the front of the car and support it on firmly based axle-stands. Suitably support the suspension unit under the rear lower arm.

2 Remove the wheel trim and the road wheel.

3 Undo and remove the two anti-roll bar link securing nuts and washers, then pull the link away from the lower suspension arm (Fig. 11.4).

4 Undo and remove the nut and spring washer from the eye bolt pin.

5 Undo and remove the front nut and spring washers from the swivel lower link pin.

6 Undo and remove the nut, bolt and spring washer that retain the tie-rod to the tie-rod fork.

7 Undo and remove the nut that retains the tie-rod fork to the lower suspension arms. Lift away the fork.

8 Undo the nut, bolt and spring washer that clamp the front and rear lower arms together.

9 The front lower suspension arm may now be lifted away.

10 Refer to Section 10 and remove the torsion bar.

11 Withdraw the eyebolt pin and then undo and remove the rear nut and spring washer from the swivel lower link pin.

12 The rear lower suspension arm may now be lifted away.

13 Refitting the lower suspension arm assembly is the reverse sequence to removal, but the following additional points should be noted:

(a) Tighten the rod fork nut to a torque wrench setting of 48 to 55 lbf ft (6.6 to 7.7 kgf m)
(b) Tighten the tie-rod to fork nut to a torque wrench setting of 22 lbf ft (3.0 kgf m)

5 Front shock absorber – removal and refitting

Note: *The torque figure given in paragraph 11(b) is the true (actual) torque for the reaction pad nut. To obtain this figure, a torque wrench has to be used with a special crowfoot adaptor (Leyland part no 18G 1237) and, because the torque is applied to the adaptor rather than the nut, a formula must be used so that the indicated (metered) torque can be related to the true torque. Referring to Fig. 11.5, the metered*

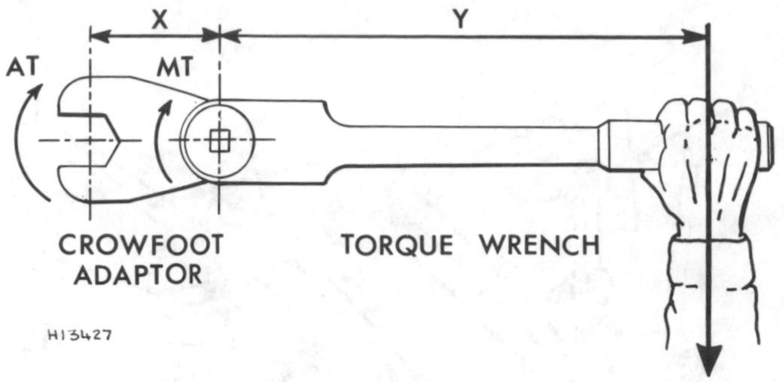

Fig. 11.5 Method of calculating metered torque when using special crowfoot adaptor (Secs 5 and 7)

MT = Metered torque
AT = Actual torque
Y = Effective length of torque wrench
X = Effective length of adaptor

torque is calculated as follows:

$$MT = \frac{AT \times Y}{Y + X}$$

If you do not have this crowfoot adaptor, an open-ended spanner ($\frac{15}{16}$ in AF) can be used, but you should arrange for your Leyland dealer to check the tightness of the nut after fitting.

1 Jack up the front of the car and support on firmly based axle-stands.
2 Remove the wheel trim and road wheel.
3 Remove the grease nipple from the swivel pin lower link and place an axle-stand beneath the lower suspension arm.
4 Unlock the reaction pad nut and remove the nut.
5 Lift away the lockwasher.
6 Free the shock absorber arm from the swivel pin balljoint using a balljoint separator tool.
7 Undo and remove the four nuts and plain washers that secure the shock absorber to its mounting.
8 Lift away the shock absorber.
9 Test the operation of the shock absorber by topping up the level if necessary and then moving the shock arm up and down. If the action is weak or jerky then either the unit is worn or air has entered the operating cylinders. Move the arm up and down ten times and if the performance has not improved, a new shock absorber must be obtained.
10 Refitting the shock absorber is the reverse sequence to removal, but the following additional points should be noted:

(a) Always use a new reaction pad lockwasher
(b) Tighten the reaction pad nut to a torque of 35 to 40 lbf ft (4.8 to 5.5 kgf m) using the crowfoot adaptor number 18G 1237
(c) Tighten the shock absorber retaining nuts to a torque wrench setting of 26 to 28 lbf ft (3.5 to 3.8 kgf m)

6 Swivel pin – removal and refitting

1 Refer to Section 5 and follow the instructions given in paragraphs 1 to 6 inclusive.
2 Raise the shock absorber arm.
3 Wipe the top of the brake master cylinder reservoir. Remove the cap and place a piece of thick polythene over the top. Refit the cap. This is to stop syphoning of fluid during subsequent operations.
4 Wipe the area around the flexible brake hose connection at the body mounted bracket. Hold the flexible hose metal end nut and undo and remove the metal pipe union nut. Undo and remove the flexible hose securing nut and star washer and draw the flexible hose from the bracket.
5 Remove the front lower suspension arm as described in Section 4, paragraphs 3 to 9 inclusive.
6 Undo and remove the lower link pin rear nut. Mark the fitted position of the special overtravel nut and remove it.
7 The swivel pin assembly may now be lifted away.
8 Refitting is the reverse of the removal procedure. Ensure that the

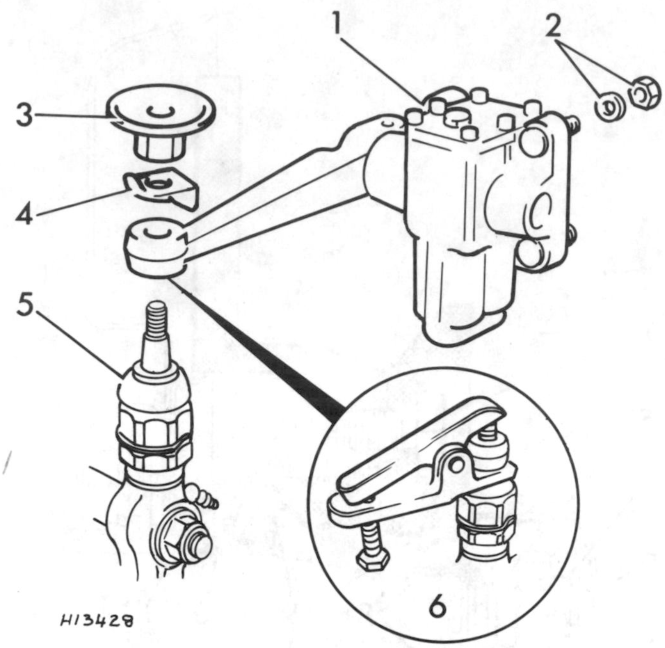

Fig. 11.6 Front shock absorber removal (Sec 5)

1 Shock absorber
2 Nut and washer
3 Reaction pad nut
4 Lockwasher
5 Swivel pin balljoint
6 Balljoint separator tool

overtravel nut is refitted in the same position.
9 Bleed the hydraulic brake system as described in Chapter 9.

7 Swivel pin balljoint – removal and refitting

Note: *The torque figure given in paragraph 9 is the true (actual) torque for the ball retainer locknut. To obtain this figure, a torque wrench has to be used with a special crowfoot adaptor (Leyland part no 18G 1192) and, because the torque is applied to the adaptor rather than the nut, a formula must be used so that the indicated (metered) torque can be related to the true torque. Referring to Fig. 11.5, the metered torque is calculated as follows:*

$$MT = \frac{AT \times Y}{Y + X}$$

If you do not have this crowfoot adaptor, an open-ended spanner ($1\frac{1}{2}$ in AF) can be used, but you should arrange for your Leyland dealer to check the tightness of the nut after refitting.

1 Refer to Section 5 and follow the instructions given in paragraphs 1 to 6 inclusive.
2 Raise the shock absorber arm.
3 Remove the dust cover and retaining clip.
4 Unlock the tab washer and using an open-ended spanner hold the locknut. With a ring spanner, undo the ball-pin retainer.
5 Lift away the ball-pin, ball seat and spring. Finally remove the tab washer and locknut.
6 To reassemble, first obtain a new tab washer. Smear the spherical surface of the ball-pin with Duckhams Q 5648 grease or an equivalent molybdenum disulphide grease prior to assembly. Pack the ball-pin retainer with a general purpose grease.
7 Fit the tab washer and locknut and then fit the ball seat and spring. Refit the ball-pin and its retainer.
8 Screw the locknut down away from the ball-pin retainer and tighten the ball retainer until the torque required to produce articulation of the ball-pin is 32 to 52 lbf in (0.38 to 0.56 kgf m).
9 Hold the ball retainer against rotation and tighten the locknut to a torque of 70 to 80 lbf ft (9.6 to 11.0 kgf m) using the crowfoot adaptor (Fig. 11.7).
10 Lock the retainer and the locknut with the tab washer.
11 Reassembly is now the reverse sequence to removal.

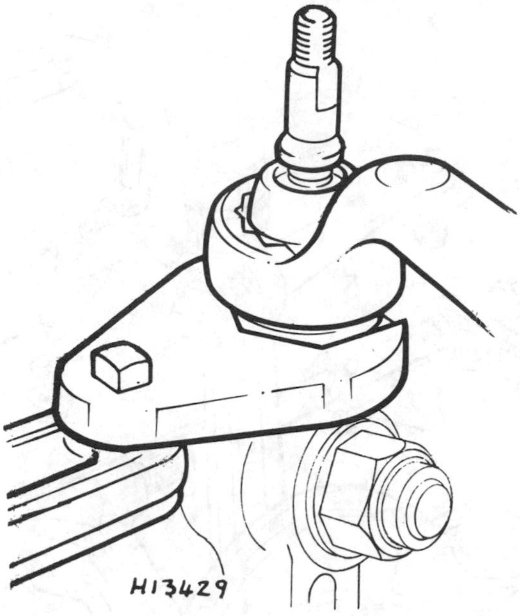

Fig. 11.7 Using the crowfoot adaptor (Sec 7)

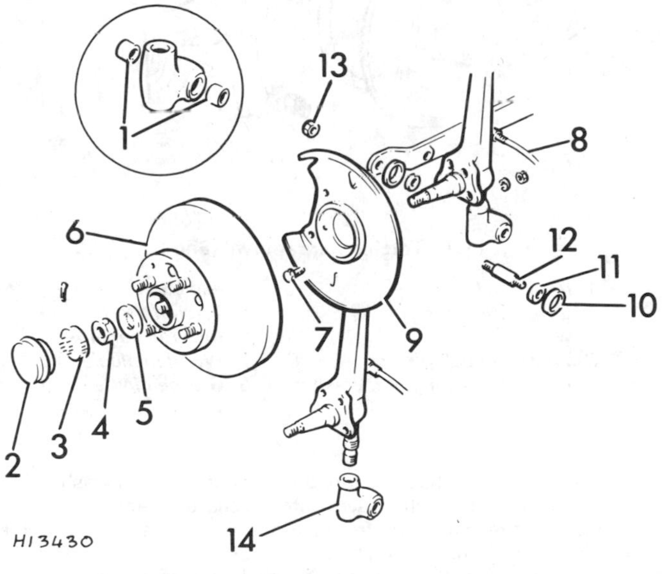

Fig. 11.8 Lower swivel pin link removal (Sec 8)

1 Lower link bushes
2 Grease cap
3 Nut retainer and split-pin
4 Nut
5 Splined washer
6 Hub
7 Mudshield retaining bolt and spring washer
8 Brake hydraulic hose
9 Disc brake mudshield
10 Sealing ring
11 Thrustwasher
12 Lower link pin
13 Link pin securing nut and spring washer
14 Lower link

8 Lower swivel pin link – removal and refitting

1 Jack up the front of the car and support it on firmly based axle-stands. Suitably support the rear lower suspension arm.
2 Remove the wheel trim and road wheel.
3 Refer to Chaper 9 and remove the disc brake caliper.
4 Using a wide blade screwdriver, carefully ease off the grease cap (Fig. 11.8).

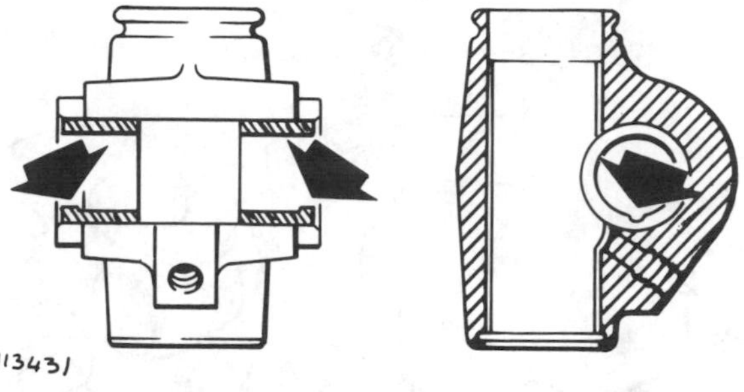

Fig. 11.9 Correct position of bush oil groove (Sec 8)

5 Straighten the split-pin legs and extract the split-pin. Lift away the nut retainer and then undo and remove the hub nut. Withdraw the splined washer.
6 The hub may now be drawn from the axle stub.
7 Wipe the top of the brake master cylinder reservoir. Remove the cap and place a piece of thin polythene over the top. Refit the cap. This is to stop syphoning of fluid during subsequent operations.
8 Wipe the area around the flexible brake hose connection at the body mounted bracket. Hold the flexible hose metal end nut and undo and remove the metal pipe union nut.
9 Undo and remove the flexible hose securing nut and star washer and draw the flexible hose from the bracket.
10 Undo and remove the four nuts, bolts and spring washers that secure the dust shield and caliper bracket to the swivel pin. Lift away the dust shield and caliper bracket.
11 Refer to Section 4, paragraphs 3 to 9, and remove the lower suspension arm.
12 Undo and remove the remaining nut and spring washer from the lower link pin.
13 Swing the swivel pin forwards and remove the rubber sealing rings and thrustwashers from the lower link.
14 Withdraw the lower link pin.
15 Unscrew and remove the lower link from the swivel pin.
16 Thoroughly wash all parts in paraffin and wipe dry using a non fluffy rag.
17 Check for excessive wear across the thrust faces and in the threaded bore. If wear is excessive, a new swivel link must be obtained.
18 Check the lower link bushes for wear and if this is evident, new bushes should be fitted by a BL garage as it has to be ream finished. If an expanding reamer and micrometer are available however, the old bushes should be drifted out. Further instructions are given in paragraphs 22 and 23.
19 Inspect the thrustwashers for signs of damage or wear which if evident, new thrustwashers must be obtained.
20 Remove the grease nipple and ensure that both it and its hole are free from obstructions.
21 Obtain a new set of rubber sealing rings.
22 If new bushes are to be fitted, these should be drifted or pressed in so that the oil groove is located as shown in Fig. 11.9. The bush oil groove blank ends should be towards the outside edge of the link.
23 Using the expanding reamer line, ream the new bushes to a finished size of 0.688 in $\pm$ 0.0005 in (17.48 mm $\pm$ 0.013 mm).
24 Pack the area between the lower link bushes and the swivel pin threads with approximately 2.5 cc of Duckhams Q5648 grease or an equivalent molybdenum disulphide grease.
25 Place the swivel pin link and seal on the swivel pin and screw on the link. Engage the seal on the recessed shoulder of the link and screw the link fully onto the swivel pin.
26 Unscrew the link one complete turn.
27 Reassembly is now the reverse sequence to removal, but the following additional points should be noted:

(a) The caliper bracket and dust shield retaining bolts should be tightened to a torque wrench setting of 35 to 42 lbf ft (4.8 to 5.8 kgf m)
(b) Refer to Section 3 and adjust the front hub bearing endfloat
(c) Bleed the brake hydraulic system, refer to Chapter 9

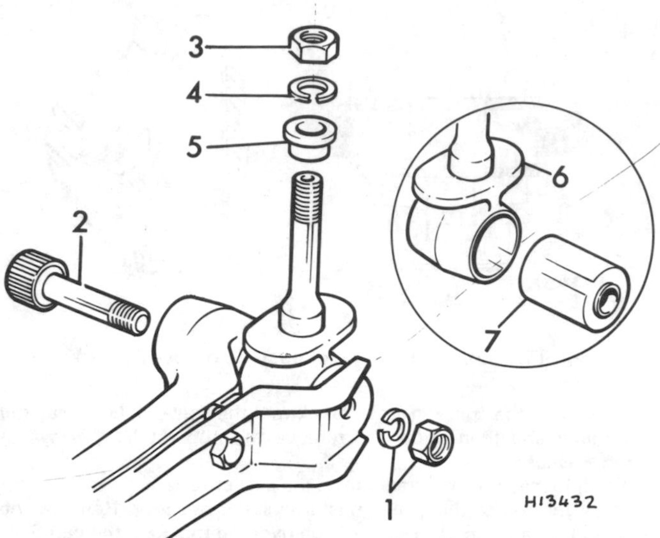

Fig. 11.10 Eyebolt bush removal (Sec 9)

1 Eyebolt pin nut and spring washer
2 Eyebolt pin
3 Nut
4 Spring washer
5 Spacer
6 Eyebolt
7 Bush

9 Eyebolt bush – removal and refitting

1 Refer to Section 10 and remove the torsion bar.
2 Undo and remove the nut and spring washer from the eyebolt.
3 Withdraw the eyebolt pin (Fig. 11.10).
4 Draw the suspension assembly clear of the eyebolt.
5 Undo and remove the nut, spring washer and spacer from the eyebolt.
6 Lift away the eyebolt, and, if fitted, the reinforcement plate.
7 The bush may be removed using pieces of suitable diameter tube and pressing it out in a bench vice.
8 To fit a new bush, lubricate its outer surface with a little soapy water and press it in using the reverse procedure to removal.
9 Refitting the eyebolt is the reverse sequence to removal.
10 Tighten the eyebolt retaining nut to a torque wrench setting of 50 to 54 lbf ft (6.9 to 7.4 kgf m).

10 Torsion bar – removal and refitting

1 Unscrew and remove the grease nipple from the swivel pin lower link.
2 Place a wooden block 8 in (200 mm) thick on the floor under the lower suspension arm as near as possible to the disc brake dust shield as shown in Fig. 11.11.
3 Jack up the front of the car. Remove the wheel trim and road wheel.
4 Carefully lower the car until the weight of the suspension is placed on the wooden block.
5 Undo and remove the two anti-roll bar link retaining nuts and washers and pull the link away from the anti-roll bar and lower suspension arm.
6 Unlock and remove the reaction pad.
7 Free the shock absorber arm from the swivel pin balljoint using a balljoint separator tool.
8 Undo and remove the steering track rod ball-pin nut.
9 Using a universal balljoint separator, release the ball-pin from the steering lever.
10 Jack up the front of the car so as to relieve the torsion bar load. Ensure the lower suspension arm is still just resting on the wooden block.

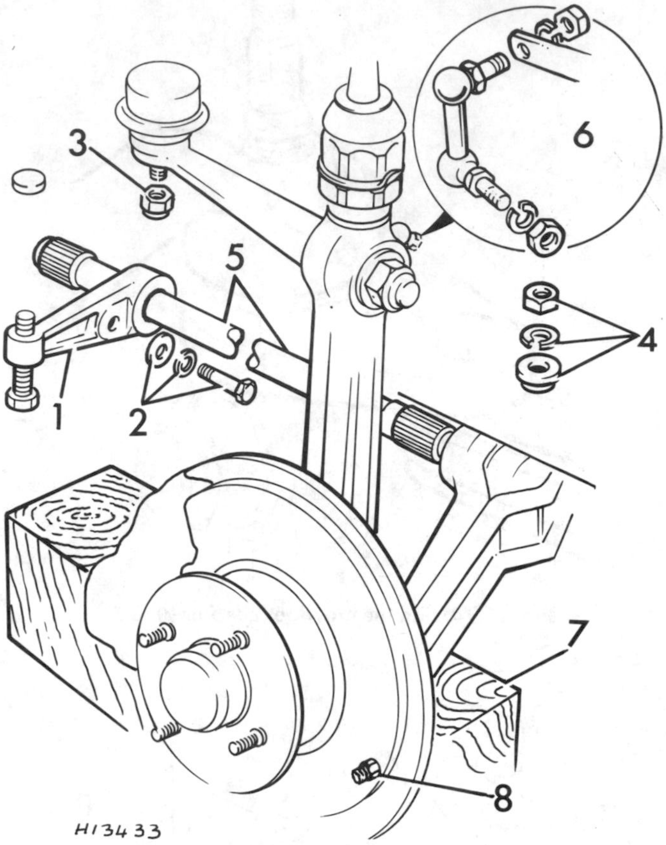

Fig. 11.11 Torsion bar removal (Sec 10)

1 Reaction lever
2 Bolt, spring and special washer
3 Track-rod ball-pin nut
4 Eyebolt retaining nut, washer and bush
5 Torsion bar
6 Anti-roll bar link
7 Wooden block
8 Grease nipple

11 Undo and remove the bolt, spring washer and special washer that secure the torsion bar reaction lever onto the chassis member.
12 Remove the reaction lever from the chassis member and move the lever forwards along the torsion bar.
13 Release the nut that retains the eyebolt through the chassis member and make sure that the suspension lowers itself by $\frac{1}{2}$ in (12 mm).
14 Ease the torsion bar forwards until it clears the shoulder from the chassis housing. Lower the torsion bar and remove it in a rearwards direction.
15 Using a pair of circlip pliers, remove the torsion bar circlip.
16 Slide off the reaction lever from the torsion bar.
17 Refitting the torsion bar is the reverse sequence to removal but the following additional points should be noted:

- *(a) Once a torsion bar has been fitted and used on one side of the car it must not under any circumstances be used on the other side. This is because a torsion bar becomes handed once it has been in use. Torsion bars are only interchangeable when new*
- *(b) Do not fit a torsion bar that is corroded or deeply scored as this will affect its reliability and in bad cases cause premature failure*

18 When refitting a torsion bar that has been removed without using a wooden block of the correct size, or when fitting a new torsion bar, it should be initially set up using the following procedure before fitting the reaction lever:
19 First measure the height of the eyebolt centre above the floor

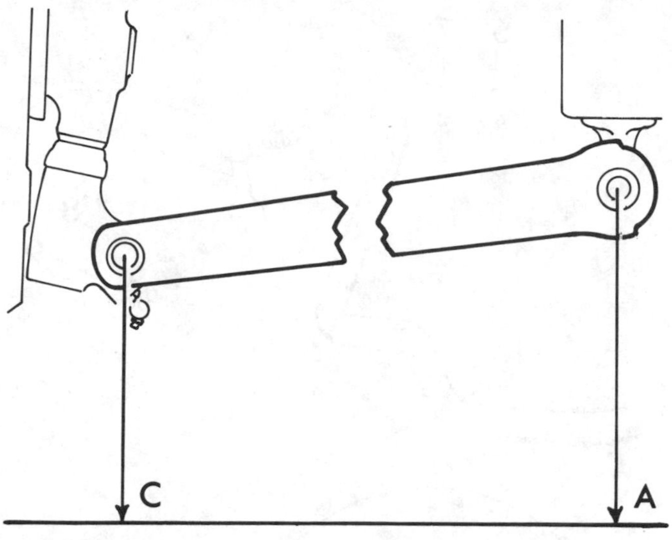

Fig. 11.12 Torsion bar setting dimensions (Sec 10)

(dimension A in Fig. 11.12). Then deduct the dimension given in the following table to obtain the dimension C in Fig. 11.12.

Torsion bar settled – 7.37 in (187.20 mm)
New unsettled bar – 7.66 in (194.56 mm)

20 Adjust the swivel link pin centre to dimension C and retain it in this position. Fit the reaction lever with the adjusting screw set in the mid-way position, making sure that the eyebolt centre to floor dimension is not altered.
21 Tighten the eyebolt nut to a torque wrench setting of 50 to 54 lbf ft (6.9 to 7.4 kgf m).
22 Tighten the reaction lever to chassis member bolt to 22 lbf ft (3.0 kgf m).
23 Tighten the track-rod ball-pin to a torque wrench setting of 20 to 24 lbf ft (2.7 to 3.3 kgf m).
24 Tighten the reaction pad nut to 35 to 40 lbf ft (4.8 to 5.5 kgf m).
25 Refer to Section 13 and adjust the front suspension trim height if necessary.

11 Tie-rod – removal and refitting

1 Jack up the front of the car and support it on firmly based axle-stands.
2 Remove the wheel trim and road wheel.
3 Using a pair of pliers, remove the tie-rod spring clip from the end of the tie-rod (Fig. 11.13).

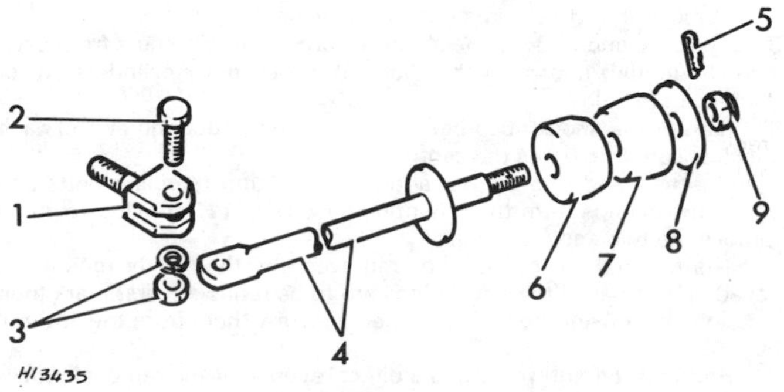

Fig. 11.13 Tie-rod removal (Sec 11)

1 Tie-rod fork
2 Tie-rod retaining bolt
3 Nut and spring washer
4 Tie-rod
5 Spring clip
6 Inner pad
7 Outer pad
8 Plain washer
9 Nut

4 Undo and remove the locking nut and large plain washer.
5 Slide off the rubber outer pad.
6 Undo and remove the nut, spring washer and bolt that secure the tie-rod to the fork end.
7 Remove the rubber inner pad from the tie-rod.
8 Undo and remove the nut that secures the rod fork and lift away the fork from the lower suspension arm.
9 Refitting the tie-rod is the reverse sequence to removal but the following additional points should be noted:

(a) Inspect the two rubber pads and if they show signs of oil contamination, cracking or perishing, obtain and fit a new pair of pads
(b) The tie-rod to fork nut should be tightened to a torque wrench setting of 22 lbf ft (3.0 kgf m)
(c) Tighten the rod fork nut to a torque wrench setting of 48 to 55 lbf ft (6.6 to 7.6 kgf m)

12 Front anti-roll bar – removal and refitting

1 Jack up the front of the car and support it on firmly based axle-stands.
2 Unscrew and remove the link retaining nuts and spring washers from both ends of the anti-roll bar and detach the link ends (Fig. 11.14).
3 Mark the anti-roll bar mounting carriers 'left' and 'right' so that they can be refitted in their original positions, then unscrew and remove the four retaining screws.
4 Note that the front left-hand face of the anti-roll bar, near the link position, is marked L and then lower the complete bar.
5 Detach the rubber bushes from the anti-roll bar and unscrew and remove the anti-roll bar links from the lower suspension arms.
6 Renew any bushes or links which show any indication of wear or deterioration.
7 Refitting is a reversal of the removal procedure but soak the rubber bushes in soapy water prior to tightening the retaining carriers.

13 Front suspension trim height – checking and adjustment

1 Before checking the front trim height of the car, it must be prepared by removing the contents of the boot with the exception of the spare wheel. Ideally there should be two gallons of petrol in the tank. Check and, if necessary, adjust the tyre pressures.
2 Stand the car on a level surface and measure the vertical distance from the centre of the hub to the underside of the front wheel arch. This height measurement should be 14.63 $\pm$ 0.25 in (371.6 $\pm$ 6.35 mm).
3 To adjust the trim height, slacken the adjustment lever lockbolt and turn the adjuster bolt as necessary to obtain the correct height (Fig. 11.15). Turning the adjuster bolt clockwise increases the trim height and turning anti-clockwise decreases the trim height. Maximum movement of the adjuster bolt from the mid-way position increases the trim height by 0.75 in (19.05 mm) or decreases the trim height by 1.25 in (31.75 mm).
4 Tighten the adjusting lever lockbolt.

14 Rear hub assembly – removal and refitting

1 Chock the front wheel, jack up the rear of the car and place it on firmly based axle-stands.
2 Remove the wheel trim and road wheel. Apply the handbrake.
3 Undo and remove the axle shaft nut and washer.
4 Release the handbrake and referring to Chapter 9, back off the brake adjuster. Remove the two countersunk screws that retain the brake drum and pull off the brake drum. If it is tight, tap the circumference with a soft-faced hammer.
5 Using a heavy duty puller and suitable thrust block, pull the hub from the end of the axleshaft.
6 Remove the axleshaft key.
7 Refitting the rear hub assembly is the reverse of the removal procedure. Clean the threads of the axleshaft and nut and apply a thread locking compound before tightening the nut to a torque wrench setting of 105 lbf ft (14.5 kgf m).

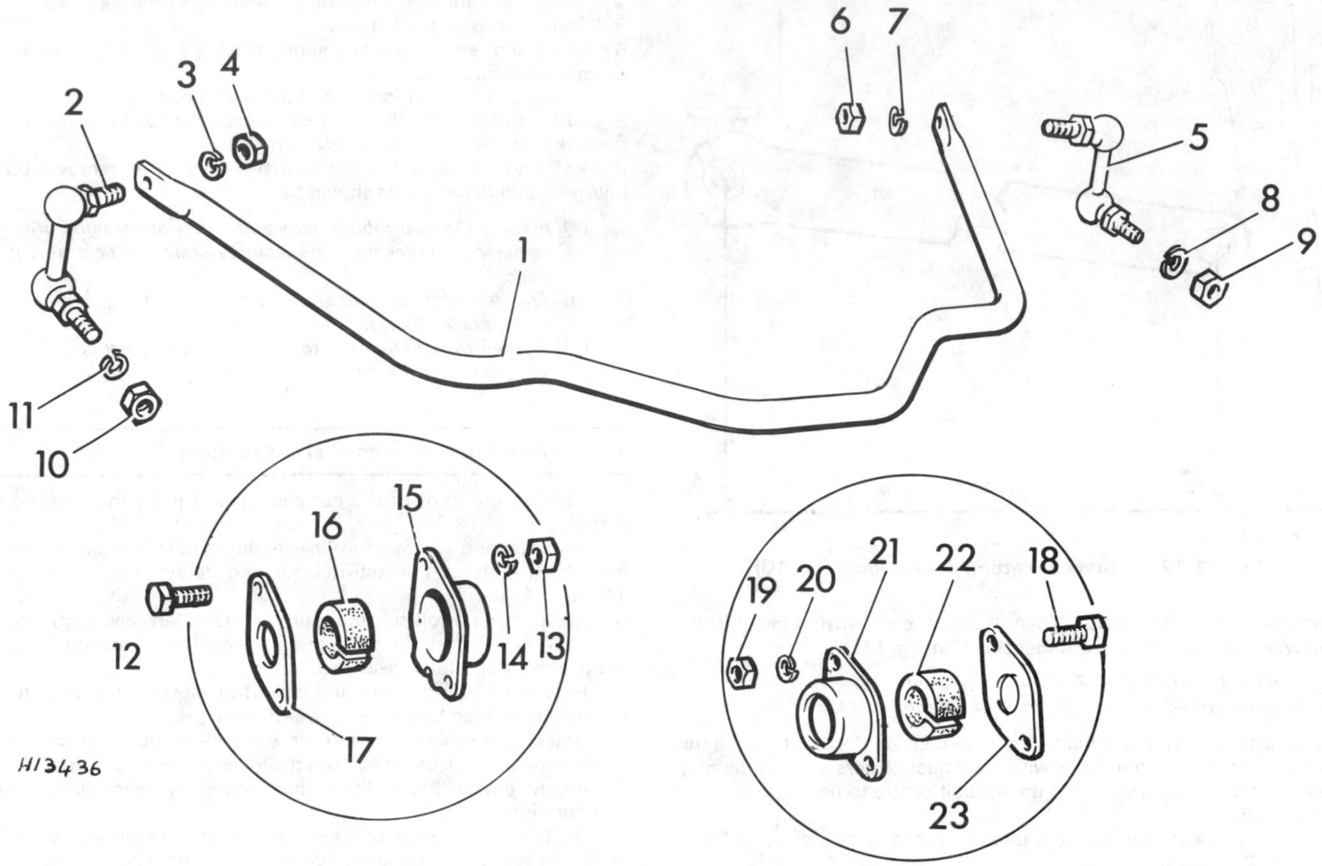

Fig. 11.14 Front anti-roll bar components (Sec 12)

1 *Front anti-roll bar*
2 *Right-hand link*
3 *Spring washer*
4 *Nut*
5 *Left-hand link*
6 *Nut*
7 *Spring washer*
8 *Spring washer*
9 *Nut*
10 *Nut*
11 *Spring washer*
12 *Bolt*
13 *Nut*
14 *Spring washer*
15 *Right-hand bearing carrier*
16 *Bearing*
17 *Bearing retainer*
18 *Bolt*
19 *Nut*
20 *Spring washer*
21 *Left-hand bearing carrier*
22 *Bearing*
23 *Bearing retainer*

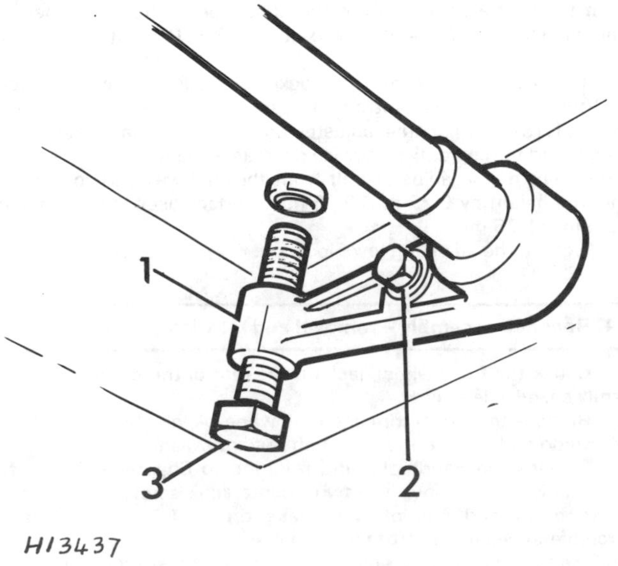

Fig. 11.15 Front suspension trim height adjustment (Sec 13)

1 *Adjustment lever*
2 *Lockbolt*
3 *Adjuster bolt*

15 Rear anti-roll bar – removal and refitting

1 Jack up the rear of the car and support it on firmly based axle-stands.
2 Remove the wheel trims and road wheels.
3 Unscrew and remove the shock absorber retaining nuts from each lower mounting and push the shock absorber lower cylinders out of their locating holes.
4 Loosen the shock absorber upper mounting nuts and swivel each shock absorber towards the centre of the car.
5 Unscrew and remove the anti-roll bar clamp retaining bolts and unclip the clamps from the mounting base (Fig. 11.16). Unscrew and remove the two end pivot bolts.
6 The anti-roll bar can now be removed and the rubber mountings eased off the bar. If the end fittings are to be renewed, first mark their location and loosen the locknuts, then unscrew them from the anti-roll bar.
7 Refitting the anti-roll bar is a direct reversal of the removal procedure but the following additional points should be noted:

(a) Make sure that the end fitting bushes line up with the chassis number bolt holes and adjust them on the anti-roll bar if necessary. It is important not to distort the rubber bushes when inserting the pivot bolt
(b) Tighten the anti-roll bar pivot bolts to 48 to 55 lbf ft (6.6 to 7.6 kgf m) with a torque wrench
(c) Make sure that the anti-roll bar locknuts are tightened securely

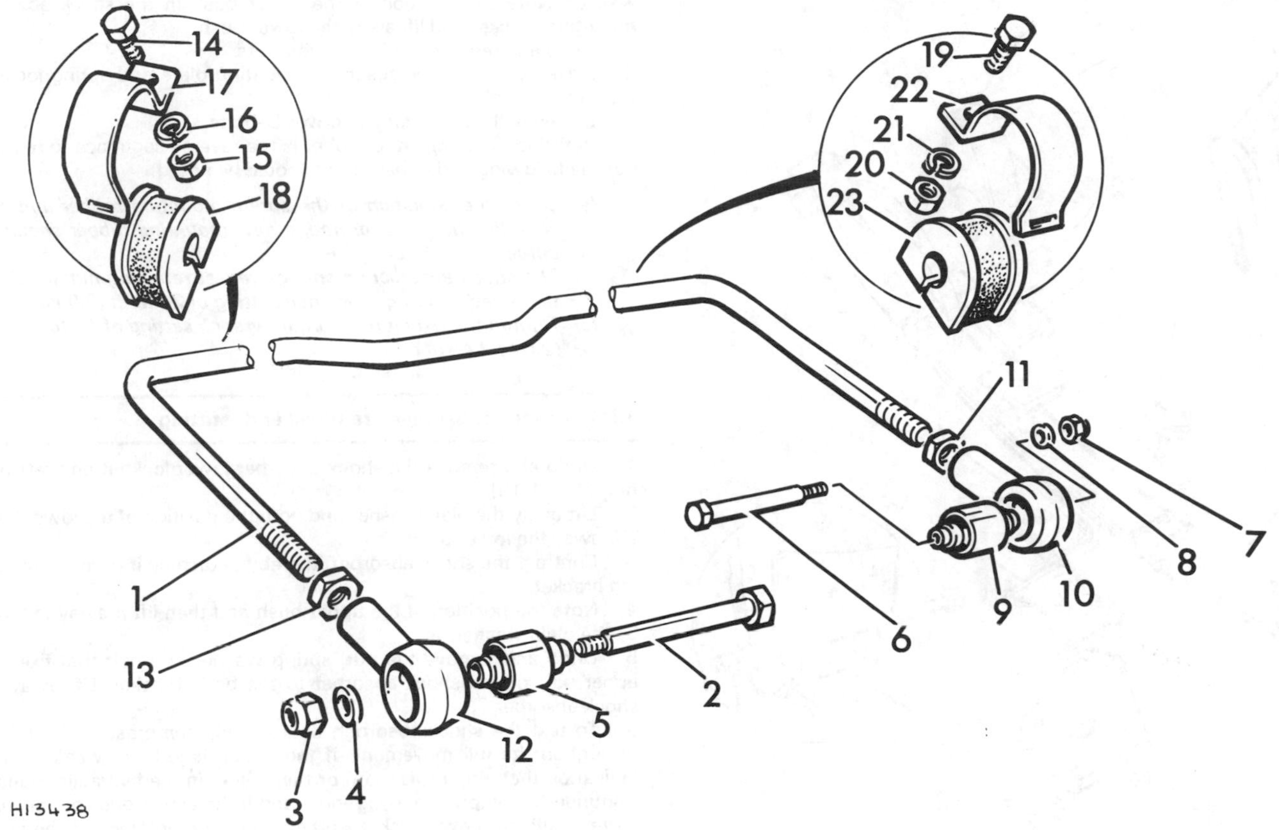

Fig. 11.16 Rear anti-roll bar components (Sec 15)

1 *Rear anti-roll bar*
2 *Bolt*
3 *Locknut*
4 *Washer*
5 *Slotted bush*
6 *Bolt*
7 *Locknut*
8 *Washer*
9 *Slotted bush*
10 *End fitting*
11 *Locknut*
12 *End fitting*
13 *Locknut*
14 *Bolt*
15 *Nut*
16 *Spring washer*
17 *Clamp*
18 *Bearing*
19 *Bolt*
20 *Nut*
21 *Spring washer*
22 *Clamp*
23 *Bearing*

16 Rear road spring – removal and refitting

1 Refer to Section 17 and remove the road spring shackle plate.
2 Jack up the rear of the car and support it on firmly based axle-stands located under the main longitudinal chassis members. Support the weight of the axle on the side from which the spring is to be removed.
3 Undo and remove the shock absorber locknut, plain nut and plain washer. Note the location of the lower bush in the shock absorber lower mounting plate and remove the lower bush.
4 Undo and remove the nut, spring washer and bolt that secure the front spring eye to the body mounted brackets (Fig. 11.17).
5 Undo and remove the four nuts from the two U-bolts.
6 Carefully lower the spring and its mountings.
7 Remove the shock absorber mounting plate followed by the spring mounting plates and mounting rubbers. Note the fitted location of the spring mounting wedge.
8 Lift away the two U-bolts and packing plate (if fitted).
9 If the spring bushes are worn or have deteriorated, they should be pressed out using suitable diameter tubes and a large bench vice.
10 Should the spring have considerably weakened or failed, necessitating the fitting of a new one, rear springs must be renewed in pairs and not singly as the remaining spring will have settled slightly. Unless the springs have the same performance and characteristics, road holding can be adversely affected.
11 Refitting the road spring is the reverse sequence to removal, but the following additional points should be noted:

(a) Tighten the upper shackle pin nuts to a torque wrench setting of 28 lbf ft (3.9 kgf m)
(b) Tighten the spring eye bush bolt nuts to a torque wrench setting of 40 lbf ft (5.5 kgf m)
(c) Tighten the U-bolt nuts to a torque wrench setting of 15 to 18 lbf ft (2.0 to 2.4 kgf m)
(d) Tighten the shock absorber to spring bracket retaining nut to a torque wrench setting of 45 lbf ft (6.2 kgf m) and then secure by tightening the locknut

17 Rear road spring shackles – removal and refitting

1 Chock the front wheels, jack up the rear of the car and place it on firmly based axle-stands located under the main longitudinal chassis members.
2 Remove the wheel trim and road wheel.
3 Undo and remove the nut and spring washer on each side of the upper shackle pin.
4 Undo and remove the nut and spring washer from the spring bush bolt.
5 Lift away the inner shackle plate.
6 Using a suitable diameter parallel pin punch, partially drift out the spring bolt and then release the outer plate from the upper pin (Fig. 11.18).

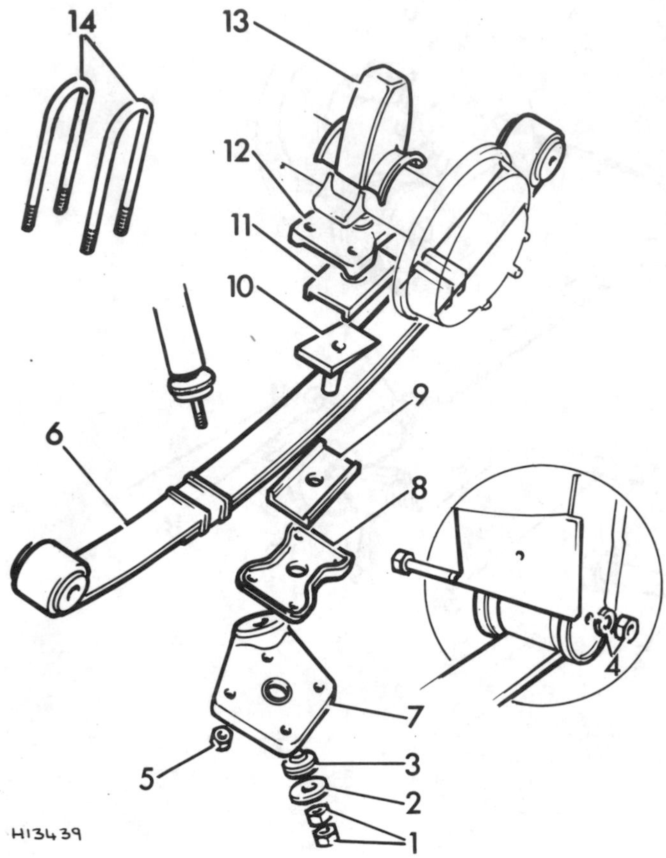

Fig. 11.17 Rear road spring removal (Sec 16)

1 *Shock absorber retaining nut and locknut*
2 *Plain washer*
3 *Lower bush*
4 *Forward spring eyebolt securing nut and spring washer*
5 *U-bolt nut*
6 *Spring assembly*
7 *Shock absorber mounting plate*
8 *Spring mounting plate (lower)*
9 *Rubber pad*
10 *Wedge*
11 *Rubber pad*
12 *Spring mounting plate (upper)*
13 *Rubber bump stop*
14 *U-bolts*

7 Remove the upper shackle pin and lift away the two half bushes.
8 Inspect the bushes for signs of deterioration or oil contamination which, if evident, new bushes should be obtained.
9 Refitting is the reverse sequence to removal, but the following additional points should be noted:

(a) Tighten the upper shackle pin nuts to a torque wrench setting of 28 lbf ft (3.9 kgf m)
(b) Tighten the spring eye bush bolt nut to a torque wrench setting of 40 lbf ft (5.5 kgf m)

18 Bump stop – removal and refitting

1 Chock the front wheels, jack up the rear of the car and place it on firmly based axle-stands located under the main logitudinal chassis members.
2 Remove the wheel trim and road wheel.
3 Support the weight of the axle on the side to be worked on.
4 Undo and remove the shock absorber locknut, plain nut and plain washer. Note the location of the lower bush in the shock absorber mounting bracket and lift away the lower bush.
5 Undo and remove the four U-bolt nuts.
6 Lift away the shock absorber mounting plate and spring locating bracket and rubber.
7 Lift away the bump stop and two U-bolts.
8 Refitting the bump stop rubber is the reverse sequence to removal but the following additional points should be noted:

(a) Check the condition of the spring mounting rubber and if its condition has deteriorated, a new mounting rubber should be obtained and fitted
(b) The shock absorber to spring bracket retaining nut should be tightened to a torque wrench setting of 28 lbf ft (3.9 kgf m)
(c) Tighten the U-bolt to a torque wrench setting of 15 to 18 lbf ft (2.0 to 2.4 kgf m)

19 Rear shock absorber – removal and refitting

1 Undo and remove the shock absorber lower locknut and retaining nut (Fig. 11.19).
2 Lift away the plain washer and note the position of the lower bush. Lift away the lower bush.
3 Contract the shock absorber, thereby detaching it from the mounting bracket.
4 Note the position of the upper bush and then lift it away followed by the plain washer.
5 Undo and remove the nut, spring washer and bolt that fixes the upper part of the shock absorber to the body bracket. Lift away the shock absorber.
6 To test the shock absorber, alternatively compress and extend it throughout its full movement. If the action is jerky or weak, it is an indication that either it is worn or there is air in the hydraulic cylinder. Continue to compress and extend it and if the action does not become more positive a new shock absorber should be obtained. If the shock absorber is showing signs of leaking it should be discarded as it is not possible to overhaul it.
7 Check the bushes and if they show signs of deterioration a new set of rubbers should be obtained.
8 Refitting the shock absorber is the reverse sequence to removal but the following additional points should be noted:

(a) Tighten the shock absorber to body bracket retaining bolt nut to a torque wrench setting of 45 lbf ft (6.2 kgf m)
(b) The shock absorber to spring bracket should be tightened to a torque wrench setting of 28 lbf ft (3.9 kgf m) and then locked with the locknut

20 Steering wheel – removal and refitting

1 Prise the motif pad from the centre of the steering wheel.
2 Unscrew and remove the steering wheel retaining nut and washer (Fig. 11.20).
3 With the palms of the hands behind the spokes and near to the centre hub, knock the steering wheel from the inner column splines. If it is tight, it will be necessary to use a universal puller and suitable thrust block. The steering column cowl will also need to be removed as follows:
4 Disconnect the choke cable at the carburettor.
5 Unscrew and remove the right-hand cowl securing screw and withdraw the cowl over the switch lever.
6 Unscrew and remove the left-hand cowl securing screws and withdraw the cowl, at the same time pulling the choke cable through the bulkhead panel until the cowl is clear of the steering wheel.
7 Refitting the steering wheel is the reverse of the removal procedure but the following additional points should be noted:

(a) Make sure that the arrow on the cancelling ring is facing towards the indicator switch when fitting the steering wheel, see Fig. 11.19
(b) With the road wheels in the straight-ahead position, fit the steering wheel on the inner column so that the spokes are in a horizontal plane
(c) The steering wheel securing nut should be tightened to a torque wrench setting of 32 to 37 lbf ft (4.43 to 5.13 kgf m)

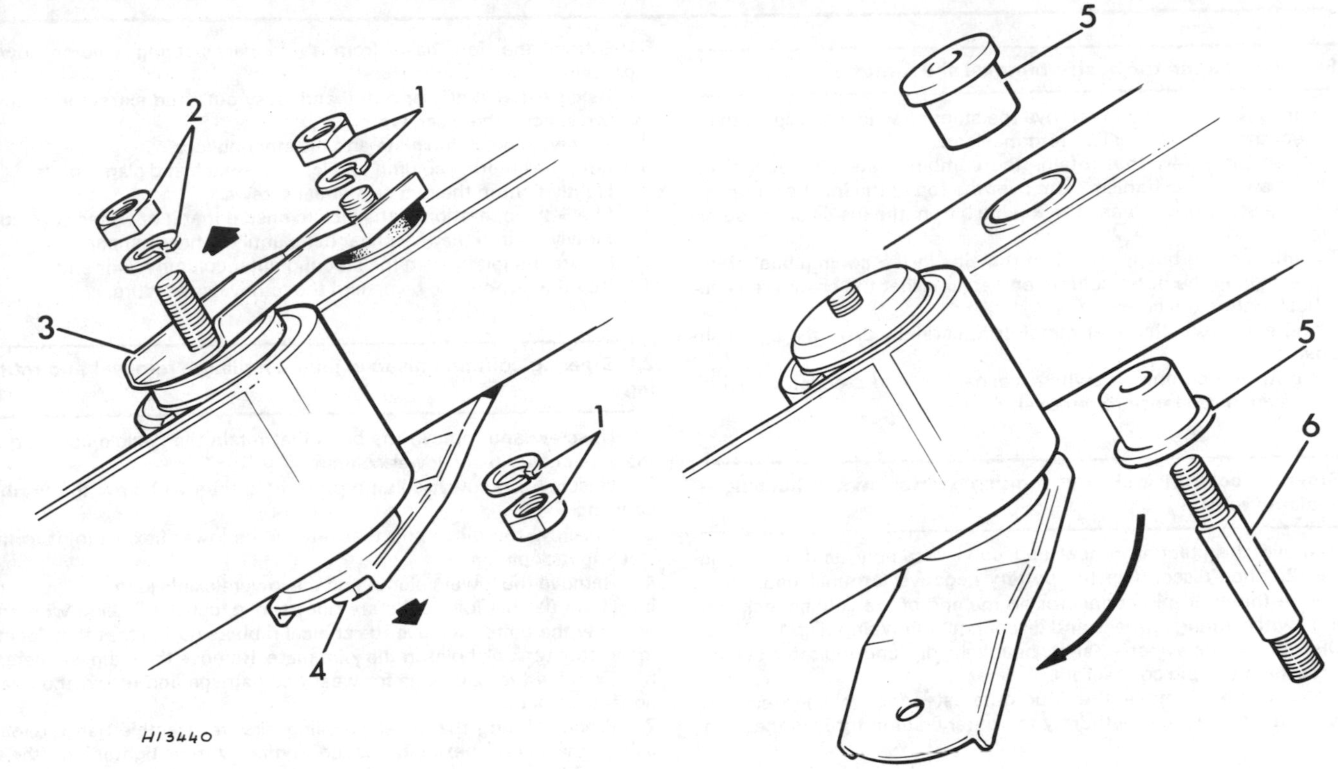

Fig. 11.18 Rear spring shackle removal (Sec 17)

1 Upper shackle pin securing nut and spring washer
2 Lower spring shackle bolt securing nut and spring washer
3 Inner shackle plate
4 Shackle bolt
5 Upper shackle bushes
6 Upper shackle pin

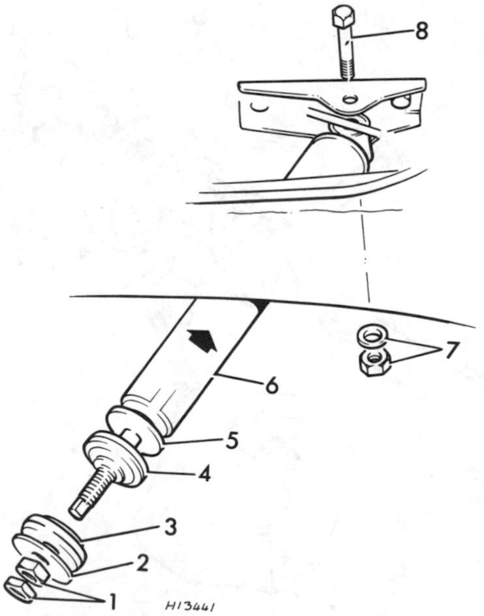

Fig. 11.19 Rear shock absorber removal (Sec 19)

1 Shock absorber retaining nut and locknut
2 Plain washer
3 Lower bush
4 Upper bush
5 Plate washer
6 Shock absorber
7 Upper mounting bolt securing nut and spring washer
8 Upper mounting bolt

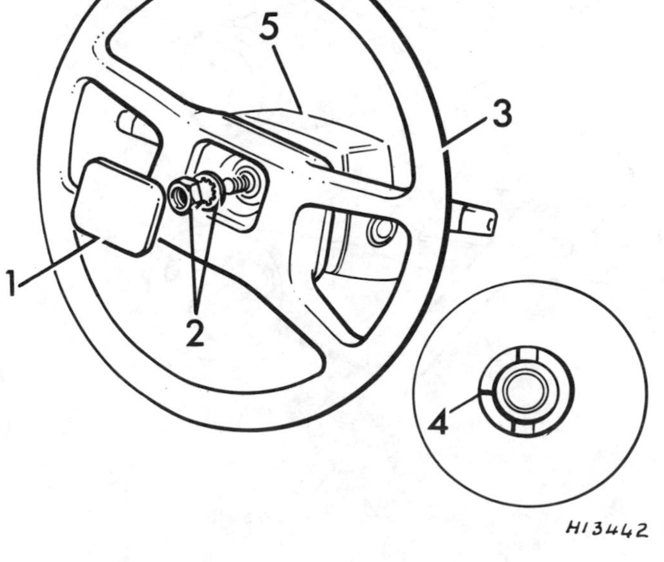

Fig. 11.20 Steering wheel removal (Sec 20)

1 Motif-pad
2 Securing nut and washer
3 Steering wheel
4 Position of cancelling ring
5 Steering column cowl

21 Steering column top bush – removal and refitting

1 Refer to Section 20 and remove the steering wheel and cowl, then disconnect the battery negative terminal lead.
2 Slacken the screw that retains the combined switch mechanism and lift the switch mechanism from over the top of the inner column.
3 Using a screwdriver, ease the top bush from the inside of the outer column.
4 To refit the top bush, first align the slits in the column bush with the depression in the outer column and ensure that the chamfered end of the bush enters the column first.
5 Using a suitable diameter metal drift, carefully drive the top bush into position.
6 Refitting the combined switch mechanism and steering wheel is now the reverse sequence to removal.

22 Steering column lock and ignition starter switch housing – removal and refitting

1 Remove the steering wheel and cowl assembly as described in Section 20 then disconnect the battery negative terminal lead.
2 Locate the multi-pin connector on the end of the column lock and ignition switch wiring harness and disconnect the wiring (Fig. 11.21).
3 Disconnect the wiper/washer, headlight dip and indicator switch wiring at the multi-pin connectors.
4 Unscrew and remove the four outer steering column securing screws and washers and withdraw the outer column through the facia panel.

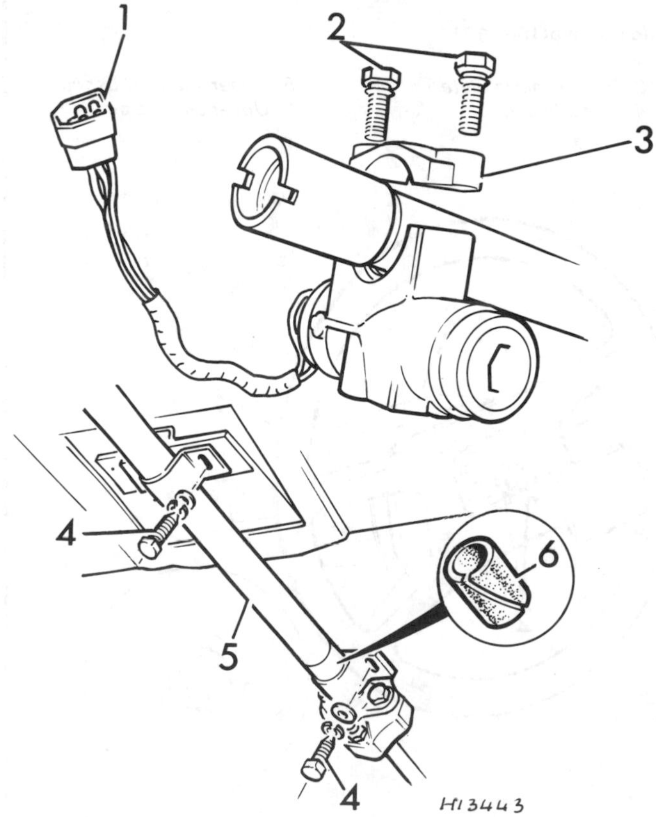

Fig. 11.21 Steering column lock and ignition starter switch housing removal (Sec 22)

1 Multi-pin connector
2 Shear bolts
3 Clamp plate
4 Mounting bolts
5 Outer column
6 Felt bush

5 Extract the felt bush from the inner steering column lower mounting.
6 Using either a drill or a drill and 'easy out' stud extractor remove the two special shear screws.
7 Lift away the clamp plate and the steering lock.
8 Refit the steering column and ignition switch and clamp plate.
9 Lightly tighten the two new shear screws.
10 Check the operation of the lock to ensure that it operates correctly.
11 Slowly tighten the two shear bolts until the heads shear.
12 Locate the felt bush in the steering outer column lower end.
13 Refitting is now the reverse of the removal procedure.

23 Steering column universal joint coupling – removal and refitting

1 Unscrew and remove the bolts that retain the flexible coupling to the upper column and lower column.
2 Unscrew the lower column pinch bolt, then withdraw the flexible coupling.
3 Unscrew the pinch bolt that retains the lower flexible joint to the steering rack pinion.
4 Remove the lower column and the lower flexible joint.
5 If the flexible joint bolts are not peened, cut the locking wire and unscrew the bolts. Remove the conical rubbers noting that they locate in the countersunk holes in the joint plate. Remove the plain washers.
6 Examine the couplings for wear and damage and renew the components as necessary.
7 When refitting the lower coupling, first reassemble half the joint using the two hexagon-headed bolts. After tightening them, reassemble the remaining half using the two Pozidrive or slotted head bolts. The hexagon type bolts should be tightened to 4 to 7 lbf ft (0.5 to 1.0 kgf m), the slotted head type to 8 to 12 lbf ft (1.1 to 1.6 kgf m).
8 Wire lock each pair of bolts.
9 Refitting is now the reverse of the removal procedure. Tighten the flexible coupling bolts to a torque of 20 to 22 lbf ft (2.77 to 3.04 kgf m) and the pinch bolts to 17 to 20 lbf ft (2.35 to 2.77 kgf m).

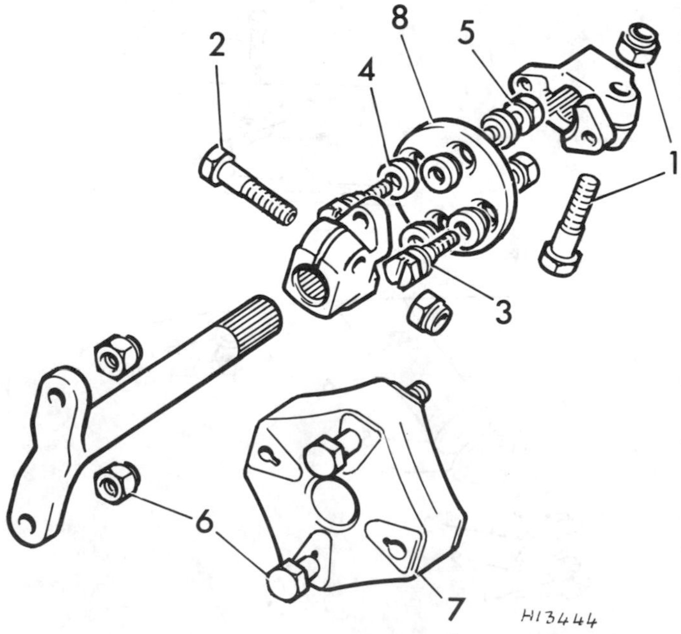

Fig. 11.22 Steering column universal joint couplings (Sec 23)

1 Pinch-bolt and locknut
2 Pinch-bolt and locknut
3 Slotted head bolt
4 Conical rubber washers
5 Plain washer and nut
6 Hexagon headed bolt and locknut
7 Flexible coupling (upper)
8 Joint plate

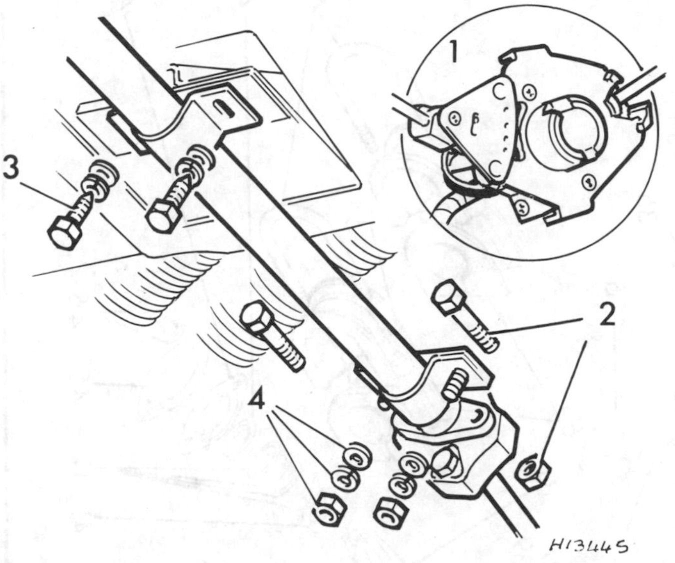

Fig. 11.23 Upper steering column removal (Sec 24)

1 Combination switch
2 Upper column to flexible coupling bolt and locknut
3 Column to upper support bracket, bolt, plain and spring washer
4 Column to lower support bracket securing nut spring and plain washer

24 Upper steering column – removal and refitting

1 Refer to Section 20 and remove the steering wheel, then disconnect the battery negative terminal lead.
2 Disconnect the multi-pin connector on the end of the wiring harness to the switch mechanism at the harness connector.
3 Slacken the screw that retains the combined switch mechanism and lift the switch mechanism from over the top of the inner column (Fig. 11.23).
4 Disconnect the multi-pin connector on the end of the wiring harness to the ignition switch at the harness connection.
5 Undo and remove the two nuts and bolts that secure the upper column to the flexible coupling.
6 Undo and remove the two screws, plain and spring washers that secure the column to the upper support bracket.
7 Undo and remove the two locknuts, plain and spring washers that attach the column to the lower support bracket bolts.
8 Lift away the upper steering column.
9 To refit the upper steering column, first engage the steering lock.
10 Centralise the steering rack by removing the rubber sealing plug and inserting a 0.25 in (6.35 mm) diameter rod so that it locates in the hole provided in the rack gear.
11 Refitting is now the reverse of the removal procedure but the following additional points should be noted:

(a) Do not forget to remove the centralising rod from the rack gear
(b) The steering column mounting bolts should be tightened to a torque wrench setting of 14 to 18 lbf ft (1.94 to 2.49 kgf m)
(c) The flexible joint coupling bolts should be tightened to a torque wrench setting of 20 to 22 lbf ft (2.77 to 3.04 kgf m)

25 Steering rack and pinion assembly – removal and refitting

1 Jack up the front of the car and support it with axle-stands placed beneath the front chassis members.
2 Unscrew and remove the steering tie-rod ball-pin retaining nuts and, using a suitable universal balljoint separator, disconnect the tie-rods from the steering levers (Fig. 11.24).
3 Unscrew and remove the lower pinch-bolt that retains the flexible joint to the rack pinion.

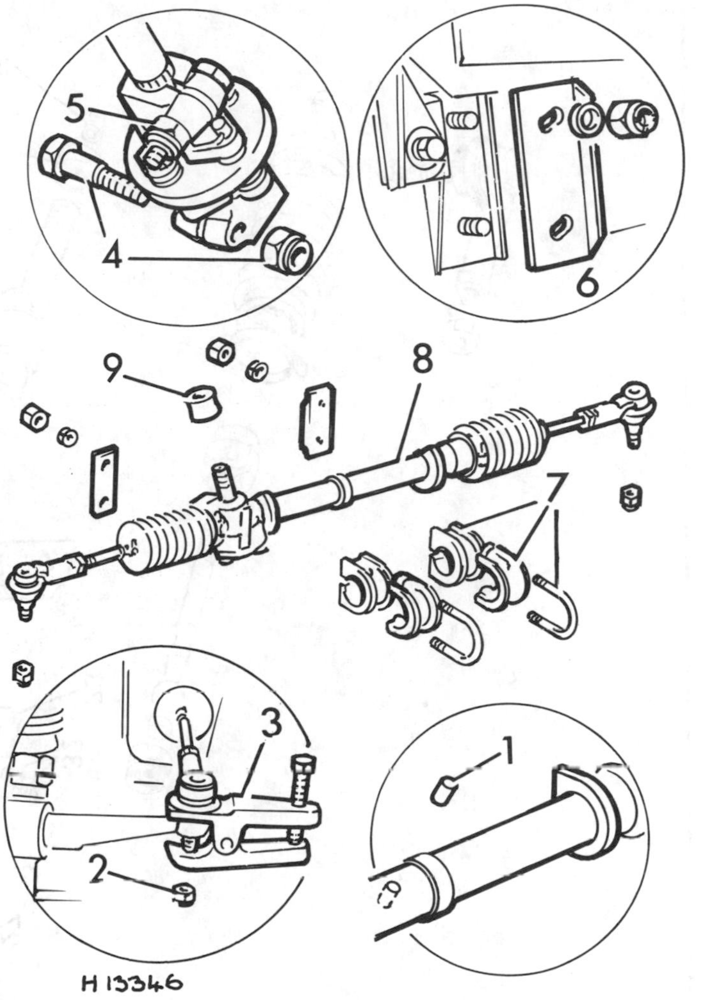

Fig. 11.24 Steering rack and pinion assembly removal (Sec 25)

1 Rubber sealing plug
2 Tie-rod ball-pin nut
3 Balljoint separator tool
4 Lower pinch-bolt and locknut
5 Upper pinch-bolt
6 Rack clamp plate
7 Rack clamps, rubber inserts and U-bolts
8 Rack and pinion assembly
9 Pinion seal

4 Loosen the lower column pinch-bolt and move the joint upwards away from the pinion.
5 Unscrew and remove the rack U-bolt retaining nuts and washers and withdraw the U-bolts, clamp brackets and rubbers from the bulkhead.
6 The rack assembly may now be lifted from the car through the wheel arch opening.
7 Lift away the pinion seal from over the end of the pinion.
8 Refitting the steering rack and pinion assembly is the reverse sequence to removal, the following additional points should be noted:

(a) Centralise the steering rack as described in Section 24
(b) Check the pinion seal and the two clamp bracket rubber inserts for signs of oil contamination or deterioration. If evident, new rubbers must be obtained
(c) The rack clamp nuts should be tightened to a torque wrench setting of 15 to 18 lbf ft (2.0 to 2.4 kgf m)
(d) The tie-rod ball-pin nuts should be tightened to a torque wrench setting of 24 lbf ft (3.3 kgf m)
(e) Tighten the lower flexible joint pinch-bolt to a torque wrench setting of 17 to 20 lbf ft (2.35 to 2.77 kgf m)

9 It will now be necessary to check and reset the front wheel alignment. Further information will be found in Section 27.

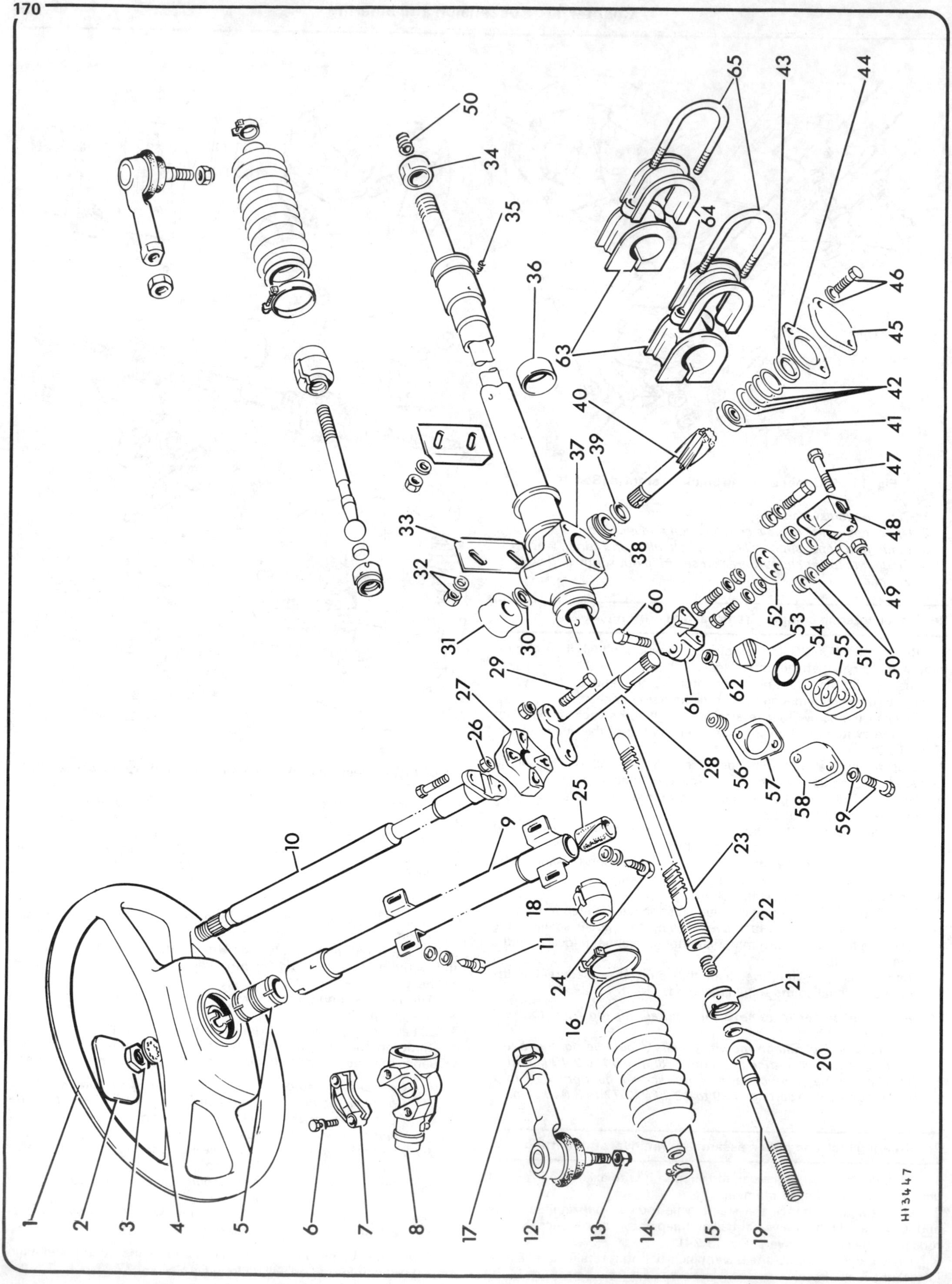
1
2
3
4
5
6
7
8
9
10
11
12
13
14
15
16
17
18
19
20
21
22
23
24
25
26
27
28
29
30
31
32
33
34
35
36
37
38
39
40
41
42
43
44
45
46
47
48
49
50
51
52
53
54
55
56
57
58
59
60
61
62
63
64
65
H13447

Fig. 11.25 Exploded view of steering mechanism (Sec 26)

1 *Steering wheel*
2 *Motif pad*
3 *Nut*
4 *Shakeproof washer*
5 *Upper bush*
6 *Shear bolt*
7 *Clamp plate*
8 *Steering lock*
9 *Upper column (outer)*
10 *Upper column (inner)*
11 *Screw*
12 *Tie-rod end*
13 *Self-locking nut*
14 *Clip (small)*
15 *Rack seal*
16 *Clip (large)*
17 *Locknut*
18 *Ball housing*
19 *Tie-rod*
20 *Ball seat*
21 *Locknut (ball housing)*
22 *Thrust spring*
23 *Rack*
24 *Bolt*
25 *Lower bush*
26 *Nut*
27 *Flexible coupling*
28 *Column (lower)*
29 *Bolt*
30 *Pinion oil seal*
31 *Sealing washer*
32 *Nut*
33 *Locating plate*
34 *Rack bearing*
35 *Rack bearing screw*
36 *Sealing rubber*
37 *Pinion housing*
38 *Pinion bearing*
39 *Washer*
40 *Pinion*
41 *Pinion bearing*
42 *Shim*
43 *Shim – 0.60 in (1.524 in)*
44 *Shim gasket – 0.010 in (0.254 mm)*
45 *End cover*
46 *Bolt and spring washer*
47 *Pinch bolt*
48 *Flexible joint (half)*
49 *Nut*
50 *Shouldered bolt*
51 *Rubber bush*
52 *Joint plate*
53 *Support yoke*
54 *O-ring*
55 *Shim*
56 *Thrust spring*
57 *Joint gasket*
58 *End cover*
59 *Bolt and spring washer*
60 *Pinch bolt*
61 *Flexible joint (half)*
62 *Nut*
63 *Rack mounting rubbers*
64 *Rack clamps*
65 *Rack U-bolts*

26 Steering rack and pinion assembly – dismantling, overhaul and reassembly

1 Wash the outside of the rack and pinion assembly in paraffin and wipe dry with a non-fluffy rag.

2 Slacken off the two tie-rod end locknuts and unscrew the two tie-rod ends as complete assemblies (Fig. 11.25).

3 Unscrew and remove the two locknuts from the ends of the tie-rods.

4 Slacken the rack seal clips at either end of the rack assembly body. Remove the clips and two rack seals.

5 Using a small chisel, carefully ease out the locknut indent from each of the balljoint housings.

6 Using two mole wrenches or one mole wrench and a soft metal drift, hold the locknut and unscrew the balljoint housing from each end of the rack. Lift away the tie-rods.

7 Recover the ball cup and spring from each end of the rack.

8 Using a small chisel, carefully ease out the locknut indent from the rack. Unscrew the locknuts.

9 Undo and remove the rack bearing pan head retaining screw located in the rack tube end as opposed to the pinion housing.

10 The bearing may now be removed from the rack housing.

11 Undo and remove the two bolts and spring washers that secure the rack yoke cover plate.

12 Lift away the cover plate, shims and joint washer.

13 Recover the rack support yoke from the pinion housing.

14 Remove the O-ring and thrust spring from the support yoke.

15 Undo and remove the two bolts and spring washers that secure the pinion end cover plate. Lift away the cover plate, shims and joint washers.

16 Carefully push out the pinion and the lower bearing. Note which way round the bearing is fitted.

17 The steering rack may now be withdrawn from the rack tube. Note which way round the rack is fitted in the rack tube.

18 Using a soft metal drift, tap out the upper pinion bearing and its washer. Note which way round the bearing is fitted.

19 Recover the pinion shaft oil seal from the pinion housing.

20 The steering rack assembly is now fully dismantled. Clean all parts in paraffin and wipe dry with a non-fluffy rag.

21 Thoroughly inspect the rack and pinion teeth for signs of wear, cracks or damage. Check the ends of the rack for wear, especially where it moves in the bushes.

22 Examine the rubber gaiters for signs of cracking, perishing or other damage which if evident, new gaiters must be obtained.

23 Inspect the ball ends and housing for wear which if evident, new parts will be necessary. Any other parts that show wear or damage must be renewed.

24 To reassemble, first fit a new rack bearing into the rack housing so that the flats of the bearing are positioned offset to the bearing retaining screw hole.

25 Using a 0.119 in (3.00mm) diameter drill located in the retaining screw hole, drill through the bearing. Clear away any swarf from the bearing and the housing.

26 Apply some non-hardening oil resistant sealing compound to the bush retaining screw and refit the screw.

27 It is very important that the screw does not protrude into the bore of the bearing. Should this condition exist, the end of the screw must be filed flat.

28 Fit the pinion washer to the pinion followed by the upper bearing. The thrust face must face towards the pinion washer.

29 Carefully fit the rack into the rack housing the correct way round as noted during dismantling.

30 Insert the pinion into the housing and then centralise the rack relative to the rack housing. Fit a peg into the centre locating hole.

31 Position the pinion, making sure the groove in the pinion serrations is facing and also parallel with the rack teeth. Remove the centralising peg.

32 Refit the lower bearing with the thrust face facing towards the pinion.

33 Refit the bearing shims and make sure that the bearing shim pack stands proud of the pinion housing. If necessary, add new shims to achieve this condition.

34 Refit the pinion housing end cover but without the paper gasket. Secure in position with the two bolts and spring washers. The two bolts should only be tightened sufficiently to nip the end cover.

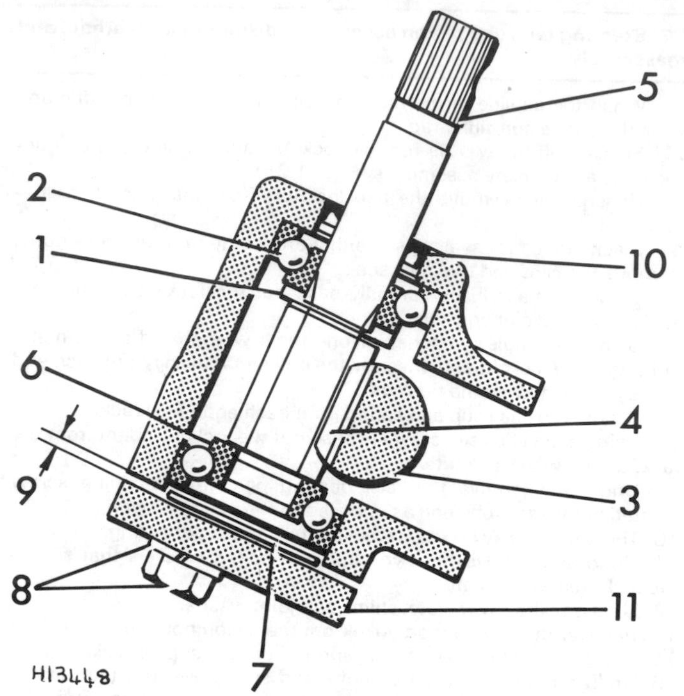

Fig. 11.26 Pinion end cross-section (Sec 26)

1 Pinion washer
2 Upper bearing
3 Rack
4 Pinion teeth
5 Pinion
6 Lower bearing
7 Shims
8 Bolt and spring washer
9 Gap measurement
10 Pinion shaft oil seal
11 End cover

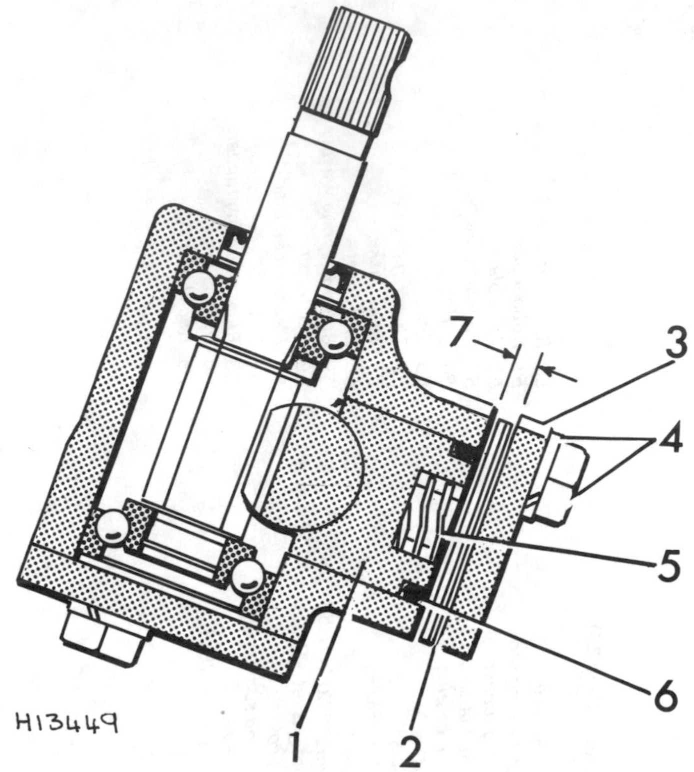

Fig. 11.27 Damper cover shim thickness (Sec 26)

1 Damper yoke
2 Shims and gasket
3 Cover plate
4 Bolt and spring washer
5 Damper spring
6 O-ring seal
7 Gap measurement

35 Using feeler gauges, measure the gap between the pinion housing and the end cover. Make a note of the measurement (Fig. 11.26).
36 Undo and remove the two pinion housing end cover securing bolts and spring washers. Lift away the end cover.
37 Adjust the number of shims in the end pack so as to obtain a 0.011 to 0.013 in (0.279 to 0.330 mm) gap. A range of shims is available for this adjustment. Details may be found in the specifications at the beginning of this chapter.
38 It is important that the 0.060 in (1.524 mm) shim is positioned next to the joint washer. Refit the shim pack, joint washer and end cover.
39 Apply a little non-hardening oil resistant sealing compound to the end cover securing bolts. Fit the two bolts and spring washers and tighten to a torque wrench setting of 12 to 15 lbf ft (1.6 to 2.0 kgf m).
40 Carefully fit a new pinion oil seal.
41 Refit the damper yoke, cover plate gasket and cover plate.
42 Refit the cover bolts and spring washers and gradually tighten these in a progressive manner whilst turning the pinion to and fro through 180° until it is just possible to rotate the pinion between the finger and thumb.
43 Using feeler gauges, measure the gap between the cover and the housing (Fig. 11.27).
44 Remove the cover and reassemble, this time including the damper spring, a new O-ring oil seal and shims to the previous determined measurement plus 0.002 to 0.005 in (0.05 to 0.13mm).
45 Tighten the bolts that secure the yoke cover to a torque wrench setting of 12 to 15 lbf ft (1.6 to 2.0 kgf m).
46 Screw a new ball housing locknut onto each end of the rack to the limits of the thread.
47 Insert the two thrust springs into the ends of the rack.
48 Fit each tie-rod into its ball housing and locate the ball cup against the thrust spring.
49 Slowly tighten the two ball housings until the tie-rod is just nipped.
50 Using a mole wrench and a soft metal drift, carefully tighten the locknut onto the ball housing. Again check that the tie-rod is still pinched.
51 Next slacken the ball housing back by $\frac{1}{8}$ turn to allow full articulation of the tie-rods.
52 Fully tighten the locking ring to the housing. Whilst this is being done, make sure the housing does not turn.
53 Using a centre punch or blunt chisel, drive the ball housing edge of the locking ring into the locking slots of the ball housing and the opposite edge into the locking slot of the rack.
54 Refit the two rack rubber seals and secure with the large clips to the housing. Position the two small clips and tighten on the pinion end.
55 Refit the tie-rod locknuts and then screw on each tie-rod end by an equal amount until the dimension between the two ball-pin centres is 43.7 in (110.9cm). Tighten the locknuts sufficiently to prevent this initial setting being lost during refitting.
56 Using a squirt type oil can, insert $\frac{1}{3}$ pint (0.19 litre) of recommended grade oil through the pinion seal. Finally position the small seal clip and lightly tighten.

27 Front wheel alignment – checking and adjustment

1 The front wheels are correctly aligned when they are turning in at the front $\frac{1}{8}$ in (3.175 mm). It is important that this measurement is taken on a centre line drawn horizontally and parallel to the ground through the centre line of the hub. The exact point should be in the centre of the sidewall of the tyre and not on the wheel rim which could be distorted and therefore give inaccurate readings.
2 The adjustment is effected by loosening the locknut on each tie-rod balljoint and also slackening the rubber gaiter clip holding it to the tie-rod; both tie-rods then being turned equally until the adjustment is correct.
3 This is a job best left to a BL garage as accurate alignment

requires the use of special equipment. If the wheels are not in alignment, tyre wear will be heavy and uneven and the steering stiff and unresponsive.

28 Wheels and tyres

1 Check the tyre pressures weekly (when they are cold).
2 Frequently inspect the tyre walls and treads for damage and pick out any large stones which have become trapped in the tread pattern.
3 If the wheels and tyres have been balanced on the car, then they should not be moved to a different axle position. If they have been balanced off the car then, in the interests of extending tread life, they can be moved between front and rear on the same side of the car and the spare incorporated in the rotational pattern.
4 Never mix tyres of different construction or very dissimilar tread patterns.
5 Always keep the roadwheels tightened to the specified torque and if the bolt holes become elongated or flattened, renew the wheel.
6 Occasionally clean the inner faces of the roadwheels and if there is any sign of rust or corrosion, paint them with metal preservative paint.
7 Before removing a roadwheel which has been balanced on the car, always mark one wheel and hub bolt hole so that the roadwheel may be refitted in the same relative position to maintain the balance.

29 Fault diagnosis – suspension and steering

Symptom	Reason/s
Steering feels vague, car wanders and floats at speed	Tyre pressures uneven Shock absorbers worn Steering gear balljoints badly worn Suspension geometry incorrect Steering mechanism free play excessive Front suspension and rear suspension pickup points out of alignment or badly worn Front suspension lacking grease
Stiff and heavy steering	Tyre pressures too low No oil in steering rack No grease in steering balljoints Front wheel toe-in incorrect Suspension geometry incorrect Steering gear incorrectly adjusted too tightly Steering column badly misaligned
Wheel wobble and vibration	Wheel nuts loose Front wheels and tyres out of balance Steering balljoints badly worn Hub bearings badly worn Steering gear free play excessive

Chapter 12 Bodywork and fittings

Contents

1 General description

The combined body and underframe is of all-welded construction. This makes a very strong and torsionally rigid shell.

The Marina 1700 saloon has four doors, with a fifth rear door on the Estate. The door hinges are securely attached to both the door and body. The front doors are locked from the outside by means of a key and all other doors may be locked from the inside.

The toughened safety glass is fitted to all windows; the windscreen has a specially toughened zone in front of the driver. In the event of the windscreen shattering, this zone breaks into much larger pieces than the rest of the screen thus giving the driver much better vision than would otherwise be possible.

The front seats are of the adjustable bucket type whilst the rear seat is a bench seat without a central arm rest.

For occupant safety, all switches and controls are suitably recessed or positioned so that they cannot cause body harm. Provision is made for the fitting of either static or inertia reel seat belts.

The instruments are contained in dials located above the steering column. A heater and ventilation system is fitted, incorporating a full-flow system with outlet ducts at instrument panel level.

2 Maintenance – bodywork and underframe

1 The general condition of a car's bodywork is the one thing that significantly affects its value. Maintenance is easy but needs to be regular and particular. Neglect, particularly after minor damage, can lead quickly to further deterioration and costly repair bills. It is important also to keep watch on those parts of the car not immediately visible, for instance the underside, inside all the wheel arches and the lower part of the engine compartment.

2 The basic maintenance routine for the bodywork is washing, preferably with a lot of water, from a hose. This will remove all the loose solids which may have stuck to the car. It is important to flush these off in such a way as to prevent grit from scratching the finish. The wheel arches and underbody need washing in the same way to remove any accumulated mud which will retain moisture and tend to encourage rust. Paradoxically enough, the best time to clean the underbody and wheel arches is in wet weather when the mud is thoroughly wet and soft. In very wet weather, the underbody is usually cleaned of large accumulations automatically and this is a good time for inspection.

3 Periodically, it is a good idea to have the whole of the underside of the car steam cleaned, engine compartment included, so that a thorough inspection can be carried out to see what minor repairs and renovations are necessary. Steam cleaning is available at many garages and is necessary for removal of accumulation of oily grime which sometimes is allowed to cake thick in certain areas near the engine, gearbox and back axle. If steam facilities are not available, there are one or two excellent grease solvents available which can be brush applied. The dirt can then be simply hosed off.

4 After washing paintwork, wipe off with a chamois leather to give an unspotted clear finish. A coat of clear protective wax polish will give added protection against chemical pollutants in the air. If the paintwork sheen has dulled or oxidised, use a cleaner/polish combination to restore the brilliance of the shine. This requires a little effort, but is usually caused because regular washing has been neglected. Always check that the door and ventilator opening drain holes and pipes are completely clear so that water can drain out. Bright work should be treated the same way as paintwork. Windscreens and windows can be kept clear of the smeary film which often appears by adding a little ammonia to the cleaning water. If they are scratched, a good rub with a proprietary metal polish will often clear them. Never use any form of wax or other body or chromium polish on glass.

3 Maintenance – upholstery and carpets

Mats and carpets should be brushed or vacuum cleaned regularly to keep them free of grit. If they are badly stained remove them from the car for scrubbing or sponging and make quite sure they are dry before replacement. Seats and interior trim panels can be kept clean by a wipe over with a damp cloth. If they do become stained (which can be more apparent on light coloured upholstery) use a little liquid detergent and a soft nail brush to scour the grime out of the grain of the material. Do not forget to keep the head lining clean in the same way as the upholstery. When using liquid cleaners inside the car do not over-wet the surfaces being cleaned. Excessive damp could get into the seams and padded interior causing stains, offensive odours or even rot. If the inside of the car gets wet accidentally it is worthwhile taking some trouble to dry it out properly, particularly where carpets are involved. **Do not** leave oil or electrical heaters inside the car for this purpose.

4 Minor body damage – repair

The colour photo sequence on pages 182 and 183 illustrates the operations detailed in the following sub-sections.

Repair of minor scratches in the car's bodywork

If the scratch is very superficial and does not penetrate to the metal of the bodywork, repair is very simple. Lightly rub the area of the scratch with a paintwork renovator or a very fine cutting paste, to remove loose paint from the scratch and to clear the surrounding bodywork of wax polish. Rinse the area with clean water.

Apply touch-up paint to the scratch using a thin paintbrush, continue to apply thin layers of paint until the surface of the paint in the scratch is level with the surrounding paintwork. Allow the new paint at least two weeks to harden; then blend it into the surrounding paintwork by rubbing the paintwork in the scratch area with a paintwork renovator or a very fine cutting paste. Finally apply wax polish.

Where the scratch has penetrated right through to the metal of the bodywork, causing the metal to rust, a different repair technique is required. Remove any loose rust from the bottom of the scratch with a penknife, then apply rust inhibiting paint to prevent the formation of rust in the future. Using a rubber nylon applicator, fill the scratch with bodystopper paste. If required, this paste can be mixed with cellulose thinners to provide a very thin paste which is ideal for filling narrow scratches. Before the stopper-paste in the scratch hardens, wrap a piece of smooth cotton rag around the top of a finger. Dip the finger in cellulose thinners and then quickly sweep it across the surface of the stopper paste in the scratch; this will ensure that the surface of the stopper paste is slightly hollowed. The scratch can now be painted over as described earlier in this Section.

Repair of dents in the car's bodywork

When deep denting of the car's bodywork has taken place, the first task is to pull the dent out until the affected bodywork almost attains its original shape. There is little point in trying to restore the original shape completely, as the metal in the damaged area will have stretched on impact and cannot be reshaped fully to its original contour. It is better to bring the level of the dent up to a point which is about $\frac{1}{8}$ in (3 mm) below the level of the surrounding bodywork. In cases where the dent is very shallow anyway, it is not worth trying to pull it out at all.

If the underside of the dent is accessible, it can be hammered out gently from behind using a mallet with a wooden or plastic head. Whilst doing this, hold a suitable block of wood firmly against the impact from the hammer blows and thus prevent a large area of bodywork from being 'belled-out.'

Should the dent be in a section of the bodywork which has a double skin or some other factor making it inaccessible from behind, a different technique is called for. Drill several small holes through the metal inside the dent area, particularly in the deeper sections. Then screw long self-tapping screws into the holes just sufficiently for them to gain a good purchase in the metal. Now the dent can be pulled out by pulling on the protruding heads of the screws with a pair of pliers.

The next stage of the repair is the removal of the paint from the damaged area, and from an inch or so of the surrounding sound bodywork. This is accomplished most easily by using a wire brush or abrasive pad on a power drill, although it can be done just as effectively by hand using sheets of abrasive paper. To complete the preparations for filling, score the surface of the bare metal with a screwdriver or the tang of a file, or alternatively, drill small holes in the affected area. This will provide a really good key for filler paste.

To complete the repair see the Section on filling and respraying.

Repair of rust holes or gashes in the car's bodywork

Remove all paint from the affected area and from an inch or so of the surrounding sound bodywork using an abrasive pad or a wire brush on a power drill. If these are not available, a few sheets of abrasive paper will do the job just as effectively. With the paint removed, you will be able to gauge the severity of the corrosion and therefore decide whether to renew the whole panel (if this is possible) or to repair the affected area. New body panels are not as expensive as most people think and it is often quicker and more satisfactory to fit a new panel than to attempt to repair large areas of corrosion.

Remove all fittings from the affected area except those which will act as a guide to the original shape of the damaged bodywork (eg, headlamp shells etc). Then, using tin snips or a hacksaw blade, remove all loose metal and any other metal badly affected by corrosion. Hammer the edges of the hole inwards in order to create a slight depression for the filler paste.

Wire brush the affected area to remove the powdery rust from the surface of the remaining metal. Paint the affected area with rust inhibiting paint. If the back of the rusted area is accessible treat this also.

Before filling can take place it will be necessary to block the hole in some way. This can be achieved by the use of zinc gauze or aluminium tape.

Zinc gauze is probably the best material to use for a large hole. Cut a piece to the approximate size and shape of the hole to be filled, then position it in the hole so that its edges are below the level of the surrounding bodywork. It can be retained in position by several blobs of filler paste around its periphery.

Aluminium tape should be used for small or very narrow holes. Pull a piece off the roll and trim it to the approximate size and shape required, then pull off the backing paper (if used) and stick the tape over the hole; it can be overlapped if the thickness of one piece is insufficient. Burnish down the edges of the tape with the handle of a screwdriver or similar, to ensure that the tape is securely attached to the metal underneath.

Bodywork repairs – filling and re-spraying

Before using this Section, see the Sections on dent, deep scratch, rust hole, and gash repairs.

Many types of bodyfiller are available, but generally speaking those proprietary kits which contain a tin of filler paste and a tube of resin hardener are best for this type of repair. A wide flexible plastic or nylon applicator will be found available for imparting a smooth and well contoured finish to the surface of the filler.

Mix up a little filler on a clean piece of card or board. Use the hardener sparingly (follow the maker's instructions on the packet) otherwise the filler will set very rapidly.

Using the applicator, apply the filler paste to the prepared area; draw the applicator across the surface of the filler to achieve the correct contour and to level the filler surface. As soon as a contour that approximates to the correct one is achieved, stop working the paste. If you carry on too long, the paste will become sticky and begin to 'pick-up' on the applicator. Continue to add thin layers of filler paste at twenty-minute intervals until the level of the filler is just proud of the surrounding bodywork.

Once the filler has hardened, excess can be removed using a Surform plane or Dreadnought file. From then on, progressively finer grades of abrasive paper should be used, starting with 40 grade production paper and finishing with 400 grade 'wet-and-dry' paper. Always wrap the abrasive paper around a flat rubber, cork, or wooden block, otherwise the surface of the filler will not be completely flat. During the smoothing of the filler surface, the 'wet-and-dry' paper should be periodically rinsed in water. This will ensure that a very smooth finish is imparted to the filler at the final stage.

At this stage the dent should be surrounded by a ring of bare metal, which in turn should be encircled by the finely 'feathered' edge of the good paintwork. Rinse the repair area with clean water until all of the dust produced by the rubbing-down operation is gone.

Spray the whole repair area with a light coat of grey primer, this will show up any imperfections in the surface of the filler. Repair these imperfections with fresh filler paste or bodystopper and once more smooth the surface with abrasive paper. If bodystopper is used, it can be mixed with cellulose thinners to form a really thin paste which is ideal for filling small holes. Repeat this spray and repair procedure until you are satisfied that the surface of the filler and the feathered edge of the paintwork are perfect. Clean the repair area with clean water and allow to dry fully.

The repair area is now ready for spraying. Paint spraying must be carried out in a warm, dry, windless and dust free atmosphere. This condition can be created artificially if you have access to a large indoor working area, but if you are forced to work in the open, you will have to pick your day very carefully. If you are working indoors, dousing the floor in the work area with water will lay the dust which would otherwise be in the atmosphere. If the repair area is confined to one body panel, mask off the surrounding panels; this will help to minimise the effects of a slight mis-match in paint colours. Bodywork fittings (eg chrome strips, door handles etc) will also need to be masked off. Use genuine masking tape and several thicknesses of newspaper for the masking operation.

Before commencing to spray, agitate the aerosol can thoroughly then spray a test area (an old tin, or similar) until the technique is mastered. Cover the repair area with a thick coat of primer; the thickness should be built up using several thin layers of paint rather than one thick one. Using 400 grade 'wet-and-dry' paper, rub down the surface of the primer until it is really smooth. Whilst doing this, the work area should be thoroughly doused with water and the 'wet-and-dry' paper periodically rinsed in water. Allow to dry before spraying on more paint.

Spray on the top coat, again building up the thickness by using several thin layers of paint. Start spraying in the centre of the repair area and then, using a circular motion, work outwards until the whole repair area and about 2 in of the surrounding original paintwork is covered. Remove all masking material 10 to 15 minutes after spraying on the final coat of paint.

Allow the new paint at least 2 weeks to harden fully; then, using a paintwork renovator or a very fine cutting paste, blend the edges of the new paint into the existing paintwork. Finally, apply wax polish.

5 Major body damage – repair

1 Because the body is built on the monocoque principle and is integral with the underframe, major damage must be repaired by specialists with the necessary welding and hydraulic straightening equipment.

2 If the damage is severe, it is vital than on completion of the repair the chassis is in correct alignment. Less severe damage may also have twisted or distorted the chassis although this may not be visible immediately. It is therefore always best on completion of repair to check for twist and squareness to make sure all is well.

3 To check for twist, position the car on a clean level floor, place a jack under each jacking point, raise the car and take off the wheels. Raise or lower the jacks until the sills are parallel with the ground. Depending where the damage occurred, using an accurate scale, take measurements at the suspension mounting points and if comparable readings are not obtained it is an indication that the body is twisted.

4 After checking for twist, check for squareness by taking a series of measurements on the floor. Drop a plumb line and bob weight from various mounting points on the underside of the body and mark these points on the floor with chalk. Draw a straight line between each point and measure and mark the middle of each line. A line drawn on the floor starting at the front and finishing at the rear should be quite straight and pass through the centres of the other lines. Diagonal measurements can also be made as a check for squareness.

6 Maintenance – locks and hinges

Once every 6000 miles (10 000 km) or 6 months, the door, bonnet and boot or tailgate hinges should be oiled with a few drops of engine oil from an oil can. The door striker plates can be given a thin smear of grease to reduce wear and ensure free movement.

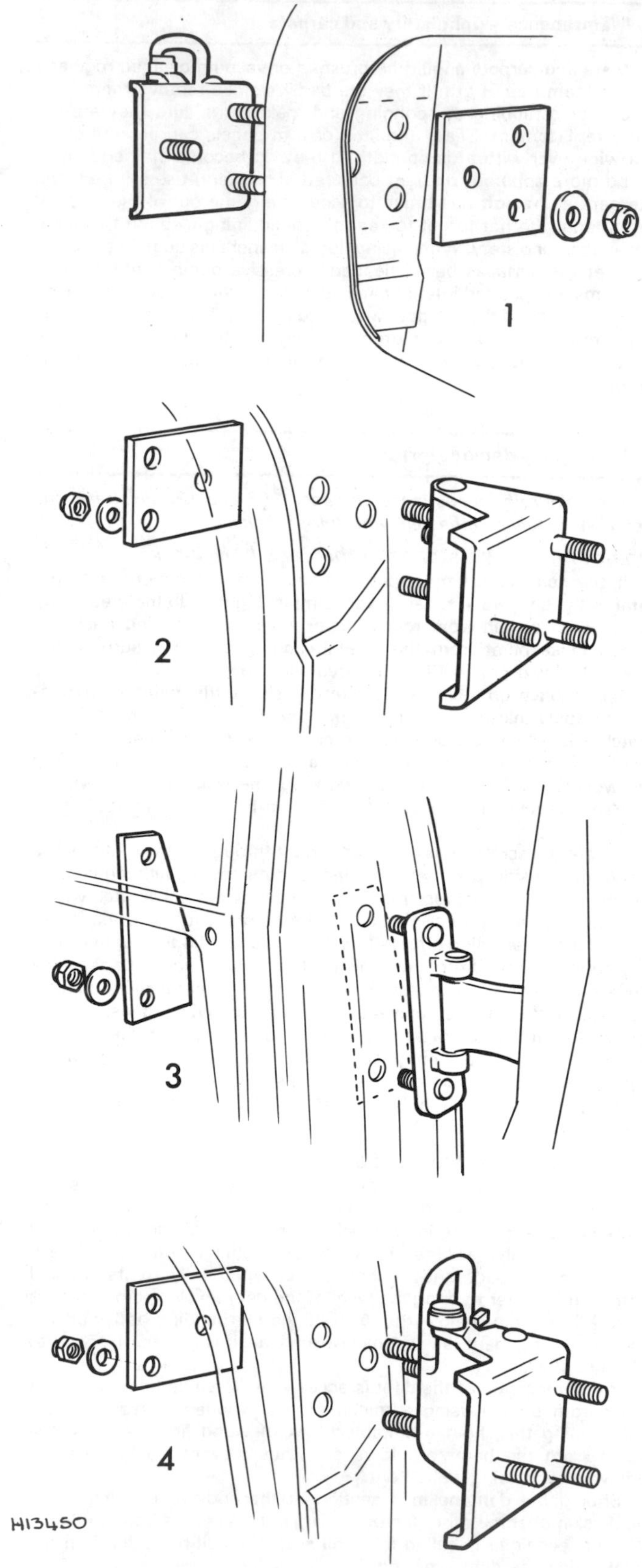

Fig. 12.1 Front and rear door hinge assemblies (Secs 8 and 9)

1 Top, front door
2 Bottom, front door
3 Top, rear door
4 Bottom, rear door

7 Door rattles – tracing and rectification

1 The most common cause of door rattles is a misaligned, loose or worn striker plate, but other causes may be:

(a) Loose door handles, window winder handles or door hinges
(b) Loose, worn or misaligned door lock components
(c) Loose or worn remote control mechanism

Or a combination of these.
2 If the striker catch is worn as a result of door rattles, renew it and adjust as described later in this Chapter.
3 Should the hinges be badly worn then they must be renewed.

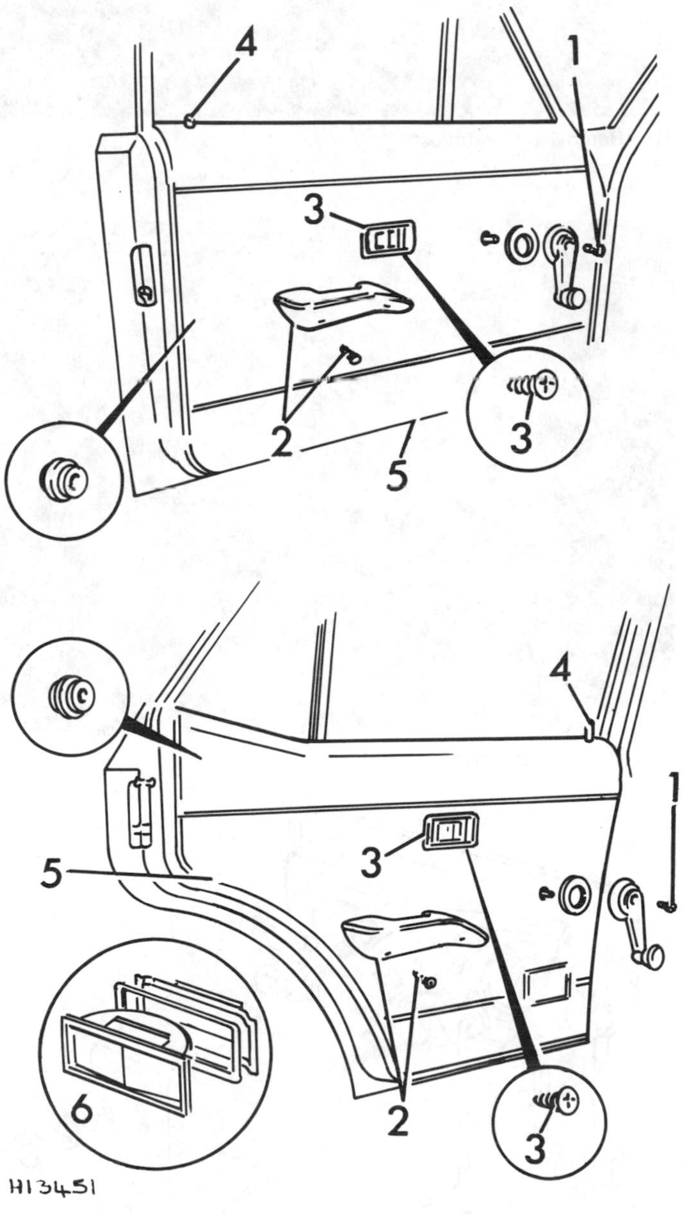

Fig. 12.2 Exploded view of door trim panel (Sec 10)

1 Screw
2 Armrest and screw
3 Bezel and screw
4 Locking button
5 Trim panel clips
6 Ashtray

8 Doors – removal and refitting

1 Refer to Section 10 and remove the door trim panel.
2 Working inside the door, mark the outline of the stiffener plate at each hinge position (Fig. 12.1). An assistant should now take the weight of the door.
3 Undo and remove the locknuts and plain washers that secure the door to the hinge.
4 Lift away the stiffener plates and finally the door.
5 Refitting the door is the reverse sequence to removal. Should it be necessary to adjust the position of the door in the aperture, leave the locknuts slightly loose and reposition the door by trial and error. Fully tighten the locknuts.

9 Door hinges – removal and refitting

1 Remove the door as described in Section 8.
2 Using a wide blade screwdriver or similar tool, carefully ease back the side trim panel door seal and then the trim panel.
3 If the rear door hinges are to be removed, use a wide blade screwdriver to ease back the B post door seals. Undo and remove the carpet finisher retaining screw, slide the front seat forward and ease the trim panel retaining clips from the B post. Hinge the trim panel up at the PVC lining crease. This will provide access to the door hinge securing nuts.
4 Undo and remove the locknuts and plain washers that secure each hinge (Fig. 12.1). Lift away the stiffener plates and finally the door hinges.
5 Refitting the door hinges is the reverse of the removal procedure.

10 Door trim panel and capping – removal and refitting

1 Wind the window up fully and note the position of the handle.
2 Undo and remove the screw and spacer that secure the window regulator handle. Lift away the handle (Fig. 12.2) (photos).
3 Undo and remove the two screws that retain the arm rest. Lift away the arm rest (photo).
4 Unscrew the remote door pull bezel retaining screw and withdraw the bezel (photo).
5 Unscrew the locking button. The ashtray must also be removed.
6 With a screwdriver, carefully slide the upper and lower bezels from the remote control door handle.

10.2a Unscrew the window regulator handle securing screw ...

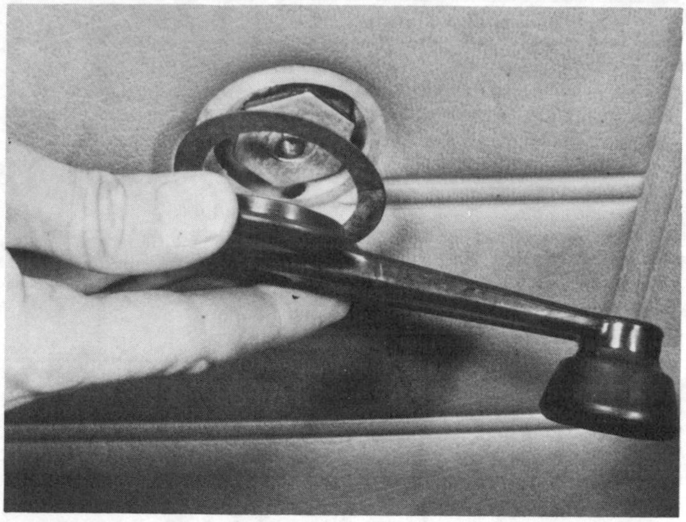

10.2b ... and lift away the handle

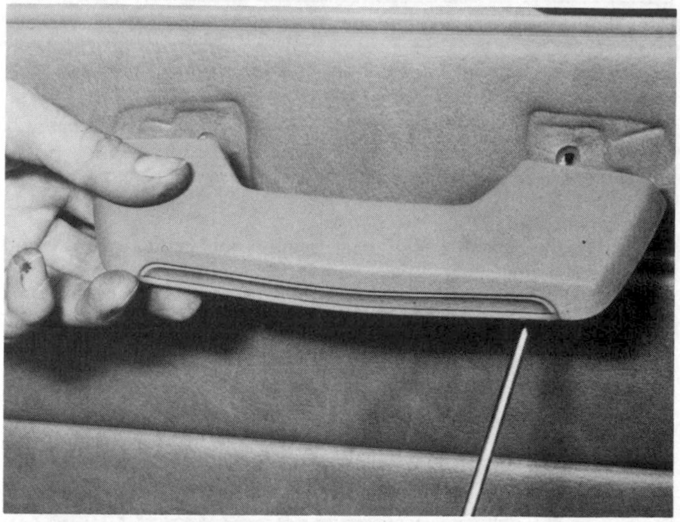

10.3 Removing the armrest

10.4 Undo the door pull retaining screw

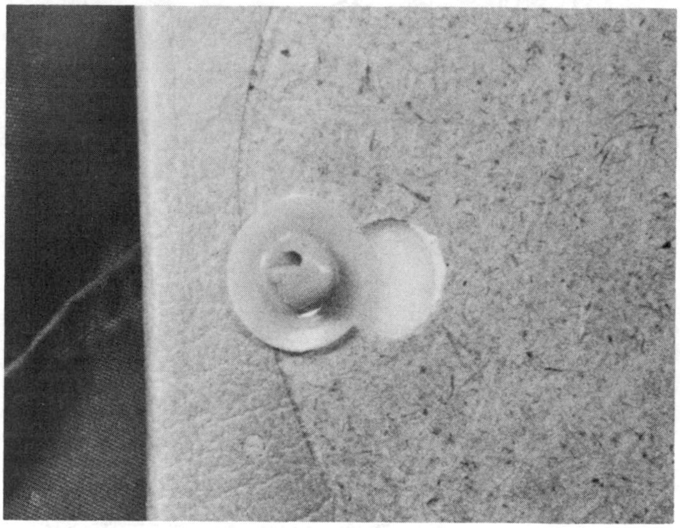

10.7 Door trim panel retaining clip

7 With a wide blade screwdriver or knife inserted between the trim panel and door, carefully ease the trim panel clips from the door. Lift away the trim panel (photo).

8 Should it be necessary to remove the trim capping, undo and remove the screws and shaped washers and unclip the trim capping from the door.

9 Refitting is the reverse sequence to removal. If the capping has been removed, make sure that the door glass seal and wiper strip are correctly positioned.

11 Door lock – removal and refitting

1 Remove the door trim panel as described in Section 10.

2 Peel back the polythene covering from the inside door panel. With the door glass in the fully raised position, remove the screw that secures the door glass rear channel to the door panel (front doors).

3 Prise off the retaining clip and detach the remote control operating link from the lock mechanism.

4 Remove the retainers and detach the private lock operating link and exterior handle link from the lock mechanism (front door). On rear doors, unscrew the knob from the end of the child safety operating link.

5 Using a pencil, mark the position of the lock on the door.

6 Undo the four screws that secure the lock and lock mechanism to

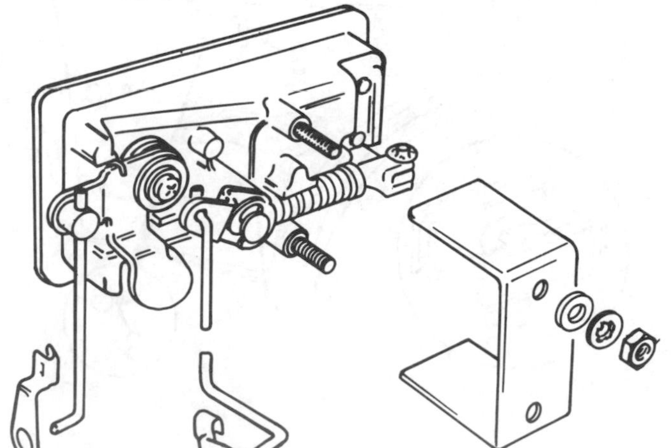

Fig. 12.3 Front door handle mechanism (Sec 11)

the door and then remove the locking mechanism and locking link through the opening in the door panel. On rear doors, release the retaining clip and disconnect the child safety operating link from the lock mechanism.

7 Refitting is the reverse of the removal procedure but before reconnecting the exterior handle operating link, move the handle to the open position and check that the cranked end of the link is in line with the bush in the operating lever of the lock mechanism; screw it up or down, as necessary, until the alignment is correct.

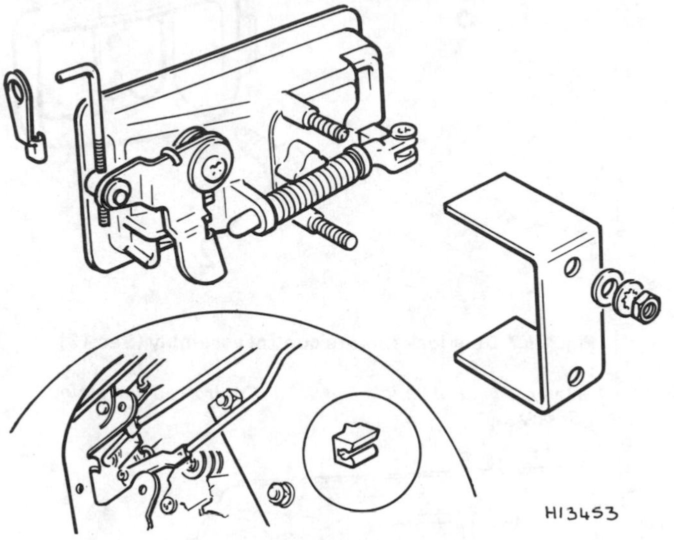

Fig. 12.4 Rear door handle mechanism (Sec 11)

12 Door lock – adjustment

Four adjustments may be made to the door locks and it will usually be found that any malfunction of a lock is caused by incorrect adjustment.

Exterior handle

1 Refer to Section 10 and remove the trim panel.
2 Close the door and partially operate the exterior release lever. Check that there is free movement of the lever before the point is reached where the transfer lever and its screwed rod move.
3 Operate the exterior release lever fully and check that the latch disc is released from the door striker before the lever is fully open.
4 To adjust, disconnect the screwed rod and screw in or out to achieve the correct setting.

Remote control

1 Refer to Section 10 and remove the trim panel.
2 Undo and remove the screw and slacken the control retaining screws.
3 Move the remote control assembly towards the latch unit. Retighten the retaining screws and make sure that the operating lever is against its stop A (Fig. 12.5 or 12.6).

Safety locking lever (front)

1 Refer to Section 10 and remove the trim panel.
2 Disconnect the long lockrod from the safety locking lever and then the short rod from the locking bar. Push the locking bar against its stop B (Fig.12.5) and move the safety locking lever to the locked position.
3 Refit the long rod in the safety locking lever and then press in the

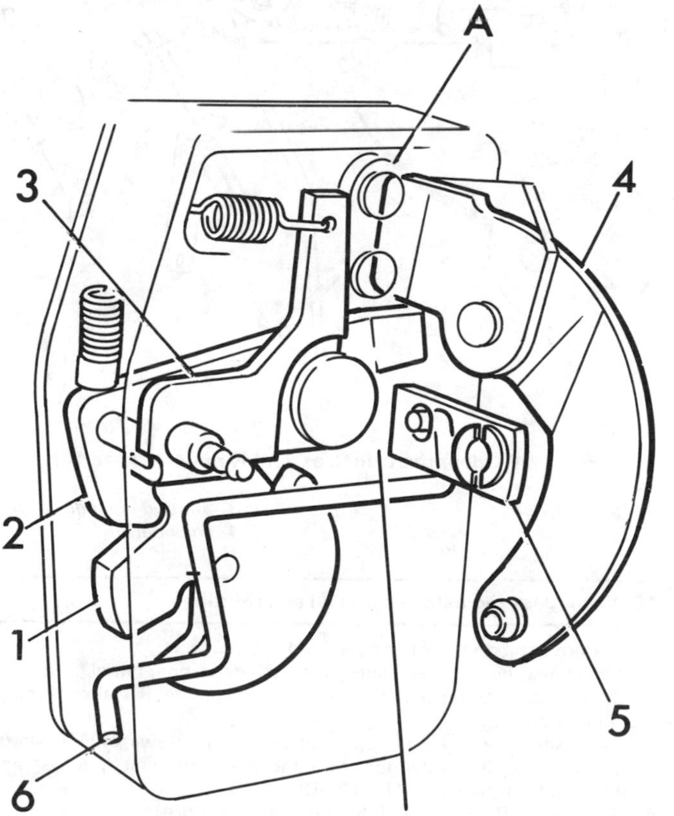

Fig. 12.5 Front door lock adjustment (Sec 12)

1 Latch disc
2 Latch disc release lever
3 Cross control lever
4 Operating lever
5 Locking bar
6 Locking bar cross-shaft
Positive stop A
Positive stop B

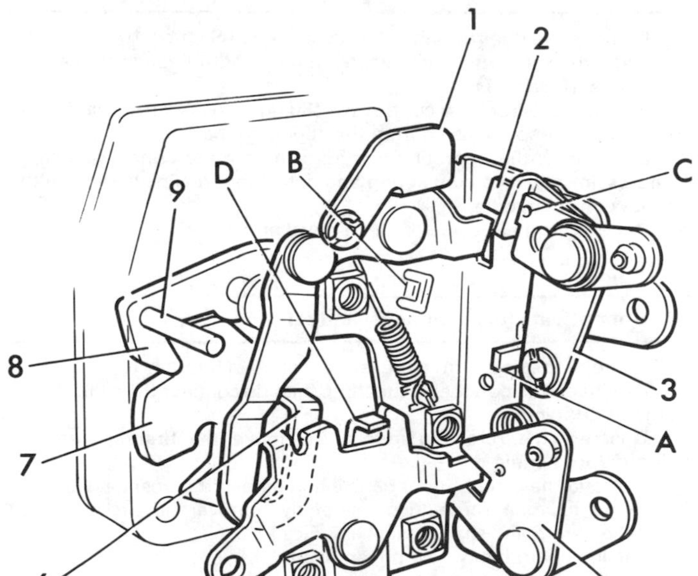

Fig. 12.6 Rear door lock adjustment (Sec 12)

1 Cross lever
2 Child safety intermediate lever
3 Operating lever
4 Locking lever
5 Freewheel actuating lever
6 Operating tab
7 Latch disc
8 Latch disc release lever
9 Striker pin
Positive stop A
Positive stop B
Positive stop C
Positive stop D

legs of the clip. Adjust the short rod so as to fit into the rod bush in the locking bar.
4 Release the safety locking lever and make sure that the operating lever is quite free to operate.

Safety locking lever (rear)

1 Refer to Section 10 and remove the trim panel.
2 Disconnect the long lockrod from the safety locking lever and then the short lockrod from the locking lever.
3 Press the free wheeling operating lever against the stop D (Fig.12.6) and position the safety locking lever in the locked position.
4 Reconnect the long lockrod to the safety locking lever. Push in the legs of the clip and adjust the short rod to fit into the rod bush in the locking lever.
5 Release the safety locking lever and ensure that the operating tab aligns with the striker pin of the latch disc release lever.

Door striker

1 It is very important that the latch disc is in the open position. **Note:** *Do not slam the door whilst any adjustment is being made otherwise damage may result.*
2 Slacken the striker plate retaining screws until it is just sufficient to allow the door to close and latch.
3 Push the door inwards or pull it outwards without operating the release lever until the door is level with the body and aperture.
4 Open the door carefully and mark with a pencil round the striker plate to act as a datum.
5 Place the striker accurately by trial and error until the door can be closed easily without signs of lifting, dropping or rattling.
6 Close the door and make sure that the striker is not positioned too far in by pressing on the door. It should be possible to press the door in slightly as the seals are compressed.
7 Finally tighten the striker plate retaining screws.

13 Door lock remote control – removal and refitting

1 Remove the door trim panel as described in Section 10.
2 Undo the screw that secures the remote control handle to the door inner panel (Fig. 12.7).
3 Release the remote control by sliding it rearwards against the spring pressure to release it from the door inner panel.
4 Disconnect the operating link from the remote control handle by compressing the retaining spring and detaching the link bush from the remote control handle.
5 Refitting is the reverse of the removal procedure.

14 Door private lock – removal and refitting

1 Remove the door trim panel as described in Section 10.
2 Peel back the polythene sheeting, then disconnect the control links from the door lock.
3 Unscrew the retaining nuts and remove the bracket from the exterior door handle.
4 Remove the exterior door handle and detach the operating links.
5 Using a small screwdriver, carefully remove the circlip and lift away the spring and special washers (Fig. 12.8).
6 The lock barrel may now be removed.
7 Refitting is the reverse sequence to removal.

15 Door glass – removal and refitting

1 Remove the door trim as described in Section 10 and the glass regulator as described in Section 16.
2 Release the outer weather strip by undoing the five retaining clips and then remove the weather strip.
3 The window glass may now be lifted out of the door.
4 After noting the position of the glass channel, remove the channel and channel rubber.
5 Refitting is the reverse of the removal procedure. When positioning the glass, make sure that dimension X in Fig. 12.9 is 6 in (152 mm).

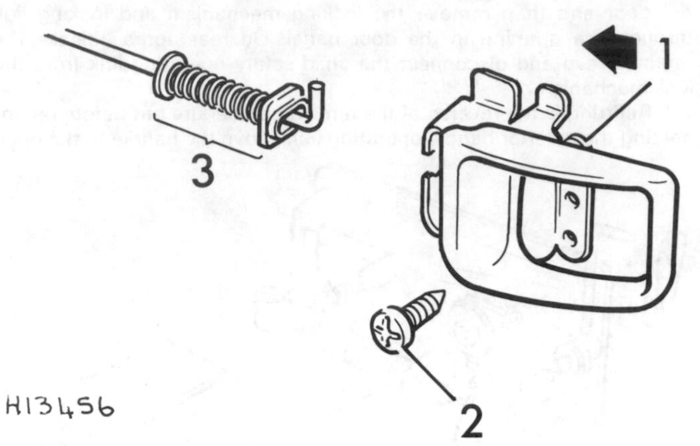

Fig. 12.7 Door lock remote control assembly (Sec 13)

1 Remote control handle *2 Screw* *3 Control rod and spring*

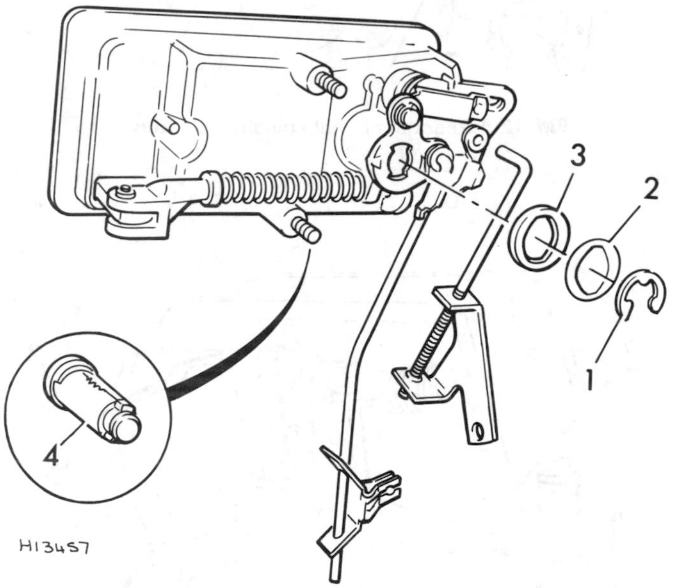

Fig. 12.8 Door private lock assembly removal (Sec 14)

1 Circlip *2 Spring washer* *3 Shaped washer* *4 Lock barrel*

16 Door glass regulator – removal and refitting

1 Remove the door trim as described in Section 10.
2 Peel back the polythene sheet from the door inner panel (photo).
3 Lower the door glass until the regulator channel is visible through the door aperture.
4 Undo and remove the regulator securing screws and washers; then disengage the regulator from the glass channel (photos) and withdraw it from the door (Fig. 12.10).
5 Refitting is the reverse of the removal procedure.

17 Bonnet – removal and refitting

1 Open the bonnet and support it on its stay.
2 With a pencil, mark the outline of the hinge on the bonnet to assist correct refitting. If the hinge is to be removed, also mark the inner panel.
3 An assistant should now take the weight of the bonnet. Undo and

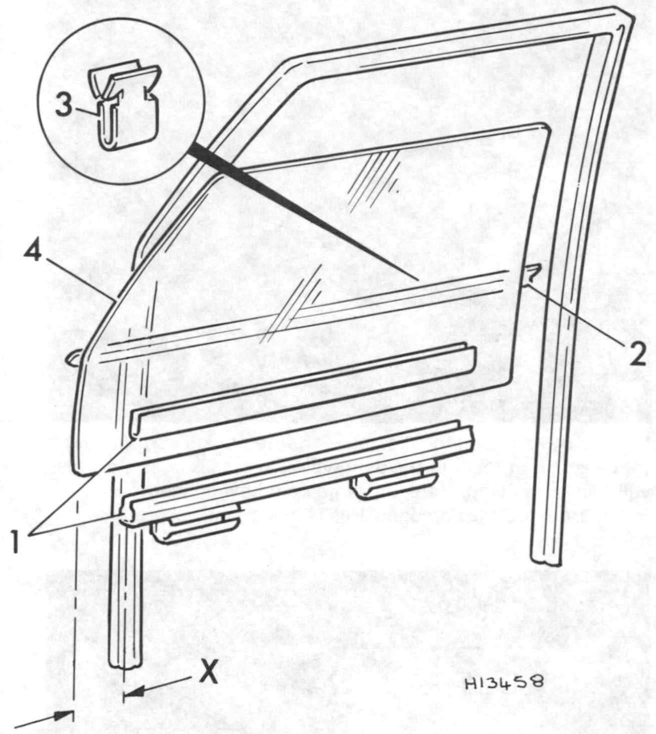

Fig. 12.9 Door glass removal (Sec 15)

1 Channel and rubber
2 Weatherstrip
3 Clip
4 Door glass
Dimension X = 6 in (152 mm)

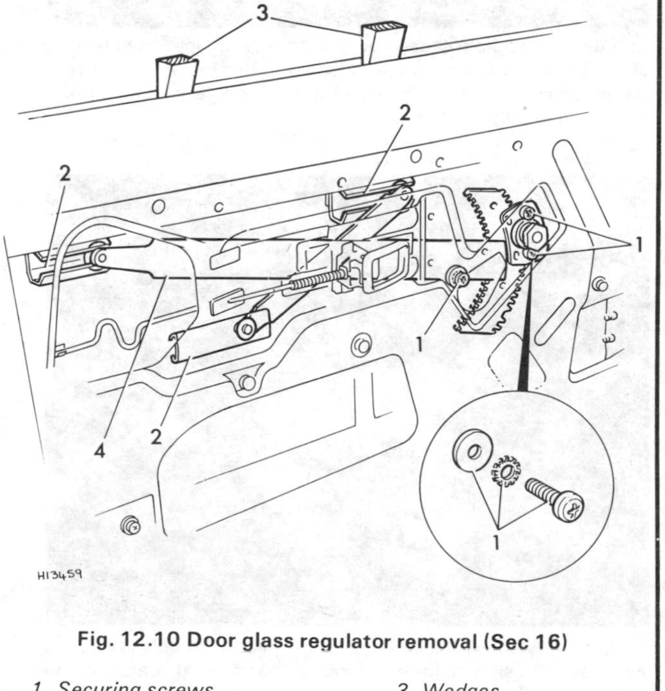

Fig. 12.10 Door glass regulator removal (Sec 16)

1 Securing screws
2 Door glass channel and door channel
3 Wedges
4 Regulator assembly

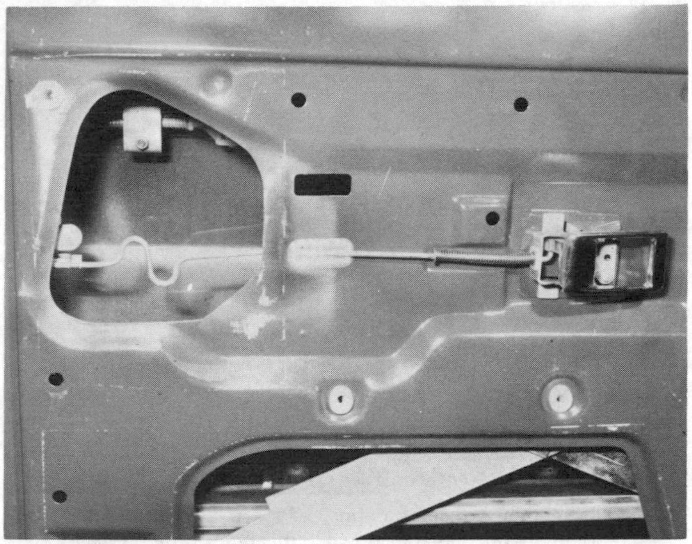

16.2 Door with trim panel and polythene sheet removed

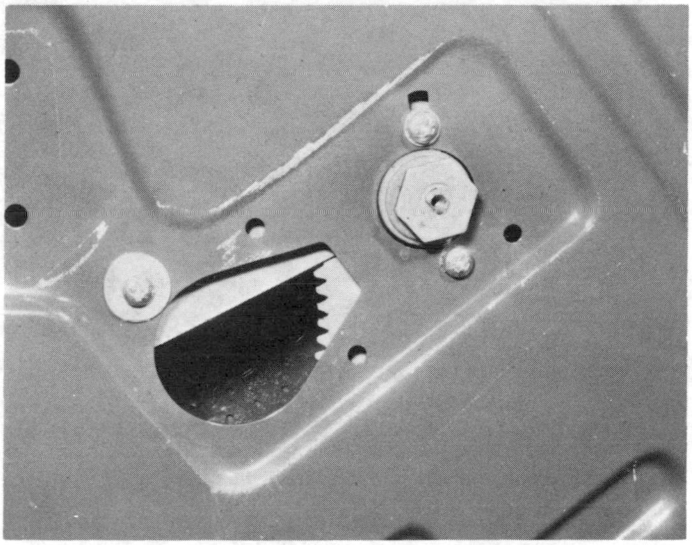

16.4a Remove the window regulator securing screws

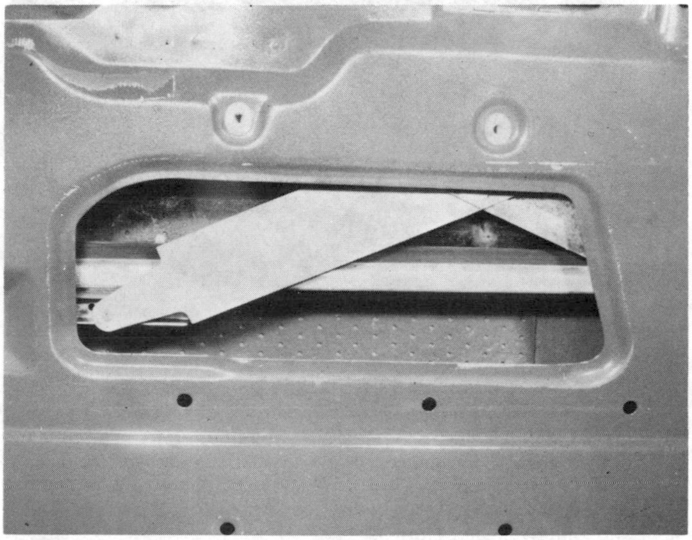

16.4b Disengage the regulator arm from the glass channel

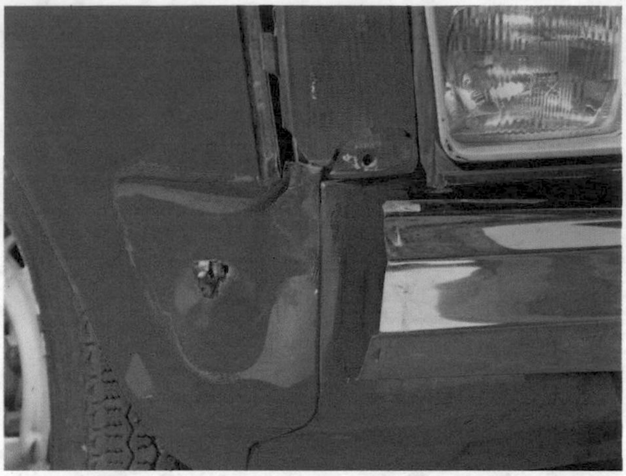

This sequence of photographs deals with the repair of the dent and paintwork damage shown in this photo. The procedure will be similar for the repair of a hole. It should be noted that the procedures given here are simplified – more explicit instructions will be found in the text

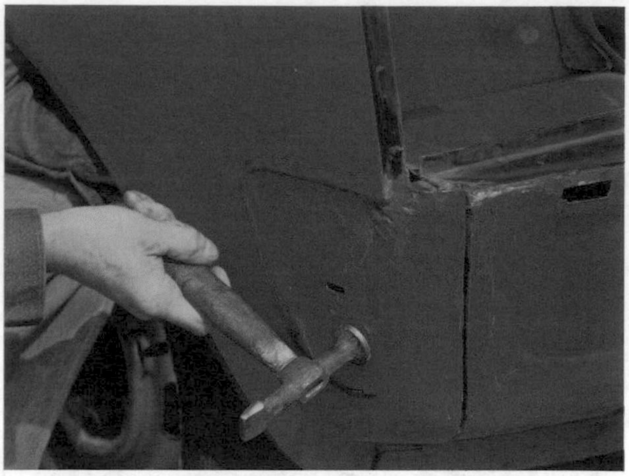

In the case of a dent the first job – after removing surrounding trim – is to hammer out the dent where access is possible. This will minimise filling. Here, the large dent having been hammered out, the damaged area is being made slightly concave

Now all paint must be removed from the damaged area, by rubbing with coarse abrasive paper. Alternatively, a wire brush or abrasive pad can be used in a power drill. Where the repair area meets good paintwork, the edge of the paintwork should be 'feathered', using a finer grade of abrasive paper

In the case of a hole caused by rusting, all damaged sheet-metal should be cut away before proceeding to this stage. Here, the damaged area is being treated with rust remover and inhibitor before being filled

Mix the body filler according to its manufacturer's instructions. In the case of corrosion damage, it will be necessary to block off any large holes before filling – this can be done with aluminium or plastic mesh, or aluminium tape. Make sure the area is absolutely clean before ...

... applying the filler. Filler should be applied with a flexible applicator, as shown, for best results; the wooden spatula being used for confined areas. Apply thin layers of filler at 20-minute intervals, until the surface of the filler is slightly proud of the surrounding bodywork

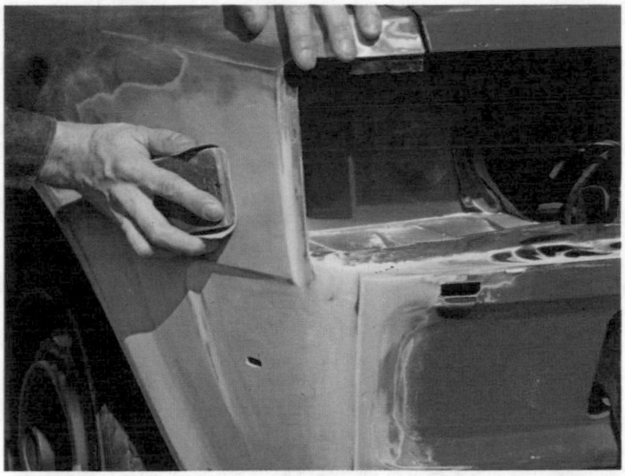

Initial shaping can be done with a Surform plane or Dreadnought file. Then, using progressively finer grades of wet-and-dry paper, wrapped around a sanding block, and copious amounts of clean water, rub down the filler until really smooth and flat. Again, feather the edges of adjoining paintwork

The whole repair area can now be sprayed or brush-painted with primer. If spraying, ensure adjoining areas are protected from over-spray. Note that at least one inch of the surrounding sound paintwork should be coated with primer. Primer has a 'thick' consistency, so will find small imperfections

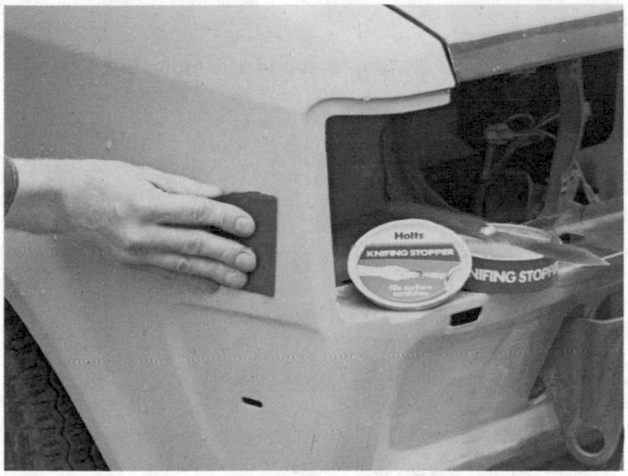

Again, using plenty of water, rub down the primer with a fine grade wet-and-dry paper (400 grade is probably best) until it is really smooth and well blended into the surrounding paintwork. Any remaining imperfections can now be filled by carefully applied knifing stopper paste

When the stopper has hardened, rub down the repair area again before applying the final coat of primer. Before rubbing down this last coat of primer, ensure the repair area is blemish-free – use more stopper if necessary. To ensure that the surface of the primer is really smooth use some finishing compound

The top coat can now be applied. When working out of doors, pick a dry, warm and wind-free day. Ensure surrounding areas are protected from over-spray. Agitate the aerosol thoroughly, then spray the centre of the repair area, working outwards with a circular motion. Apply the paint as several thin coats

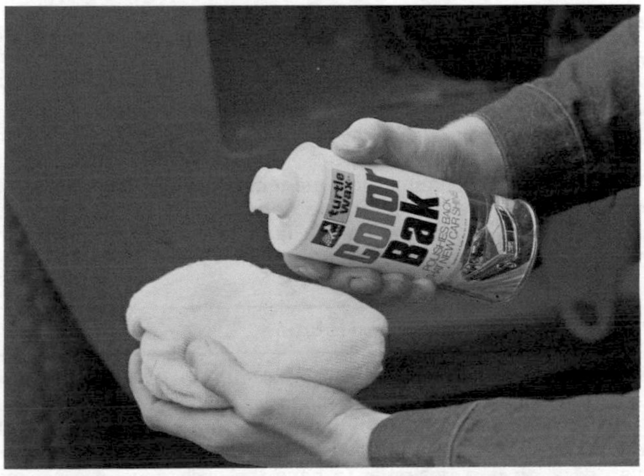

After a period of about two weeks, which the paint needs to harden fully, the surface of the repaired area can be 'cut' with a mild cutting compound prior to wax polishing. When carrying out bodywork repairs, remember that the quality of the finished job is proportional to the time and effort expended

remove the bonnet to hinge retaining bolts, spring and plain washers at both hinges. Carefully lift away the bonnet over the front of the car (Fig. 12.11).

4 Refitting is the reverse sequence to removal. Alignment in the body may be made by leaving the securing bolts slightly loose and repositioning by trial and error. The securing bolts must not be overtightened as they could damage the outer bonnet panel.

18 Bonnet lock – removal and refitting

1 Open the bonnet and support it on its stay, then unscrew and remove the two radiator support bracket retaining screws and move the top of the radiator back to gain access to the lock.

2 Slacken the nut and detach the release cable from the trunnion located at the lock lever (Fig. 12.12).

3 Detach the release cable and its clip from the bonnet lock.

4 Undo and remove the three bolts, plain and shakeproof washers that secure the bonnet lock. Lift away the bonnet lock.

5 Undo and remove the two bolts, plain and shakeproof washers that secure the locking pin assembly to the underside of the bonnet.

6 Detach the return spring and remove the rivet that secures the safety catch.

7 Refitting is the reverse sequence to removal. It is now necessary to adjust the lock pin assembly until a clearance of 2 in (50.8 mm) exists (A, Fig. 12.12) between the thimble and bonnet panel.

8 Carefully lower the bonnet and check the alignment of the pin thimble with the lock hole. If misaligned, slacken the fixing bolts and move the assembly slightly. Retighten the fixing bolts.

9 Close the bonnet and check its alignment with the body wing panels. If necessary, reposition the locking pin assembly.

10 The bonnet must contact the rubber stops. To adjust the position of the stops, screw in or out as necessary.

11 Lubricate all moving parts and finally check the bonnet release operations.

19 Bonnet lock control cable – removal and refitting

1 Open the bonnet and support it on its stay, then unscrew and remove the two radiator support bracket retaining screws and move the top of the radiator back to gain access to the lock.

2 Slacken the nut and detach the release cable from the trunnion located at the lock lever (Fig. 12.12).

3 Detach the release cable and its clip from the bonnet lock.

4 Release the outer control cable from its snap clamp and then unscrew and remove the screw that secures each clip to the wing valance. Lift away the two clips.

5 Undo and remove the nut and shakeproof washer that secure the outer cable to the body side bracket mounted below the facia panel.

6 Carefully withdraw the control cable assembly through the body grommet.

7 Refitting is the reverse sequence to removal. It is however, necessary to adjust the inner cable. Push the release knob in fully and make sure that the lock release lever is not pre-loaded by the release cable.

8 There must be a minimum movement of 0.5 in (12.7 mm) prior to the release of the bonnet. To adjust, slacken the cable trunnion nut and re-adjust the cable so that the bonnet is released with 0.5 to 2.0 in (12.7 to 50.8 mm) of cable movement.

20 Boot lid hinge – removal and refitting

1 Open the lid and using a pencil, mark the position of the hinge relative to the luggage compartment lid.

2 Undo and remove the four bolts, spring and plain washers that secure the hinges to the lid. Lift away the lid over the back of the car. For this operation it is desirable to have the assistance of a second person.

3 To remove the hinge, undo and remove the two nuts, plain and spring washers that secure each hinge to the body bracket. Lift away the hinge (Fig. 12.13).

4 Refitting is the reverse sequence to removal, but if adjustment is necessary, leave the bolts that secure the hinge to the lid slack.

5 Close the lid and adjust the position to ensure correct trim spacing.

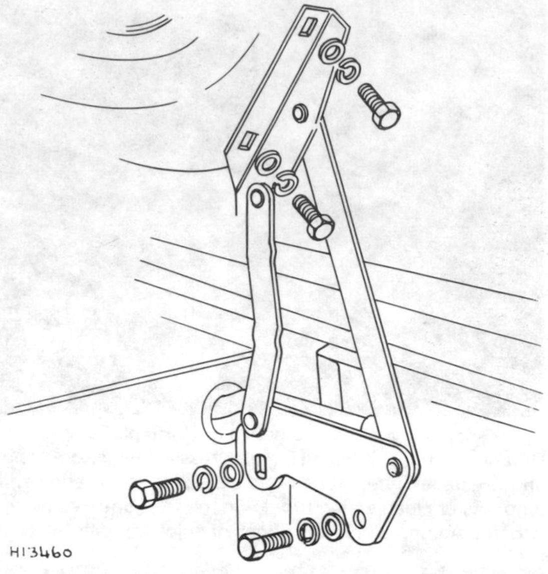

Fig. 12.11 Bonnet hinge assembly (Sec 17)

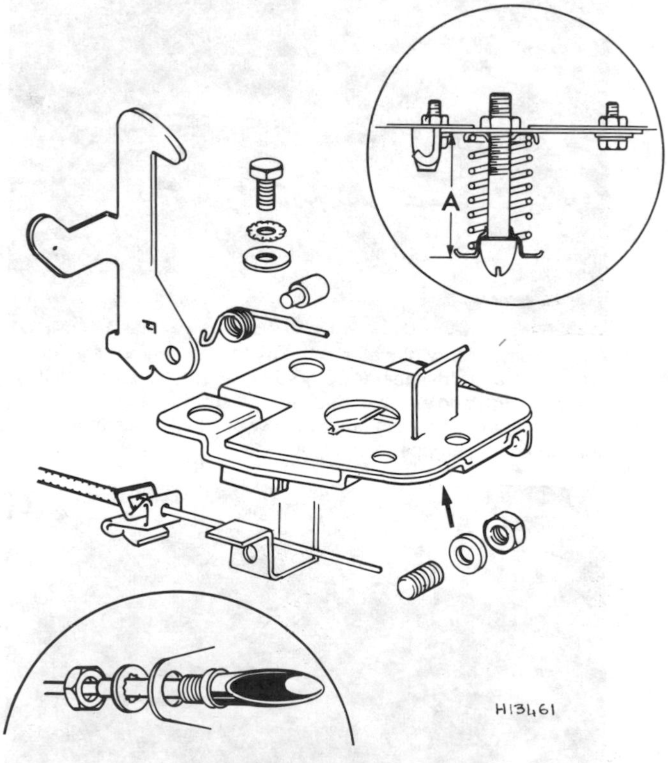

Fig. 12.12 Bonnet lock assembly (Sec 18)

Open the lid and tighten the hinge bolts. Do not overtighten as they could damage the outer lid panel.

21 Boot lid lock – removal and refitting

1 Using a pencil, mark the outline of the lock catch plate on the lid under the panel.

2 Undo and remove the three bolts, spring and plain washers that secure the lock catch (Fig. 12.14).

3 Slacken the locknut and unscrew the spindle. Lift away the shakeproof washer, spindle striker and spring.

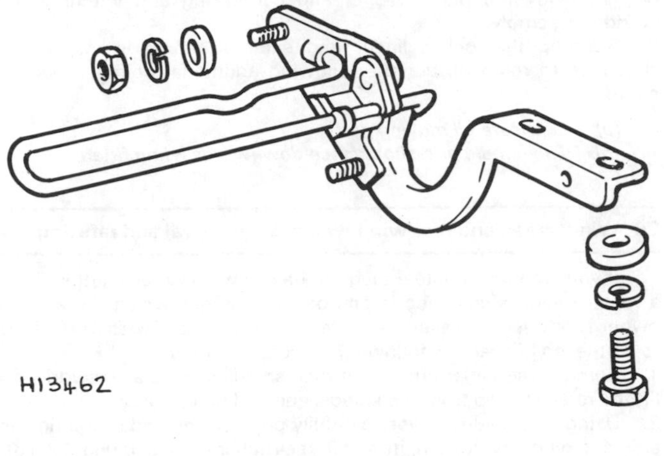

Fig. 12.13 Boot lid hinge (Sec 20)

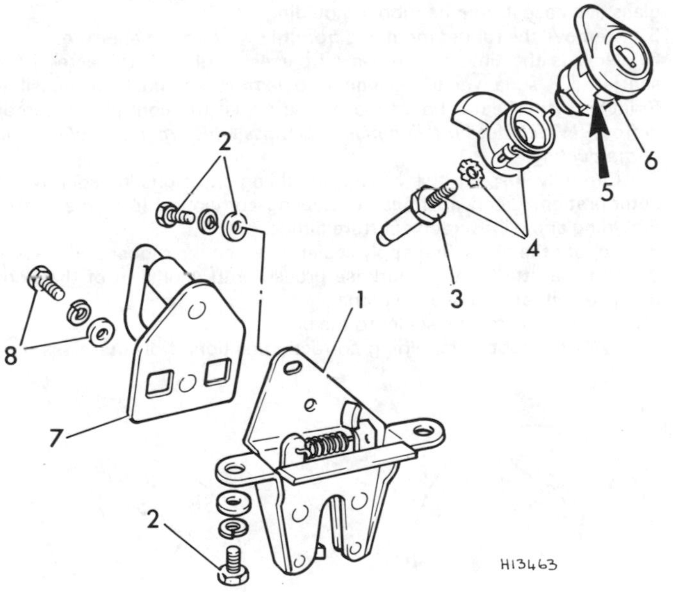

Fig. 12.14 Boot lock assembly (Sec 21)

1 Lock catch plate
2 Bolt, spring and plain washer
3 Locknut
4 Shakeproof washer, spindle and spring
5 Barrel housing spring
6 Lock barrel assembly
7 Striker
8 Bolt, spring and plain washer

4 Using a screwdriver, break off the two retaining ears of the barrel housing spring retaining clip and withdraw the barrel assembly and sealing gasket from outside the lid. A new clip will be necessary during assembly.
5 Using a pencil, mark the outline of the striker on the body panel.
6 Undo and remove the two bolts, spring and plain washers that retain the striker. Lift away the striker.
7 Refitting is the reverse sequence to removal. Lubricate all moving parts with engine oil.

22 Tailgate and tailgate hinges – removal and refitting

Tailgate

1 The tailgate of the Estate is removed by undoing the hinge bolts on the tailgate itself. Have an assistant hold the tailgate in the open position so that as the bolts are removed it will still remain supported. Do not forget to scribe round the hinges so that they can be refitted in a similar position. Refitting a new tailgate will mean that the exact positioning when closed will have to be adjusted using the same method as for a side door.

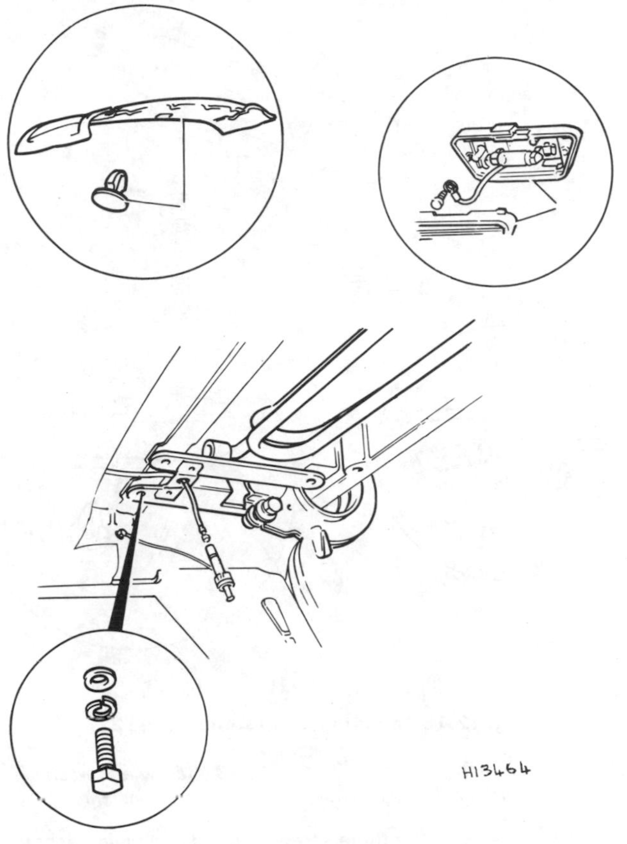

Fig. 12.15 Tailgate hinge removal (Sec 22)

Tailgate hinges

2 Remove the rear interior light; note the earth wire secured beneath one fixing screw (Fig. 12.15).
3 Carefully detach the rear compartment rear headlining and collect the press fasteners.
4 Detach the courtesy light switch from the right-hand hinge bracket. Disconnect the wire from the rear of the switch.
5 Using a pencil, mark the position of the hinges relative to the body to act as a datum for refitting.
6 Undo and remove the four screws, spring and plain washers that secure the hinge and torsion bar to each side of the body.
7 Lift away the hinge and torsion bar assembly.
8 Refitting the hinge and torsion bar assembly is the reverse sequence to removal.

23 Tailgate lock – removal and refitting

1 Remove the tailgate trim pad (Fig. 12.16).
2 Carefully unclip the retainer and detach the operating rod from the outside handle assembly.
3 Undo and remove the three screws and spring washers that secure the lock to the tailgate.
4 Lift away the tailgate lock assembly.
5 Refitting the tailgate lock assembly is the reverse sequence to removal. Lubricate all moving parts.

24 Tailgate lock striker plate – removal and refitting

1 Lift up the tailgate.
2 Using a pencil , mark the position of the striker plate relative to the body to act as a datum for refitting.

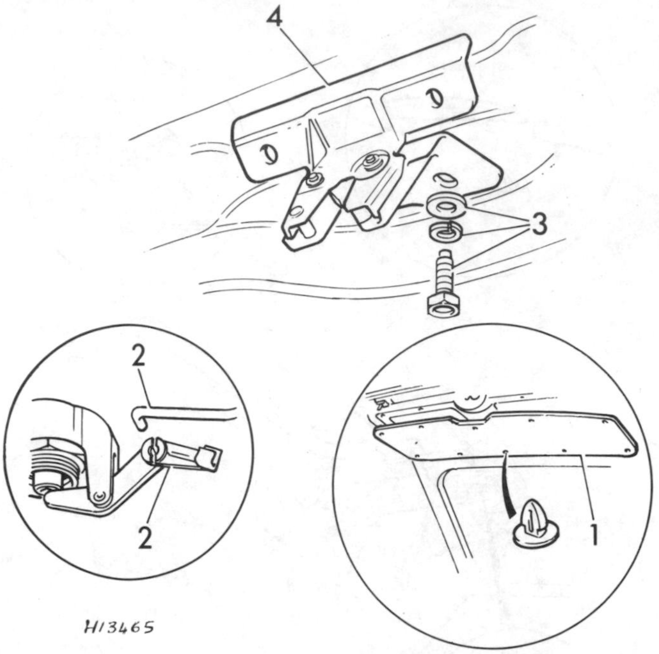

Fig. 12.16 Tailgate lock assembly (Sec 23)

1 Trim panel
2 Operating rod and retainer
3 Screw and washer
4 Lock assembly

3 Undo and remove the three screws, spring and plain washers that secure the striker plate to the body (Fig. 12.17).
4 Lift away the striker plate.
5 Refitting the striker plate is the reverse sequence to removal.

25 Tailgate exterior handle and lock – removal and refitting

1 Lift up the tailgate and remove the trim pad.
2 Carefully unclip the retainer and detach the lock operating rod from the operating lever.
3 Undo and remove the nut that secures the handle assembly to the tailgate.
4 Detach the operating lever assembly and remove the handle assembly and seal.
5 To remove the lock cylinder from the outside handle first detach the retaining circlip.

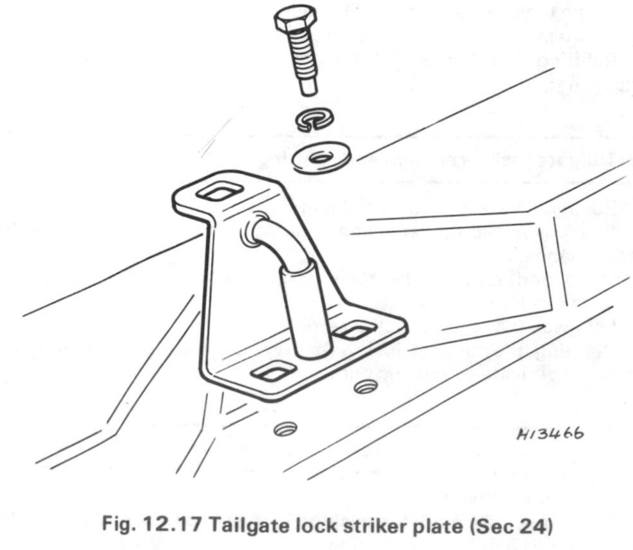

Fig. 12.17 Tailgate lock striker plate (Sec 24)

6 Lift away the plain washer and coil spring and withdraw the cylinder assembly.
7 Refitting the lock cylinder and exterior handle is the reverse sequence to removal, but the following additional points should be noted:

(a) Lubricate all moving parts
(b) The handle grip should face downwards when fitted

26 Windscreen and rear window glass – removal and refitting

If you are unfortunate enough to have a windscreen shatter, fitting a replacement windscreen is one of the few jobs which the average owner is advised to leave to a professional. For the owner who wishes to do the job himself the following instructions are given:
1 Remove the wiper arms from their spindles using a screwdriver to lift the retaining clip from the spindle end and pull away.
2 Using a screwdriver, very carefully prise up the end of the finisher strip and withdraw it from, its slot in the rubber moulding (Fig. 12.18).
3 The assistance of a second person should now be enlisted, ready to catch the glass when it is released from its aperture.
4 Working inside the car, commencing at one top corner, press the glass and ease it from its rubber moulding.
5 Remove the rubber moulding from the windscreen aperture.
6 Now is the time to remove all pieces of glass if the screen has shattered. Use a vacuum cleaner to extract as much as possible. Switch on the heater boost motor and adjust the controls to 'screen defrost'. *Watch out for flying pieces of glass which might be blown out of the ducting.*
7 Carefully inspect the rubber moulding for signs of splitting or deterioration. Clean all traces of sealing compound from the rubber moulding and windscreen aperture flange.
8 To refit the glass, first apply sealer between the rubber and glass.
9 Press a little general purpose grease onto 4 or 5 in of the body flange on either side of each corner.
10 Apply some mastic sealer to the body flange.
11 With the rubber moulding correctly positioned on the glass, it is

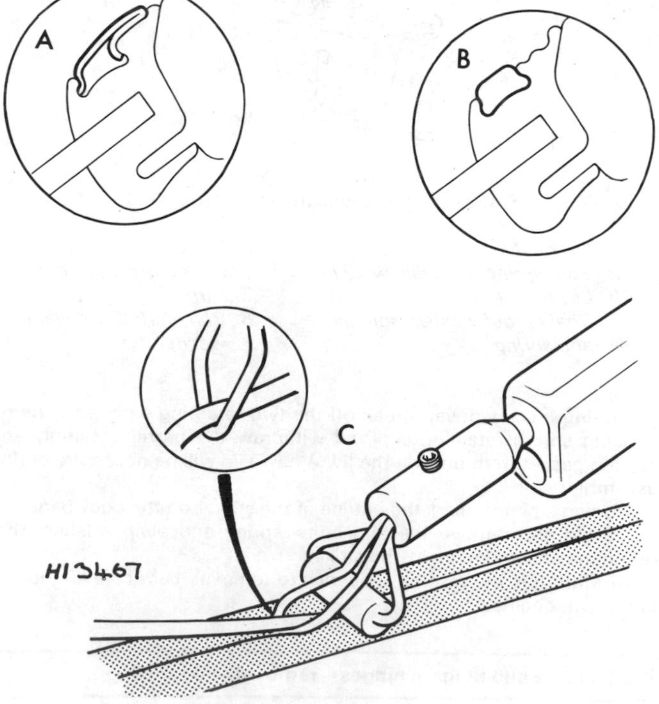

Fig. 12.18 Glass removal and refitting (Sec 26)

A and B Profiles of two types of finisher strip in place
C Type of tool necessary to fit finisher

now necessary to insert a piece of cord about 16 ft long all round the outer channel in the rubber surround which fits over the windscreen aperture flange. The two free ends of the cord should finish at either top or bottom centre and overlap each other by a minimum of 1 ft.
12 Offer the screen up to the aperture and get an assistant to press the rubber surround hard against the body flange. Slowly pull one end of the cord, moving round the windscreen so drawing the lip over the windscreen flange on the body. If necessary, use a piece of plastic or tapered wood to assist in locating the lip on the windscreen flange.
13 The finisher strip must next be fitted to the moulding and for this a special tool is required. An illustration of this tool is shown in Fig.12.18 and a handyman should be able to make up an equivalent using netting wire and a wooden file handle.
14 Fit the eye of the tool into the groove and feed in the finisher strip.
15 Push the tool around the complete length of the moulding, feeding the finisher into the channel as the eyelet opens it. The back half beds the finisher into the moulding.
16 Clean off all traces of sealer using turpentine.

27 Rear body side glass (Estate) – removal and refitting

1 With an assistant ready to catch the glass and rubber surround assembly, push outwards on the glass to release it from the aperture flange.
2 Carefully remove the finisher from the rubber surround.
3 Remove the rubber surround from the glass.
4 If the original glass and/or rubber surround are to be used again, all traces of old sealer should be removed.
5 Refit the rubber surround to the glass.
6 Using a suitable sealer, seal the rubber surround to the glass at the outside face.
7 Lubricate the finisher channel in the rubber surround with a soapy solution or washing up liquid.
8 Fit the finisher to the rubber surround.
9 Apply some sealer to the middle groove around the outside edge of the rubber surround.
10 Apply some sealer to the outside face of the window aperture in the body.
11 Fit some cord around the locating groove in the rubber surround and with the ends inside the body, position the assembly up in the aperture.
12 Pull on the cord whilst an assistant pushes hard on the glass. The retaining lip should now move over the aperture flange and hold the glass and surround assembly in position. Clean off any surplus sealer.

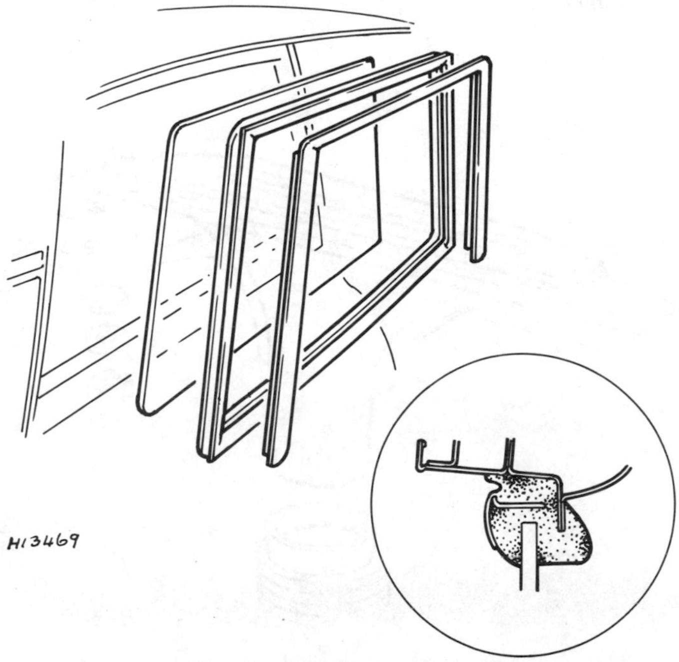

Fig. 12.19 Body side glass assembly – Estate (Sec 27)

28 Tailgate glass (Estate) – removal and refitting

The procedure is the same as described in Section 26, except that it will be necessary to disconnect the supply leads to the heated rear glass.

29 Facia panel – removal and refitting

1 Disconnect the battery negative terminal.
2 Refer to Chapter 11 and remove the steering wheel
3 Remove the instrument panel as described in Section 33 and the glovebox as described in Section 30.
4 Remove the parcel shelf by undoing the two securing nuts and washers, the screws and the clip.
5 Disconnect the face level vent tubes and detach the steering column bridge bracket (Fig. 12.20).
6 Carefully pull off the heater control knobs.
7 Unscrew and remove the heater control retaining screws and secure the heater control to one side.
8 Note the position of all the switches and multi-plug connectors and then disconnect them from the facia panel.
9 Unscrew and remove the five facia panel securing screws and the three panel retaining nuts, then press the centre of the panel downwards to release it from the retaining clip.
10 The facia panel assembly can now be removed.
11 If further dismantling is necessary, unscrew the retaining screws and detach the two face level vents and glovebox hinge brackets.
12 Remove the ashtray and cigar lighter (if fitted).
13 Remove the speaker grille and prise out the radio blank.
14 Unscrew and remove the two retaining screws that secure the instrument pack support bracket to the facia panel.
15 Finally drill out the rivets to release the facia panel stud clip fasteners and remove the vent, glovebox and steering column bracket fixing.
16 Refitting is the reverse of the removal procedure.

30 Glovebox – removal and refitting

1 Open the lid and detach the hinge stay from the lid.
2 Unscrew and remove the upper and lower hinge bracket retaining screws and withdraw the lid.
3 Unscrew and remove the two glovebox lock striker plate retaining screws then remove the two glovebox retaining screws.
4 The glovebox can now be withdrawn from the facia panel, but on HL models it will be necessary to disconnect the electrical leads from the glovebox lamp and switch.
5 Refitting is a reversal of the removal procedure.

31 Rear parcel tray – removal and refitting

1 Remove the rear seat squab and cushion.
2 Working inside the boot, carefully push up the rear parcel tray and release it from the retaining clips.
3 Refitting the rear parcel tray is the reverse sequence to removal.

32 Centre console – removal and refitting

1 Remove the knob from the gearchange lever.
2 Unclip the gearchange lever gaiter and tray and lift them away.
3 Undo the two forward housing securing screws and release the retaining clip at the rear end of the housing.
4 Disconnect the supply cable for the clock and remove the clock lamp holder.
5 Prise up the front edge of the console and then withdraw the trim panel from the handbrake lever opening. Remove the seat belt stalk seal and the ashtray.
6 Remove the console securing screws and lift out the centre console.
7 Refitting the centre console is the reverse of the removal procedure.

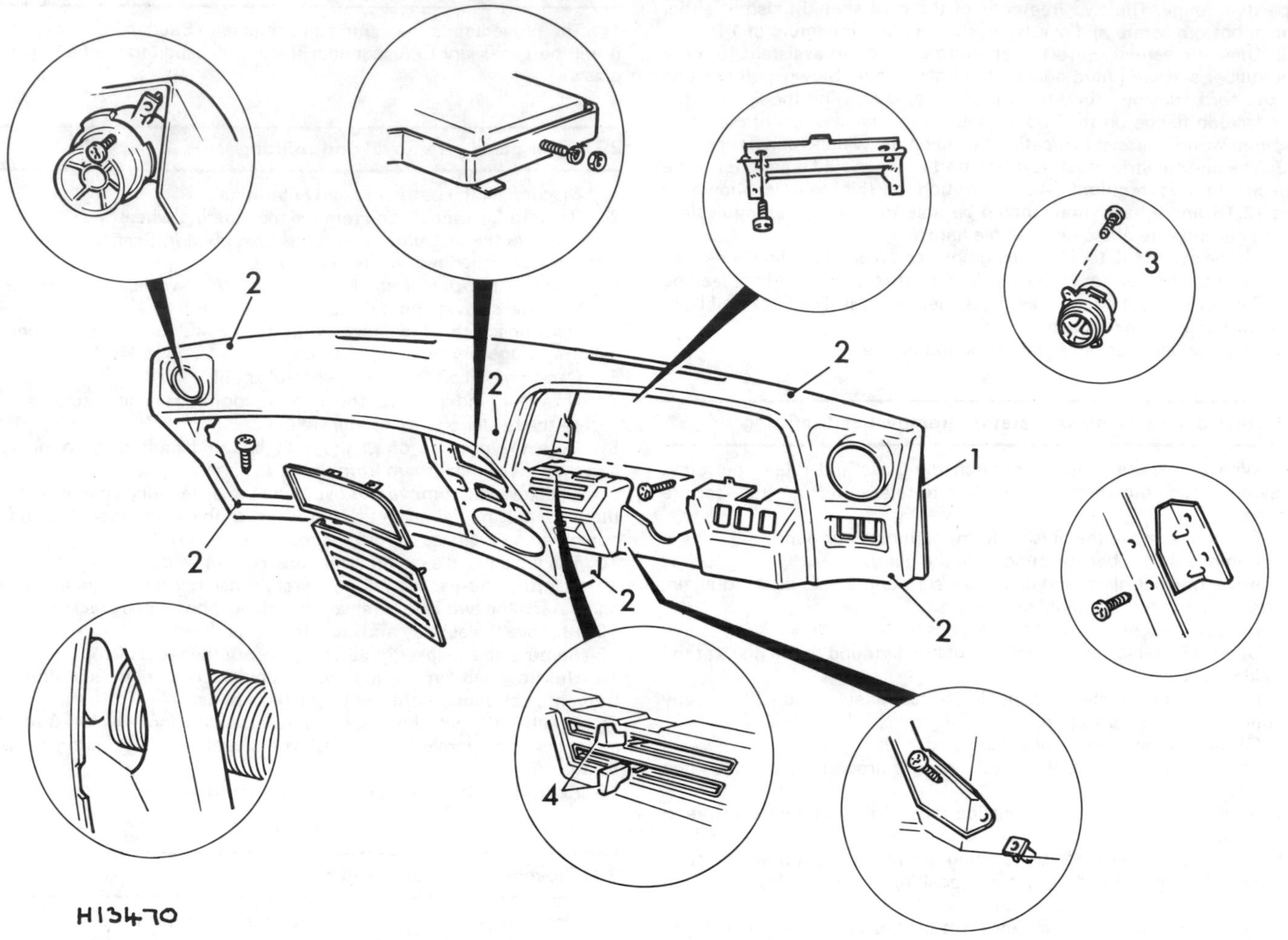

Fig. 12.20 Facia panel removal (Sec 29)

1 Facia panel
2 Retaining screw locations
3 Face level vent retaining screw
4 Heater controls

33 Instrument panel – removal and refitting

1 Disconnect the battery negative terminal.
2 Unscrew and remove the two instrument cowl securing screws and withdraw the cowl (photo).
3 Depress the speedometer cable locking clip and release the cable from the speedometer head.
4 Unscrew and remove the instrument panel securing screws and pull the assembly forward.
5 Note the location of the various leads and bulb holders, then disconnect them and withdraw the instrument panel (photos).
6 Refitting is the reverse of the removal procedure.

34 Windscreen demister duct – removal and refitting

1 Remove the instrument panel and facia panel as described in Sections 33 and 29.
2 Disconnect the demister duct tube, then unscrew and remove the demister duct securing nuts and washers and lift the duct from its location holes.
3 Refitting the duct is the reverse of the removal procedure.

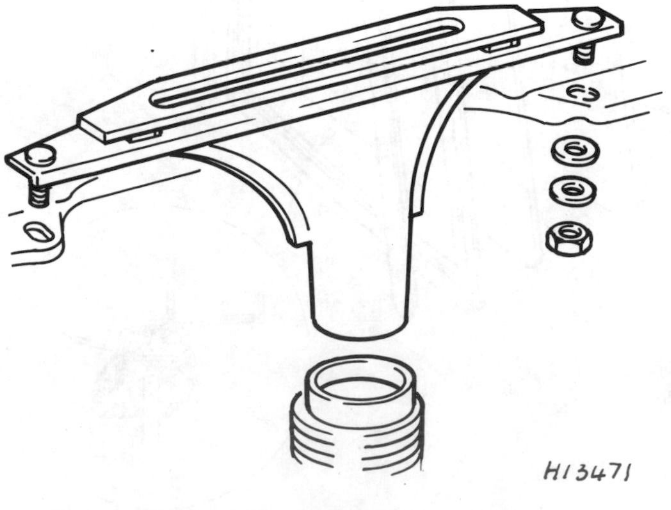

Fig. 12.21 Windscreen demister duct removal (Sec 34)

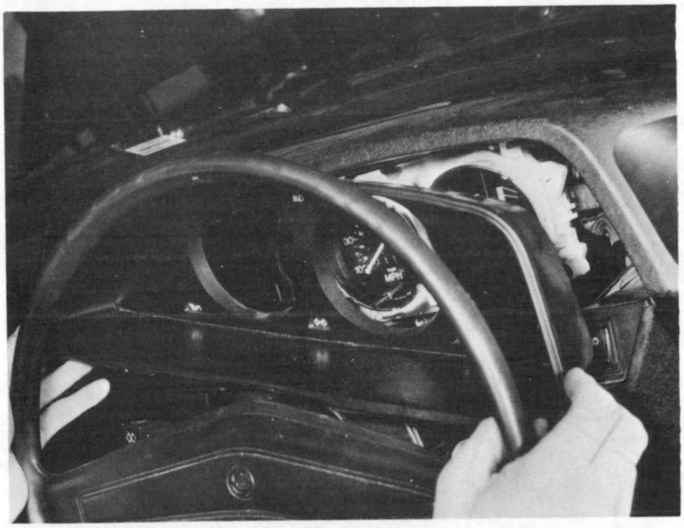

33.2 Removing the instrument cowl

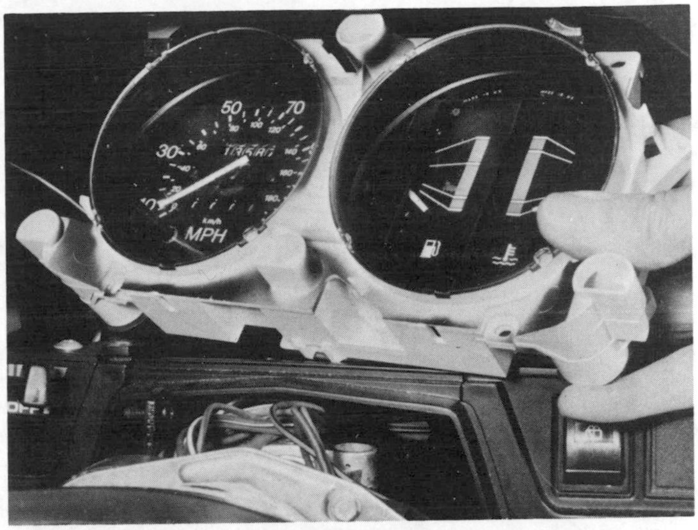

33.5a Lifting out the instrument panel

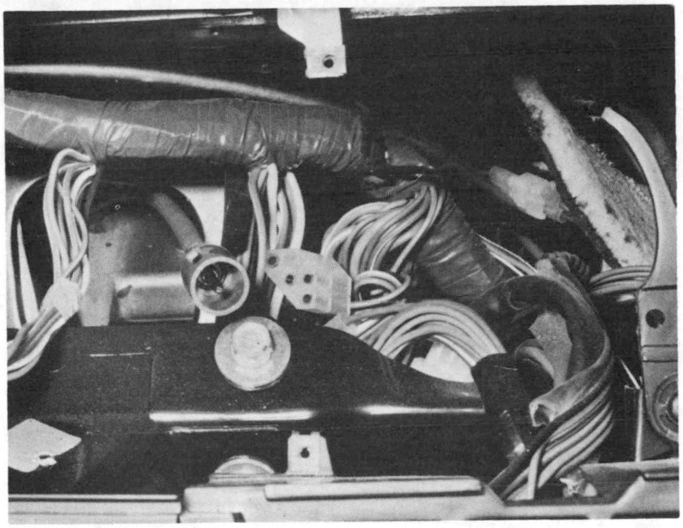

33.5b Facia panel with instrument panel removed

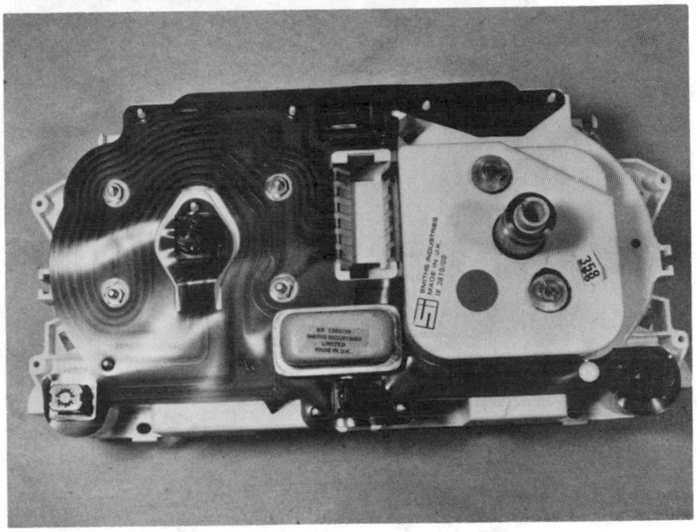

33.5c Back of the instrument panel

35 Front seat – removal and refitting

1 Remove the two screws that secure the outer seat runner (Fig. 12.22).
2 Unscrew the two locknuts that secure the inner seat runner to the car floor. Lift out the seat and lock bar assembly.
3 Refitting is the reverse of the removal procedure.

36 Rear seat squab and cushion (Saloon) – removal and refitting

1 Release the two cushion retaining clips and lift away the seat cushion.
2 Slacken the two screws and remove the arm rest.
3 Using a drill, remove the two rivet heads that secure the squab brackets. **Caution**: *Do not pass the drill through the body as it may damage or even puncture the brake pipes.*
4 Raise the squab to release the back panel from the three retaining hooks and lift away the squab.
5 Refitting the seat squab and cushion is the reverse sequence to removal. Always use pop rivets to retain the squab brackets. Do not use selftapping screws.

37 Rear seat cushion (Estate) – removal and refitting

1 Pivot the cushion assembly forwards.
2 Using a pencil, mark the position of the hinges on the seat to act as a datum for refitting.
3 Undo and remove the two screws, shakeproof and plain washers that secure the hinges to the seat (Fig. 12.24).
4 Lift the cushion from the hinges.
5 Refitting the rear seat cushion is the reverse of the removal procedure.

38 Rear seat squab (Estate) – removal and refitting

1 Release the squab from its retaining catches and pivot the squab forwards.
2 Undo and remove the two countersunk screws that secure the squab pivots to the body at each side (Fig. 12.25).
3 The squab may now be lifted away.
4 Refitting the rear seat squab is the reverse sequence to removal.

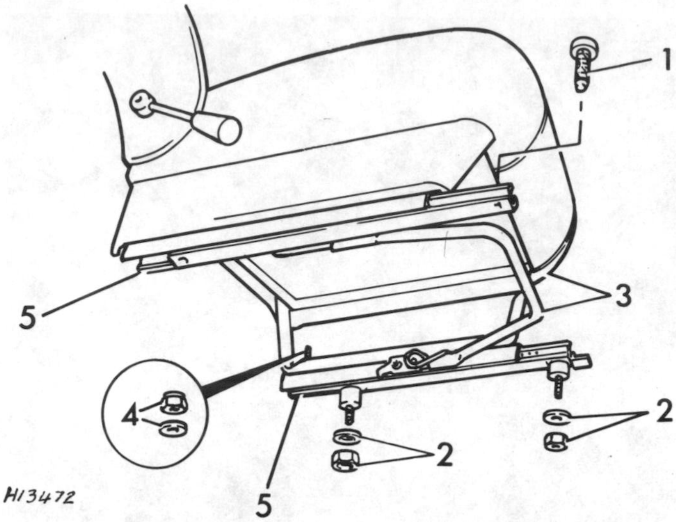

Fig. 12.22 Front seat removal (Sec 35)

1 Securing screws
2 Locknuts
3 Seat and lockbar
4 Seat runner securing nut
5 Seat runners

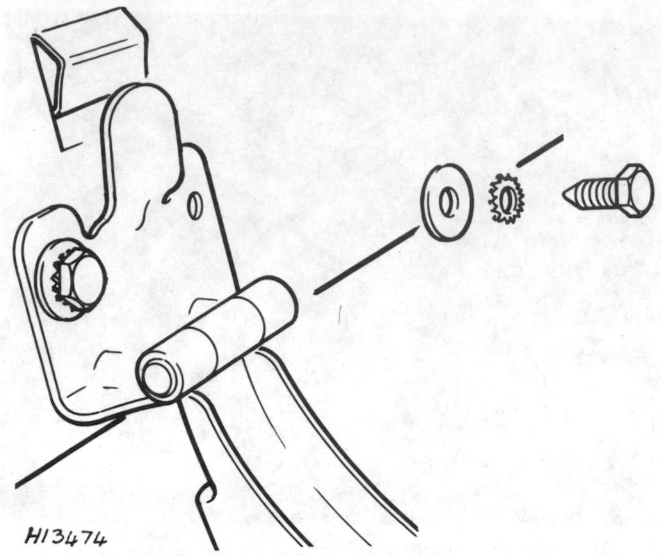

Fig. 12.24 Rear seat cushion hinge assembly – Estate (Sec 37)

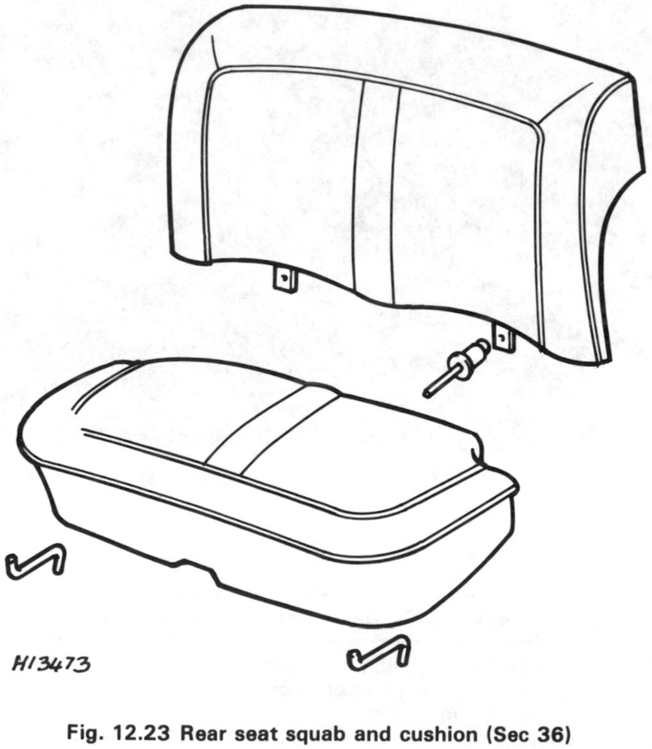

Fig. 12.23 Rear seat squab and cushion (Sec 36)

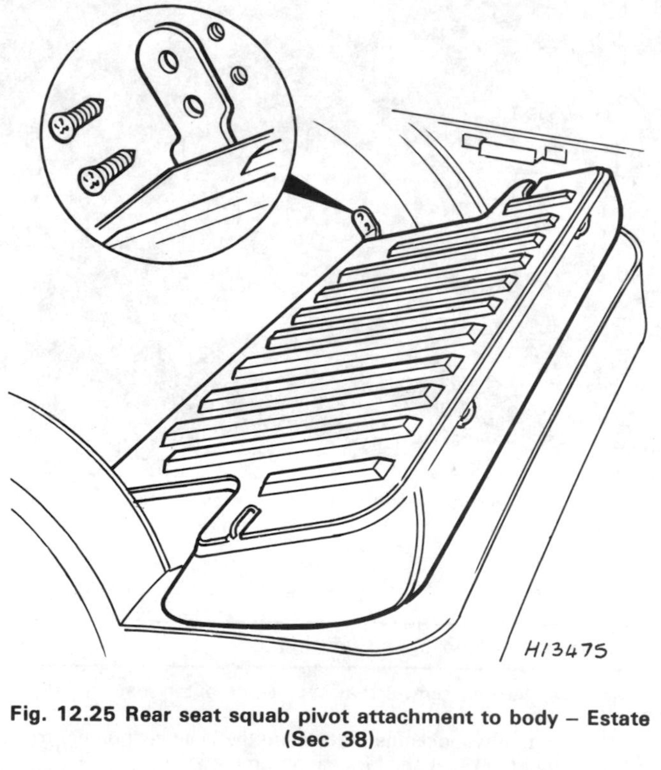

Fig. 12.25 Rear seat squab pivot attachment to body – Estate (Sec 38)

39 Radiator grille – removal and refitting

1 Refer to Fig. 12.26 and undo and remove the four selftapping screws and plain washers that secure the case to the body.
2 The grille and case may now be lifted upwards and forwards away from the front of the car.
3 To detach the grille from the case, undo and remove the selftapping screws and the plain washers.
4 Refitting is the reverse sequence to removal. Make sure that the rubber inserts are correctly positioned in the body panel cutout and the locating pegs of the case are located in the centre of the rubber inserts.

40 Sump guard – removal and refitting

1 Undo and remove the three screws and washers that secure each side of the guard to the mounting brackets (Fig. 12.27).
2 Lift the sump guard away from under the front of the car.
3 Refitting the sump guard is the reverse sequence to removal.

41 Front bumper – removal and refitting

1 Working inside the engine compartment, disconnect the wiring

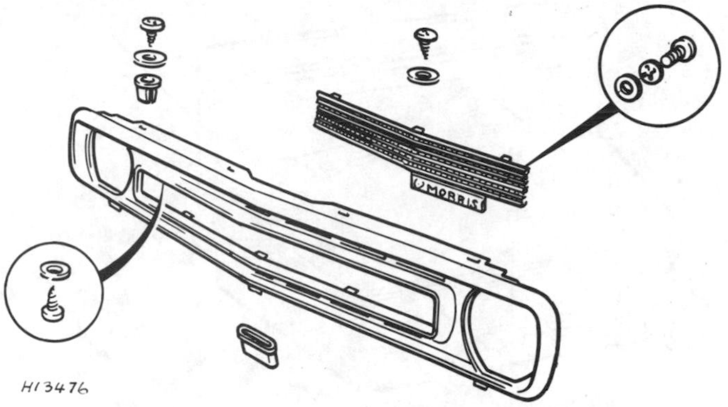

Fig. 12.26 Radiator grille assembly (Sec 39)

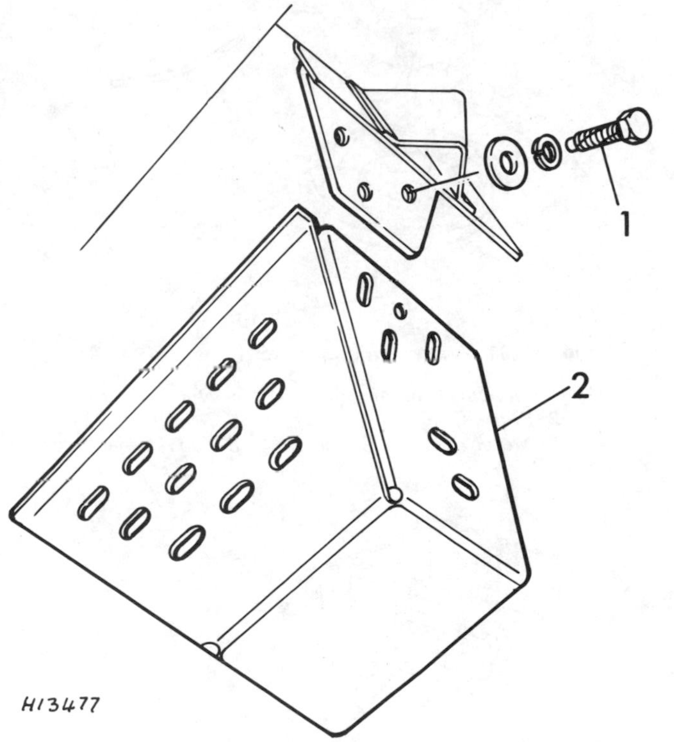

Fig. 12.27 Sump guard removal (Sec 40)

1 Retaining bolt *2 Sump guard*

from the front flasher lamps at the Lucar connectors (Fig. 12.28).
2 Undo and remove the bolt, spring, plain washer and mounting rubber from each end of the bumper.
3 Undo and remove the bolts, spring and plain washers that secure each support bracket to the body.
4 Remove the grommet that protects the flasher lamp wiring harness. Thread the harness through the front valance.
5 The bumper can now be lifted away from the car. If necessary, the flasher lamps can be removed from the bumper after removing the lamp securing nuts.
6 Refitting is the reverse of the removal procedure.

42 Rear bumper – removal and refitting

1 On Estate models, disconnect the wiring for the fog and reverse lamps at the harness located in the box section behind the rear valance.
2 The number plate lamps can be withdrawn from the bumper after pressing in the two securing clips.
3 Remove the lamp cover and lens from each lamp and pass the

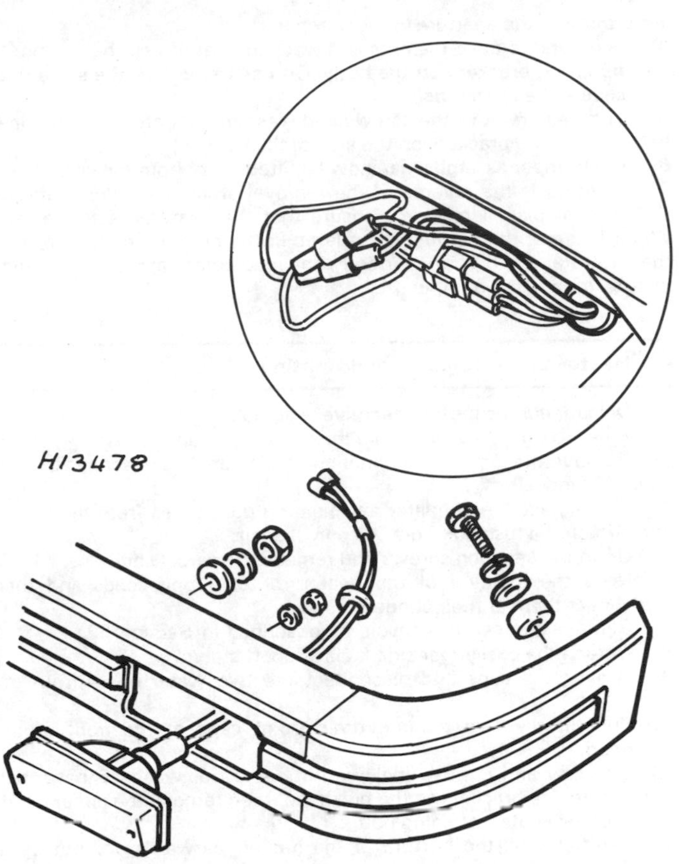

Fig. 12.28 Front bumper removal (Sec 41)

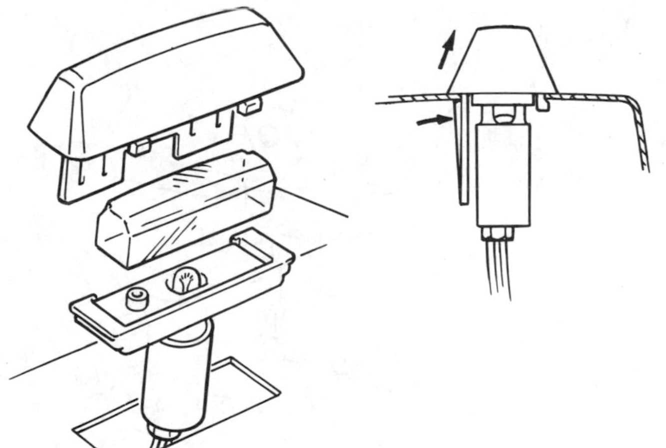

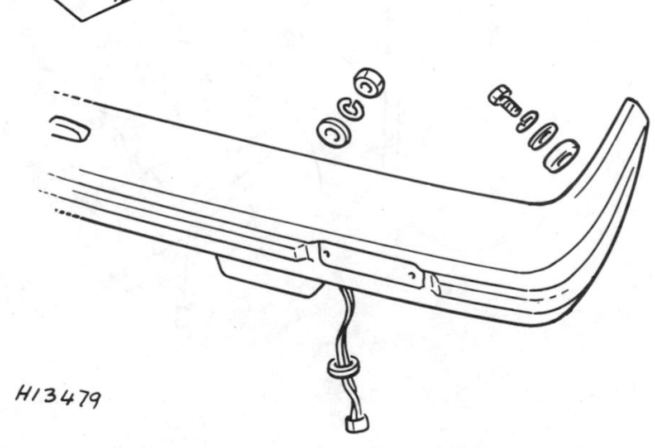

Fig. 12.29 Rear bumper removal (Sec 42)

lamp through the aperture in the bumper.
4 Undo and remove the nuts and washers that attach the bumper to the mounting brackets on the body. On Estate models, the same nuts also secure the fog lamps.
5 Undo and remove the screws and washers that attach the bumper to the mounting brackets on the side of the body.
6 The bumper assembly can now be lifted away from the car.
7 Refitting is the reverse of the removal procedure. When refitting the number plate lamp lens, ensure that the chamfer is to the rear. When locating the number plate lamp in the bumper, make sure that the two retaining lugs are correctly engaged before pressing the lamps into position in the bumper.

43 Heater unit – removal and refitting

1 Disconnect the battery negative terminal.
2 Refer to Chapter 2 and drain the cooling system.
3 Remove the instrument panel and facia panel as described in Sections 33 and 29.
4 Disconnect the demister and heating duct tubes from the heater and detach the fuse box from the parcel shelf.
5 Undo the securing screws and remove the parcel shelf.
6 Note the location of the heater motor supply leads and then disconnect them at their connectors.
7 Remove the centre console as described in Section 32.
8 Detach the passenger side facia support stay.
9 Loosen the clips and disconnect the two water hoses from the heater.
10 Pull off the two plenum drain point caps at the front of the bulkhead.
11 Unscrew and remove the two nuts and four washers that retain the heater side brackets to the bulkhead, then remove the upper windshield panel heater retaining bolt.
12 Carefully pull the bottom of the heater rearwards and out of its location.

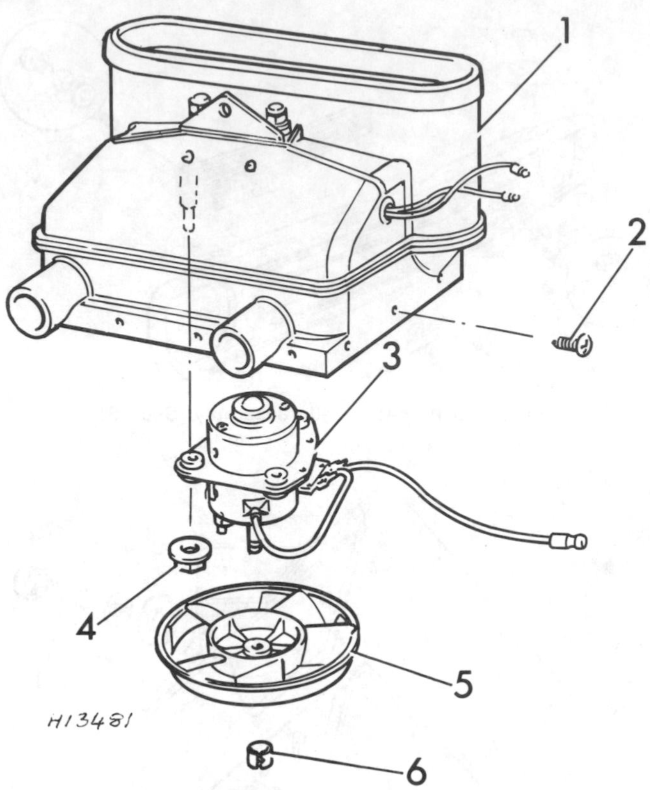

Fig. 12.31 Heater fan and motor removal (Sec 44)

1 Plenum chamber
2 Screws
3 Motor
4 Nut
5 Fan
6 Fan retainer

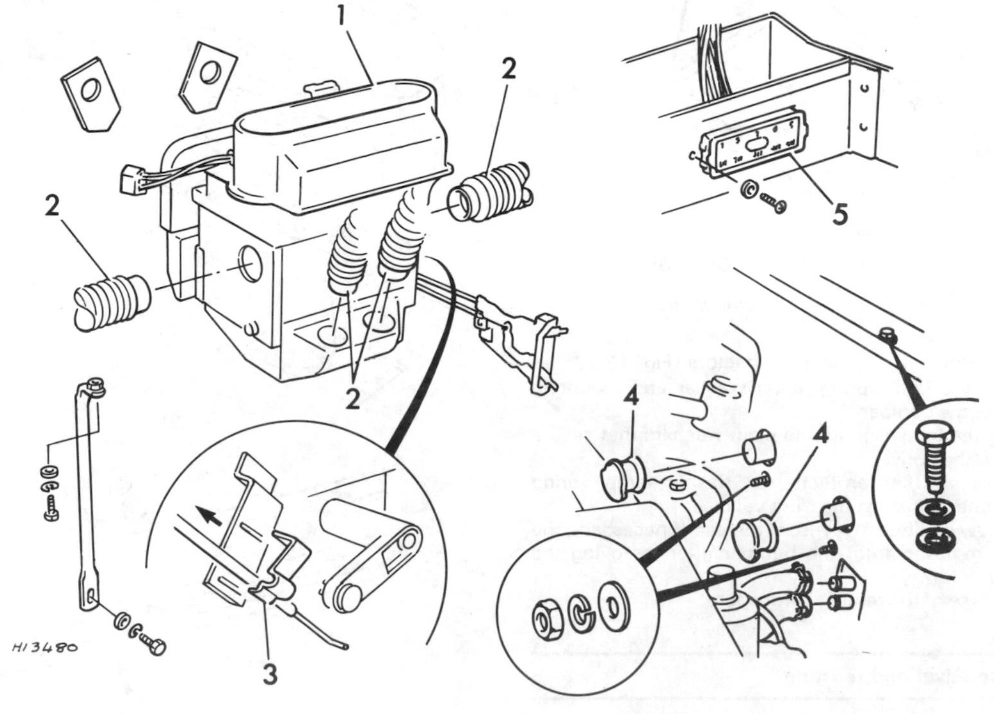

Fig. 12.30 Heater unit removal (Sec 43)

1 Heater assembly
2 Air ducts
3 Outer control cable clamp
4 Plenum drain point caps
5 Fuse box (on parcel shelf)

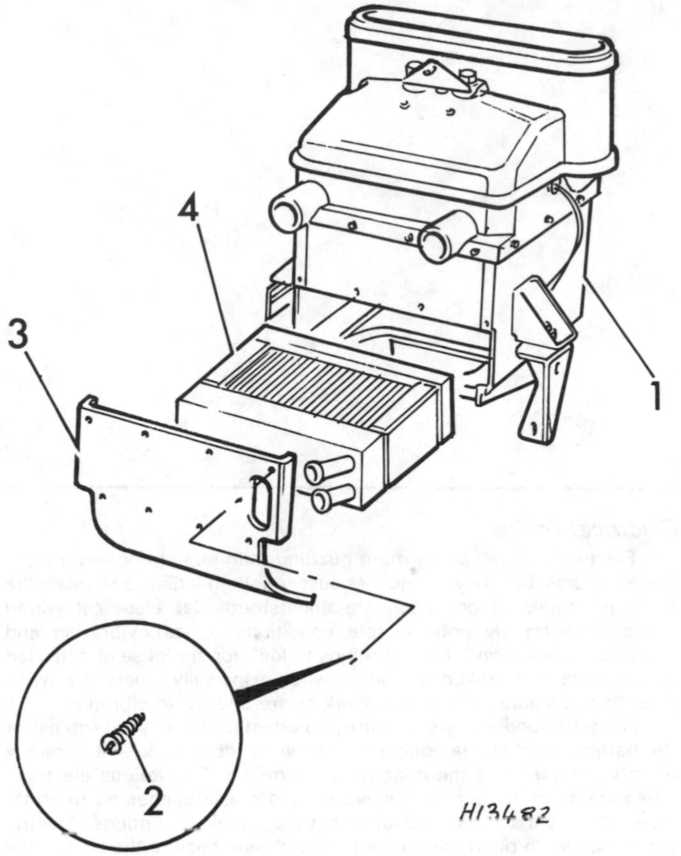

Fig. 12.32 Heater matrix removal (Sec 45)

1 *Heater casing*
2 *Screw*
3 *Cover plate*
4 *Matrix*

13 Detach the inner and outer control cables from the heater unit, withdraw the heater and remove the insulation and seal pads.
14 Refitting follows a reversal of the removal procedure but before connecting the outer control cable clamp, hold the control levers in the OFF position and slightly pull the outer cable away from the operating arm.

44 Heater fan and motor – removal and refitting

1 Refer to Section 43 and remove the heater unit.
2 Undo and remove the heater plenum chamber securing selftapping screws and lift away the plenum chamber (Fig. 12.31).
3 Undo and remove the three nuts and plain washers that secure the motor and fan assembly to the heater body. Lift away the motor.
4 If it is necessary to remove the fan, note which way round on the motor spindle it is fitted and remove the spring clip on the fan boss. Lift away the fan.
5 Refitting the heater fan and motor is the reverse sequence to removal. Before fitting a new motor, always test it by placing the cable terminals on the battery terminals.

45 Heater matrix – removal and refitting

1 Refer to Section 43 and remove the heater unit.
2 Carefully remove the packing rubber from the forward end of the heater unit.
3 Undo and remove the screws that secure the matrix cover plate to the heater body. Lift away the cover plate (Fig. 12.32).
4 The heater matrix may now be slid out from its location in the heater body.
5 If the matrix is leaking or blocked, follow the instructions given in Chapter 2, Section 6.
6 Refitting the heater matrix is the reverse sequence to removal.

Fault diagnosis

Introduction

The car owner who does his or her own maintenance according to the recommended schedules should not have to use this section of the manual very often. Modern component reliability is such that, provided those items subject to wear or deterioration are inspected or renewed at the specified intervals, sudden failure is comparatively rare. Faults do not usually just happen as a result of sudden failure, but develop over a period of time. Major mechanical failures in particular are usually preceded by characteristic symptoms over hundreds or even thousands of miles. Those components which do occasionally fail without warning are often small and easily carried in the car.

With any fault finding, the first step is to decide where to begin investigations. Sometimes this is obvious, but on other occasions a little detective work will be necessary. The owner who makes half a dozen haphazard adjustments or replacements may be successful in curing a fault (or its symptoms), but he will be none the wiser if the fault recurs and he may well have spent more time and money than was necessary. A calm and logical approach will be found to be more satisfactory in the long run. Always take into account any warning signs or abnormalities that may have been noticed in the period preceding the fault – power loss, high or low gauge readings, unusual noises or smells, etc – and remember that failure of components such as fuses or spark plugs may only be pointers to some underlying fault.

The pages which follow here are intended to help in cases of failure to start or breakdown on the road. There is also a Fault Diagnosis Section at the end of each Chapter which should be consulted if the preliminary checks prove unfruitful. Whatever the fault, certain basic principles apply. These are as follows:

Verify the fault. This is simply a matter of being sure that you know what the symptoms are before starting work. This is particularly important if you are investigating a fault for someone else who may not have described it very accurately.

Don't overlook the obvious. For example, if the car won't start, is there petrol in the tank? (Don't take anyone else's word on this particular point, and don't trust the fuel gauge either!) If an electrical fault is indicated, look for loose or broken wires before digging out the test gear.

Cure the disease, not the symptom. Substituting a flat battery with a fully charged one will get you off the hard shoulder, but if the underlying cause is not attended to, the new battery will go the same way. Similarly, changing oil-fouled spark plugs for a new set will get you moving again, but remember that the reason for the fouling (if it wasn't simply an incorrect grade of plug) will have to be established and corrected.

Don't take anything for granted. Particularly, don't forget that a 'new' component may itself be defective (especially if it's been rattling round in the boot for months), and don't leave components out of a fault diagnosis sequence just because they are new or recently fitted. When you do finally diagnose a difficult fault, you'll probably realise that all the evidence was there from the start.

Electrical faults

Electrical faults can be more puzzling than straightforward mechanical failures, but they are no less susceptible to logical analysis if the basic principles of operation are understood. Car electrical wiring exists in extremely unfavourable conditions – heat, vibration and chemical attack – and the first things to look for are loose or corroded connections and broken or chafed wires, especially where the wires pass through holes in the bodywork or are subject to vibration.

All metal-bodied cars in current production have one terminal of the battery 'earthed', ie connected to the car bodywork, and in nearly all modern cars it is the negative (–) terminal. The various electrical components, motors, bulb holders etc – are also connected to earth, either by means of a lead or directly by their mountings. Electric current flows through the component and then back to the battery via the car bodywork. If the component mounting is loose or corroded, or if a good path back to the battery is not available, the circuit will be incomplete and malfunction will result. The engine and/or gearbox are also earthed by means of flexible metal straps to the body or subframe; if these straps are loose or missing, starter motor, generator and ignition trouble may result.

Assuming the earth return to be satisfactory, electrical faults will be due either to component malfunction or to defects in the current supply. Individual components are dealt with in Chapter 10. If supply wires are broken or cracked internally this results in an open-circuit, and the easiest way to check for this is to bypass the suspect wire temporarily with a length of wire having a crocodile clip or suitable connector at each end. Alternatively, a 12V test lamp can be used to verify the presence of supply voltage at various points along the wire and the break can be thus isolated.

If a bare portion of a live wire touches the car bodywork or other earthed metal part, the electricity will take the low-resistance path thus formed back to the battery: this is known as a short-circuit. Hopefully a short-circuit will blow a fuse, but otherwise it may cause burning of the insulation (and possibly further short-circuits) or even a fire. This is why it is inadvisable to bypass persistently blowing fuses with silver foil or wire.

Spares and tool kit

Most cars are only supplied with sufficient tools for wheel changing; the *Maintenance and minor repair* tool kit detailed in *Tools and working facilities,* with the addition of a hammer, is probably sufficient for those repairs that most motorists would consider attempting at the roadside. In addition a few items which can be fitted without too much trouble in the event of a breakdown should be carried. Experience and available space will modify the list below, but the following may save having to call on professional assistance:

Spark plugs, clean and correctly gapped
HT lead and plug cap – long enough to reach the plug furthest from the distributor
Distributor rotor, condenser and contact breaker points
Drivebelts – emergency type may suffice
Spare fuses
Set of principal light bulbs

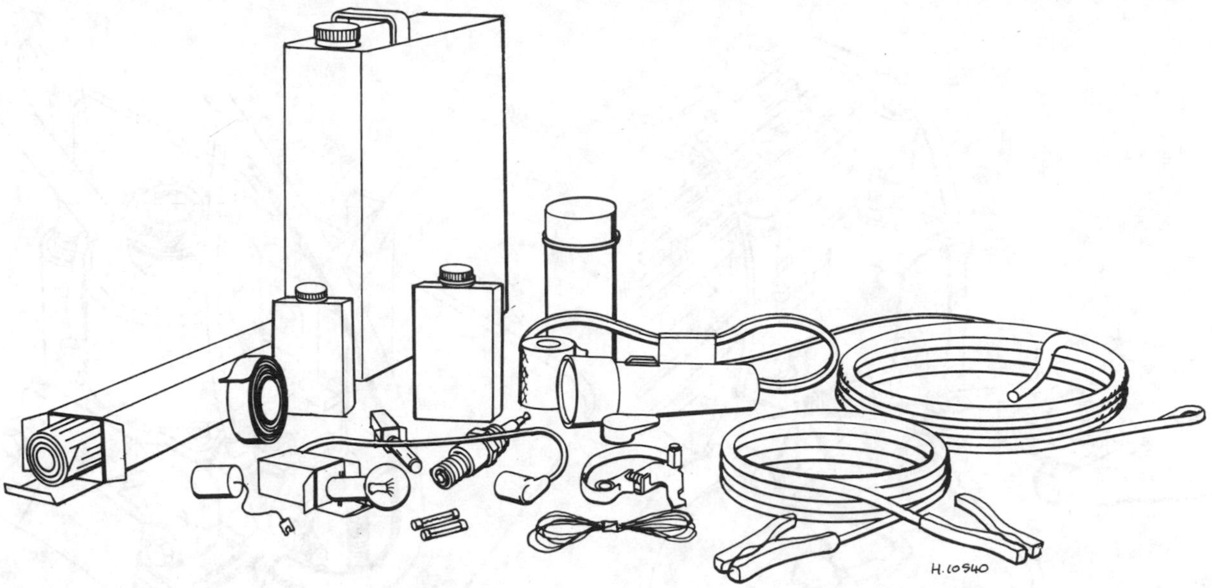

Carrying a few spares may save you a long walk!

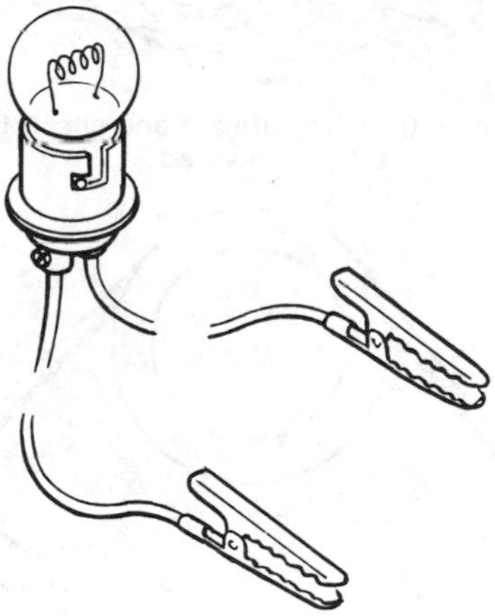

A simple test lamp is useful for investigating electrical faults

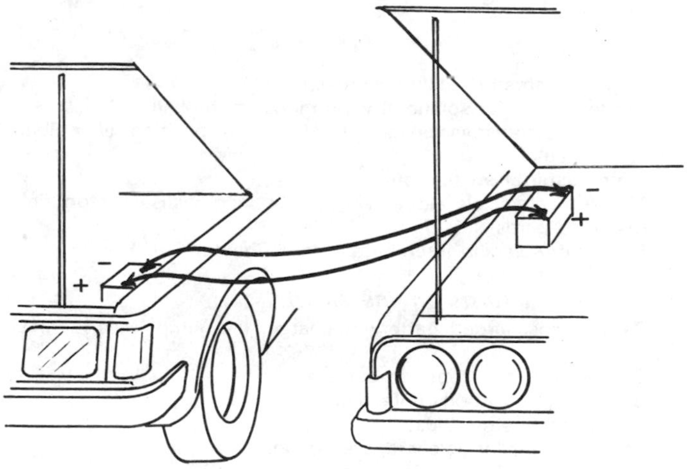

Correct way to connect jump leads. Do not allow car bodies to touch!

Tin of radiator sealer and hose bandage
Exhaust bandage
Roll of insulating tape
Length of soft iron wire
Length of electrical flex
Torch or inspection lamp (can double as test lamp)
Battery jump leads
Tow-rope
Ignition waterproofing aerosol
Litre of engine oil
Sealed can of hydraulic fluid
Emergency windscreen
Worm drive hose clips
Tube of filler paste
Tyre valve core

If spare fuel is carried, a can designed for the purpose should be used to minimise risks of leakage and collision damage. A first aid kit and a warning triangle, whilst not at present compulsory in the UK, are obviously sensible items to carry in addition to the above.

When touring abroad it may be advisable to carry additional spares which, even if you cannot fit them yourself, could save having to wait while parts are obtained. The items below may be worth considering:

Throttle cable
Cylinder head gasket
Alternator brushes
Fuel pump repair kit

One of the motoring organisations will be able to advise on availability of fuel etc in foreign countries.

Engine will not start

Engine fails to turn when starter operated

Flat battery (recharge, use jump leads, or push start)
Battery terminals loose or corroded
Battery earth to body defective

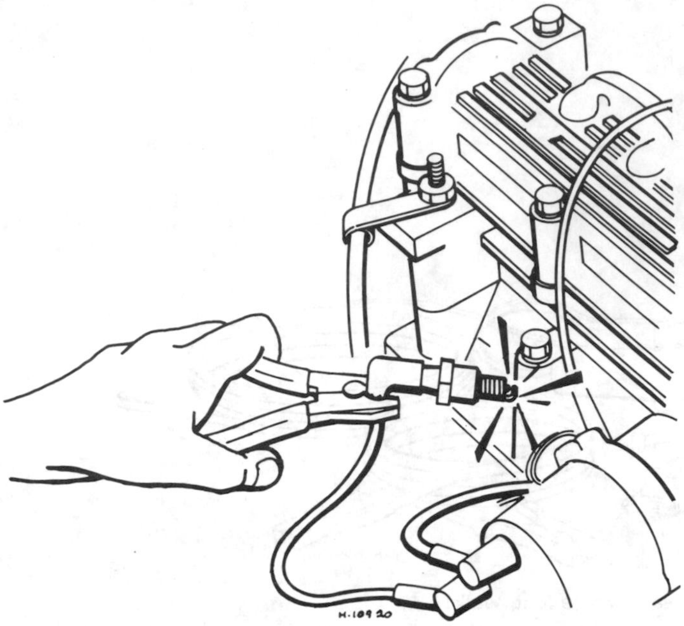

Crank engine and check for a spark. Note use of insulated pliers – dry cloth or a rubber glove will suffice

Engine earth strap loose or broken
Starter motor (or solenoid) wiring loose or broken
Automatic transmission selector in wrong position, or inhibitor switch faulty
Ignition/starter switch faulty
Major mechanical failure (seizure) or long disuse (piston rings rusted to bores)
Starter or solenoid internal fault (see Chapter 10)

Starter motor turns engine slowly

Partially discharged battery (recharge, use jump leads, or push start)
Battery terminals loose or corroded
Battery earth to body defective
Engine earth strap loose
Starter motor (or solenoid) wiring loose
Starter motor internal fault (see Chapter 10)

Starter motor spins without turning engine

Flat battery
Starter motor pinion sticking on sleeve
Flywheel gear teeth damaged or worn
Starter motor mounting bolts loose

Engine turns normally but fails to start

Damp or dirty HT leads and distributor cap (crank engine and check for spark)
Dirty or incorrectly gapped CB points
No fuel in tank (check for delivery at carburettor)
Excessive choke (hot engine) or insufficient choke (cold engine)
Fouled or incorrectly gapped spark plugs (remove, clean and regap)
Other ignition system fault (see Chapter 4)
Other fuel system fault (see Chapter 3)
Poor compression (see Chapter 1)
Major mechanical failure (eg camshaft drive)

Engine fires but will not run

Insufficient choke (cold engine)
Air leaks at carburettor or inlet manifold
Fuel starvation (see Chapter 3)
Ballast resistor defective, or other ignition fault (see Chapter 4)

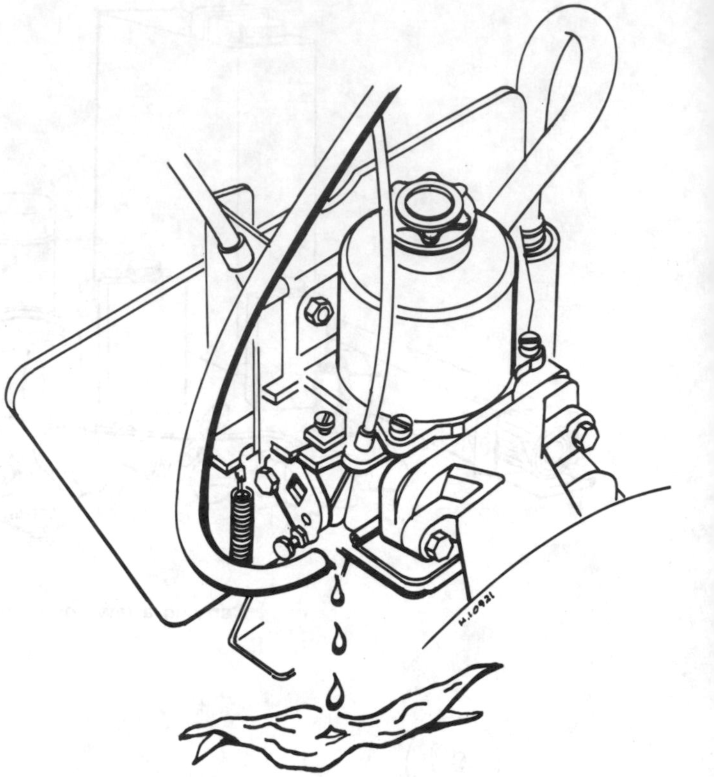

Remove fuel pipe from carburettor and check that fuel is being delivered

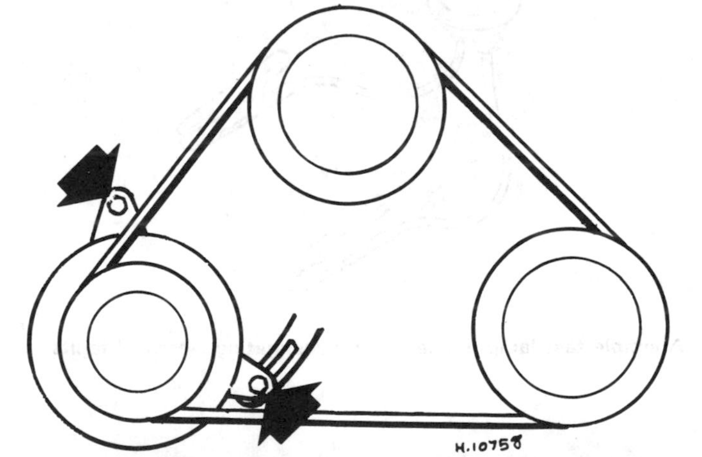

A slack drivebelt may cause overheating and battery charging problems. Slacken bolts (arrowed) to adjust

---

Engine cuts out and will not restart

Engine cuts out suddenly – ignition fault

Loose or disconnected LT wires
Wet HT leads or distributor cap (after traversing water splash)
Coil or condenser failure (check for spark)
Other ignition fault (see Chapter 4)

Engine misfires before cutting out – fuel fault

Fuel tank empty
Fuel pump defective or filter blocked (check for delivery)
Fuel tank filler vent blocked (suction will be evident on releasing cap)
Carburettor needle valve sticking
Carburettor jets blocked (fuel contaminated)
Other fuel system fault (see Chapter 3)

Engine cuts out – other causes

Serious overheating
Major mechanical failure (eg camshaft drive)

Engine overheats

Ignition (no-charge) warning light illuminated

Slack or broken drivebelt – retension or renew (Chapter 10)

Ignition warning light not illuminated

Coolant loss due to internal or external leakage (see Chapter 2)
Thermostat defective
Low oil level
Brakes binding
Radiator clogged externally or internally
Engine waterways clogged
Ignition timing incorrect or automatic advance malfunctioning
Mixture too weak

Note: *Do not add cold water to an overheated engine or damage may result*

Low engine oil pressure

Gauge reads low or warning light illuminated with engine running

Oil level low or incorrect grade
Defective gauge or sender unit
Wire to sender unit earthed
Engine overheating
Oil filter clogged or bypass valve defective
Oil pressure relief valve defective
Oil pick-up strainer clogged
Oil pump worn or mountings loose
Worn main or big-end bearings

Note: *Low oil pressure in a high-mileage engine at tickover is not necessarily a cause for concern. Sudden pressure loss at speed is far more significant. In any event, check the gauge or warning light sender before condemning the engine!*

Engine noises

Pre-ignition (pinking) on acceleration

Incorrect grade of fuel
Ignition timing incorrect
Distributor faulty or worn
Worn or maladjusted carburettor
Excessive carbon build-up in engine

Whistling or wheezing noises

Leaking vacuum hose
Leaking carburettor or manifold gasket
Blowing head gasket

Tapping or rattling

Incorrect valve clearances
Worn valve gear
Worn timing belt
Broken piston ring (ticking noise)

Knocking or thumping

Unintentional mechanical contact (eg fan blades)
Worn fanbelt
Peripheral component fault (generator, water pump etc)
Worn big-end bearings (regular heavy knocking, perhaps less under load)
Worn main bearings (rumbling and knocking, perhaps worsening under load)
Piston slap (most noticeable when cold)

Conversion factors

Length (distance)				
Inches (in)	X 25.4	= Millimetres (mm)	X 0.039	= Inches (in)
Feet (ft)	X 0.305	= Metres (m)	X 3.281	= Feet (ft)
Miles	X 1.609	= Kilometres (km)	X 0.621	= Miles
Volume (capacity)				
Cubic inches (cu in; in^3)	X 16.387	= Cubic centimetres (cc; cm^3)	X 0.061	= Cubic inches (cu in; in^3)
Imperial pints (Imp pt)	X 0.568	= Litres (l)	X 1.76	= Imperial pints (Imp pt)
Imperial quarts (Imp qt)	X 1.137	= Litres (l)	X 0.88	= Imperial quarts (Imp qt)
Imperial quarts (Imp qt)	X 1.201	= US quarts (US qt)	X 0.833	= Imperial quarts (Imp qt)
US quarts (US qt)	X 0.946	= Litres (l)	X 1.057	= US quarts (US qt)
Imperial gallons (Imp gal)	X 4.546	= Litres (l)	X 0.22	= Imperial gallons (Imp gal)
Imperial gallons (Imp gal)	X 1.201	= US gallons (US gal)	X 0.833	= Imperial gallons (Imp gal)
US gallons (US gal)	X 3.785	= Litres (l)	X 0.264	= US gallons (US gal)
Mass (weight)				
Ounces (oz)	X 28.35	= Grams (g)	X 0.035	= Ounces (oz)
Pounds (lb)	X 0.454	= Kilograms (kg)	X 2.205	= Pounds (lb)
Force				
Ounces-force (ozf; oz)	X 0.278	= Newtons (N)	X 3.6	= Ounces-force (ozf; oz)
Pounds-force (lbf; lb)	X 4.448	= Newtons (N)	X 0.225	= Pounds-force (lbf; lb)
Newtons (N)	X 0.1	= Kilograms-force (kgf; kg)	X 9.81	= Newtons (N)
Pressure				
Pounds-force per square inch (psi; lbf/in^2; lb/in^2)	X 0.070	= Kilograms-force per square centimetre (kgf/cm^2; kg/cm^2)	X 14.223	= Pounds-force per square inch (psi; lbf/in^2; lb/in^2)
Pounds-force per square inch (psi; lbf/in^2; lb/in^2)	X 0.068	= Atmospheres (atm)	X 14.696	= Pounds-force per square inch (psi; lbf/in^2; lb/in^2)
Pounds-force per square inch (psi; lbf/in^2; lb/in^2)	X 0.069	= Bars	X 14.5	= Pounds-force per square inch (psi; lbf/in^2; lb/in^2)
Pounds-force per square inch (psi; lbf/in^2; lb/in^2)	X 6.895	= Kilopascals (kPa)	X 0.145	= Pounds-force per square inch (psi; lbf/in^2; lb/in^2)
Kilopascals (kPa)	X 0.01	= Kilograms-force per square centimetre (kgf/cm^2; kg/cm^2)	X 98.1	= Kilopascals (kPa)
Torque (moment of force)				
Pounds-force inches (lbf in; lb in)	X 1.152	= Kilograms-force centimetre (kgf cm; kg cm)	X 0.868	= Pounds-force inches (lbf in; lb in)
Pounds-force inches (lbf in; lb in)	X 0.113	= Newton metres (Nm)	X 8.85	= Pounds-force inches (lbf in; lb in)
Pounds-force inches (lbf in; lb in)	X 0.083	= Pounds-force feet (lbf ft; lb ft)	X 12	= Pounds-force inches (lbf in; lb in)
Pounds-force feet (lbf ft; lb ft)	X 0.138	= Kilograms-force metres (kgf m; kg m)	X 7.233	= Pounds-force feet (lbf ft; lb ft)
Pounds-force feet (lbf ft; lb ft)	X 1.356	= Newton metres (Nm)	X 0.738	= Pounds-force feet (lbf ft; lb ft)
Newton metres (Nm)	X 0.102	= Kilograms-force metres (kgf m; kg m)	X 9.804	= Newton metres (Nm)
Power				
Horsepower (hp)	X 745.7	= Watts (W)	X 0.0013	= Horsepower (hp)
Velocity (speed)				
Miles per hour (miles/hr; mph)	X 1.609	= Kilometres per hour (km/hr; kph)	X 0.621	= Miles per hour (miles/hr; mph)
Fuel consumption*				
Miles per gallon, Imperial (mpg)	X 0.354	= Kilometres per litre (km/l)	X 2.825	= Miles per gallon, Imperial (mpg)
Miles per gallon, US (mpg)	X 0.425	= Kilometres per litre (km/l)	X 2.352	= Miles per gallon, US (mpg)
Temperature				
Degrees Fahrenheit (°F)	$= (°C \times \frac{9}{5}) + 32$			
Degrees Celsius (Degrees Centigrade; °C)	$= (°F - 32) \times \frac{5}{9}$			

**It is common practice to convert from miles per gallon (mpg) to litres/100 kilometres (l/100km), where mpg (Imperial) x l/100 km = 282 and mpg (US) x l/100 km = 235*

Index

C

D

E

Printed by
J H Haynes & Co Ltd
Sparkford Nr Yeovil
Somerset BA22 7JJ England